Chambers
Air and Space
Dictionary

GENERAL EDITOR
Professor P. M. B. Walker, CBE, FRSE

CONSULTANT EDITORS
Professor J. E. Allen
BSc (Eng), CEng, FRAeS, FIMechE

Dr D. J. Shapland
BSc, PhD, CPhys, FBIS

Chambers

Published 1990 by W & R Chambers Ltd, 43–45 Annandale Street,
Edinburgh EH7 4AZ

British Library Cataloguing in Publication Data
Chambers air and space dictionary.
  1. Aeronautics & astronautics
  I. Walker, P. M. B. (Peter Martin Brabazon) *1922 –*
  629.1

  ISBN 0 550 13242 2
  ISBN 0 550 13243 0 Pbk

Cover design by John Marshall

Printed in Great Britain by
Richard Clay Ltd, Bungay, Suffolk

Contents

How the dictionary was made

Chambers Air and Space Dictionary was compiled and designed on a COMPAQ 386 personal computer. The original database was made with the INMAGIC library retrieval software from Head Computers Ltd. The text was set using the Xerox VENTURA desktop publishing system and the drawings made with the Micrografx DESIGNER graphics program.

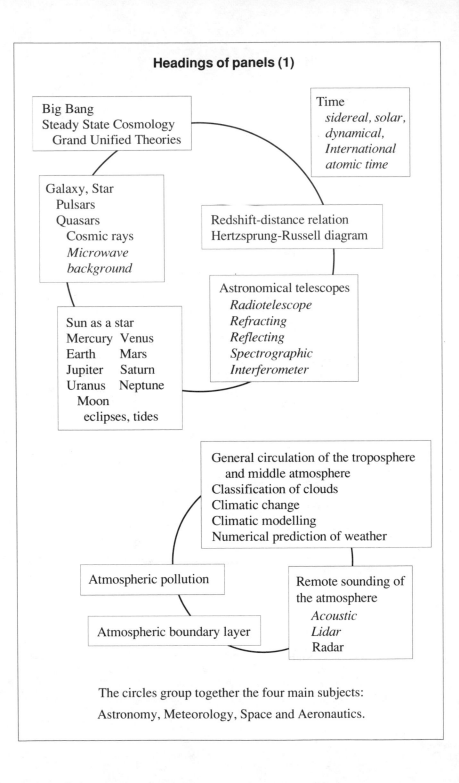

Headings of panels (1)

Big Bang
Steady State Cosmology
  Grand Unified Theories

Time
  sidereal, solar,
  dynamical,
  International
  atomic time

Galaxy, Star
  Pulsars
  Quasars
    Cosmic rays
    Microwave
    background

Redshift-distance relation
Hertzsprung-Russell diagram

Astronomical telescopes
  Radiotelescope
  Refracting
  Reflecting
  Spectrographic
  Interferometer

Sun as a star
Mercury Venus
Earth   Mars
Jupiter Saturn
Uranus   Neptune
   Moon
     eclipses, tides

General circulation of the troposphere
   and middle atmosphere
Classification of clouds
Climatic change
Climatic modelling
Numerical prediction of weather

Atmospheric pollution

Remote sounding of
the atmosphere
  Acoustic
  Lidar
  Radar

Atmospheric boundary layer

The circles group together the four main subjects:
Astronomy, Meteorology, Space and Aeronautics.

Headings of panels (2)

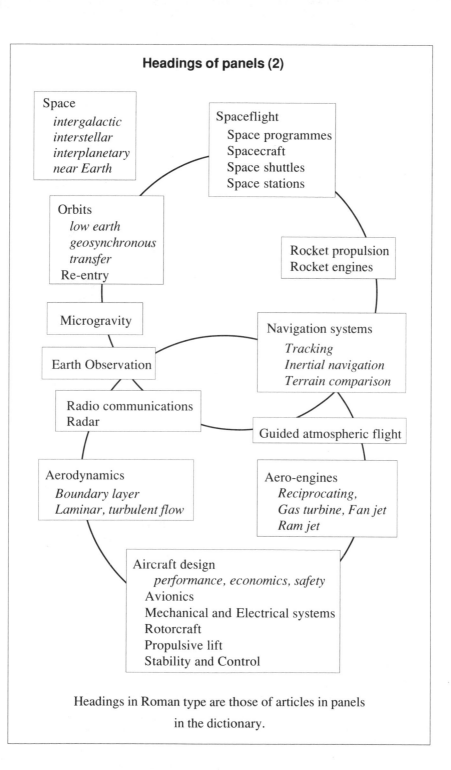

Space
intergalactic
interstellar
interplanetary
near Earth

Spaceflight
Space programmes
Spacecraft
Space shuttles
Space stations

Orbits
low earth
geosynchronous
transfer
Re-entry

Rocket propulsion
Rocket engines

Microgravity

Navigation systems
Tracking
Inertial navigation
Terrain comparison

Earth Observation

Radio communications
Radar

Guided atmospheric flight

Aerodynamics
Boundary layer
Laminar, turbulent flow

Aero-engines
Reciprocating,
Gas turbine, Fan jet
Ram jet

Aircraft design
performance, economics, safety
Avionics
Mechanical and Electrical systems
Rotorcraft
Propulsive lift
Stability and Control

Headings in Roman type are those of articles in panels
in the dictionary.

Preface

Chambers Air and Space Dictionary is a unique dictionary concerned with the sciences and technologies of everything above the Earth's surface. It has been developed from the database of the acclaimed *Chambers Science and Technology Dictionary* with many new and revised definitions in this group of fast-changing subjects.

The book contains some 6000 definitions, including 1500 in aeronautics, 600 in astronomy, 500 in meteorology and 400 for space and radar, together with a large number of relevant definitions from acoustics, physics, telecommunications and engineering. The aim has been to give a comprehensive treatment of subjects which interact with each other in many ways. Aircraft and meteorology, spacecraft and astronomy are obvious examples.

The special articles, placed in separate panels, are a major feature of *Chambers Air and Space Dictionary*. These, many of which are illustrated, supply fuller treatments of major topics within each subject. There are, for example, panels on aerodynamics, aircraft design, guided flight and propulsive lift in aeronautics; the Big Bang, cosmic rays, pulsars and the redshift-distance relation in astronomy; on atmospheric pollution, the classification of clouds, climate modelling and the circulation of the atmosphere in meteorology; launch systems, microgravity, re-entry and space stations in the subject of space.

Both astronomy and meteorology are well-established observational sciences with a range of terms, some of which stretch back to antiquity but with many others which have evolved to describe their rapid advances in recent years. Aeronautics and space, on the other hand, are essentially concerned with the engineering problems concerned with vehicle design in which many solutions have been tried, some discarded and with the remainder making up a very complex web of subjects, often difficult for the more general reader to penetrate. For example, *inertial navigation* is clearly part of *navigation* but the necessary equipment is also an *avionic* unit. *Stability and control* is a fundamental feature of an inherently unstable flying device which has to fly through the air with safety and economy and justifies a separate article. However, in the guided missile field it is customary to consider *guidance and control* which itself includes *automatic navigation*. Similarly *radio* is part of *avionics* and *air traffic control* and is also a *navigational aid*. To help overcome these problems some articles in these subjects, in particular in the older-established aeronautics, have been designed to provide an overview of a particular system like *stability and control* with cross-references to other articles and relevant individual entries in the main body of the dictionary. The diagrams on pp. iv and v illustrate some of the major interactions between the various sections of this dictionary.

Arrangement

The entries in this dictionary are strictly alphabetical with single letter entries occurring at the beginning of each letter. The panels occur either on the same page as the parent entry or on the pages immediately following, with the alphabetical entry stating 'See panel on this page' or 'See panel on p. 0'.

Italic and Bold

Italic is used for:

(1) alternative forms of, or alternative names for, the headword, usually after 'also' at the end of an entry;
(2) the expanded form of an abbreviated headword, provided that the expanded form is not found as a headword elsewhere;
(3) terms derived from the headword, often after 'adj.' or 'pl.';
(4) variables in mathematical formulae;
(5) for emphasis.

Bold is used for:

cross-references, either after 'see', 'cf.' etc. or in the **body of the entry**. It is also used after 'Abbrev. for' when the expanded form can be found as a headword elsewhere.

Bold italic is used in the panels to highlight a term explained within that article.

Appendices

The following tables will be found in the appendices:

(1) the radio and radar frequencies used in aircraft navigation;
(2) the letter designations of the frequency bands;
(3) the specifications for jet fuels;
(4) physical concepts expressed in SI units;
(5) SI conversion factors and
(6) physical constants, standard values and their equivalents in SI units.

Peter M. B. Walker, 1990

Contributors

Extensive use has been made of the entries in *Chambers Science and Technology Dictionary* and I would like to thank all those who contributed to the individual sections of that dictionary. In addition the following have revised entries and contributed longer articles used in the panels:

Professor J. E. Allen (Aeronautics)
Mr R. P. W. Lewis (Meteorology)
Dr S. Mitton (Astronomy)
Dr D. J. Shapland (Space)

P.M.B.W.

Acknowledgments

Drawings on pp. 5, part of that on 7 and 30 are redrawn from *Aerodynamics* by J. E. Allen, Allen Brothers and Father, Blythburgh, Suffolk, 1986; drawings on pp. 155 and 176 are redrawn from *Above and Beyond*, New Horizon, Chicago, 1968; the tables on navigation frequencies and letter designations on pp. 208–209 come from *Kempe's Engineer's Handbook*, Morgan-Grampian Books, London, 1989; the table on aviation fuels has been compiled by Dr Erik M. Goodyear, Cranfield Institute of Technology; diagram on p. 77 redrawn from *On the Mean Meridional Mass Motions of the Stratosphere and Mesosphere* by Dr T. J. Dunkerton in the *Journal of Atmospheric Sciences*, Vol. 35, No. 12, pp. 2325–2333, published by the American Metereological Society, 45 Beacon Street, Boston, MA 02108–3693, USA, 1978; diagram on p. 75 redrawn from *The Physics of Atmospheres* by John Y. Houghton, Cambridge University Press, 1986. Grateful thanks are due to Dr S. Mitton for the drawings on pp. 87, 148 and 182.

A

Å Symbol for Ångström.

[A] A strong absorption band in deep red of the solar spectrum (wavelength 762.128 nm) caused by oxygen in the Earth's atmosphere. The first of the Fraunhofer lines.

A & AEE Abbrev. for *Aeroplane & Armament Experimental Establishment*, at Boscombe Down, UK.

A and R display See **R-display**.

abaxial Said of rays of light which do not coincide with the optical axis of a lens system.

aberration (1) In an image-forming system, e.g. an optical or electronic lens, failure to produce a true image, i.e. a point object as a point image etc. Five geometrical aberrations are recognized by von Seidel, viz. spherical aberration, coma, astigmatism, curvature of the field and distortion. See **chromatic aberration**. (2) In astronomy an apparent change of position of a heavenly body, due to the velocity of light having a finite ratio to the relative velocity of the source and the observer.

A-bomb See **atomic bomb**.

abort The termination of a vehicle's flight either by failure or deliberate action to prevent dangerous consequences; if manned, a predetermined sequence of events is followed to ensure the safety of the crew.

absolute ceiling The height at which the rate of climb of an aircraft, in standard atmosphere, would be zero; the maximum height attainable under standard conditions.

absolute humidity Same as **vapour concentration**.

absolute magnitude See **magnitude**.

absolute temperature A temperature measured with respect to **absolute zero**, i.e. the zero of the **Kelvin thermodynamic scale of temperature**, a scale which cannot take negative values. See **kelvin scale**.

absolute zero The least possible temperature for all substances. At this temperature the molecules of any substance possess no heat energy. A figure of −273.15°C is generally accepted as the value of absolute zero.

absorber Any material which converts energy of radiation or particles into another form, generally heat. Energy transmitted is not absorbed. Scattered energy is often classed with absorbed energy.

absorption hygrometer An instrument by which the quantity of water vapour in air may be measured. A known volume of air is drawn through tubes containing a drying agent such as phosphorus pentoxide; the in-crease in weight of the tubes gives the weight of water vapour in the known volume of air.

absorption lines Dark lines in a continuous spectrum caused by absorption by a gaseous element. The positions (i.e. the wavelengths) of the dark absorption lines are identical with those of the bright lines given by the same element in emission.

absorption spectrum The system of absorption bands or lines seen when a selectively absorbing substance is placed between a source of white light and a spectroscope. See **Kirchhoff's law**.

AC-boundary layer See **Stokes layer**.

accelerate-stop distance The total distance, under specified conditions, in which an aircraft can be brought to rest after accelerating to **critical speed** for an engine failure at take-off.

acceleration The rate of change of velocity, expressed in metres (or feet) per second squared. It is a vector quantity and has both magnitude and direction.

acceleration due to gravity Acceleration with which a body would fall freely under the action of gravity in a vacuum. This varies according to the distance from the Earth's centre, but the internationally adopted value is 9.806 65 m/s² or 32.1740 ft/s². See **Helmert's formula**.

acceleration error The error in an airborne magnetic compass due to manoeuvring; caused by the vertical component of the Earth's magnetic field when the centre of gravity of the magnetic element is displaced from normal.

acceleration stress The influence of acceleration (or deceleration) on certain physiological parameters of the human body. Man can withstand transverse accelerations better than longitudinal ones, which have a profound effect on the cardiovascular system. The degree of tolerance also depends on the magnitude and duration of the acceleration.

acceleration tolerance The maximum of *g* forces an astronaut can withstand before 'blacking out' or otherwise losing control.

accelerator A device, similar to a **catapult**, but generally mounted below deck level, for assisting the acceleration of aircraft flying off aircraft carriers. Land versions have been tried experimentally.

accelerometer Transducer used to provide a signal proportional to the rate of acceleration of a vibrating or other body, usually employing the piezo-electric principle. It is carried in aircraft, guided missiles and spacecraft for measuring acceleration in a specific direction. Main types are *indicating*, *maximum-reading*, *recording* (graphical), and *counting* (digital, totalling all accelerations above a set value). See **impact accelerometer, vertical-gust recorder**.

accessory gearbox A gearbox, driven remotely from an aero-engine, on which aircraft accessories, e.g. hydraulic pump, electrical generator, are mounted.

accretion Process in which a celestial body, particularly an evolved dwarf star or a planet, is enlarged by the accumulation of extraneous matter falling in under gravity.

accumulated temperature The integrated product of the excess of air temperature above a threshold value and the period in days during which such excess is maintained.

aceval Abbrev. for *Air Combat EVALuation*.

achromatic lens A lens designed to minimize chromatic aberration. The simplest form consists of two component lenses, one convergent, the other divergent, made of glasses having different dispersive powers, the ratio of their focal lengths being equal to the ratio of the dispersive powers. Also *antispectroscopic lens*.

acoustic absorption Transfer of energy into thermal energy when sound is incident at an interface.

acoustic absorption factor, co-efficient The measure of the ratio of the acoustic energy absorbed by a surface to that which is incident on the surface. For an open window this can be 1.00, for painted plaster 0.02. The value varies with the frequency of the incident sounds e.g. for 2 cm glass fibre it is 0.04 at 125 Hz, 0.80 at 4000 Hz.

acoustical mass, inertia Given by M, where ωM is that part of the acoustical reactance which corresponds to the inductance of an electrical reactance: ω is the pulsatance $= 2\pi t \times$ frequency.

acoustical stiffness For an enclosure of volume V, given by $S = \rho \chi^2/\varsigma$, where c is velocity of propagation of sound and ρ is density. It is assumed that the dimensions of the enclosure are small compared with the sound wavelength and that the walls around the volume do not deflect.

acoustic amplifier One amplifying mechanical vibrations.

acoustic centre The effective 'source' point of the spherically divergent wave system observed at distant points in the radiation field of an acoustic transducer.

acoustic compliance Reciprocal of acoustical stiffness.

acoustic distortion Distortion in sound-reproducing systems.

acoustic filter One which uses tubes and resonating boxes in shunt and series as reactance elements, providing frequency cut-offs in acoustic wave transmission, as in an electric wave filter.

acoustic grating A diffraction grating for production of directive sound. Spacings are much larger than in optical gratings due to the longer wavelength of sound waves. Both transmission and reflection gratings are used.

acoustic impedance, reactance and resistance Impedance is given by complex ratio of sound pressure on surface to sound flux through surface, and has imaginary (reactance) and real (resistance) components, respectively. Unit is the **acoustic ohm**.

acoustic interferometer Instrument in which measurements are made by study of interference pattern set up by two sound or ultrasonic waves generated at the same source.

acoustic lens A system of slats or disks to spread or converge sound waves.

acoustic models Scale models of rooms (e.g. concert halls) or structures which are used to measure qualities important for architectural acoustics and noise control, e.g. sound distribution. The scale is typically between 1:10 and 1:20. In order to adjust the wavelength, the frequency has to be increased by a factor 10 to 20.

acoustic ohm Unit of acoustic resistance, reactance and impedance. 10^5 Pa s/m^3.

acoustic perspective The quality of depth and localization inherent in a pair of ears, which is destroyed in a single channel for sound reproduction. It is transferable with 2 microphones and 2 telephone ear-receivers with matched channels, and more adequately realized with 3 microphones and 3 radiating receivers with 3 matched channels.

acoustic pressure See **sound pressure**.

acoustic radiator Device to generate and radiate sound. The most common radiators are: (1) vibrating elastic systems (membrane, string, vocal cord) which cause a fluctuating pressure in the surrounding medium; (2) electrically driven membranes and plates (loudspeaker, sonar transducer); (3) vortices in turbulent fluid flow.

acoustic ratio The ratio between the directly radiated sound intensity from a source, at the ear of a listener (or a microphone), and the intensity of the reverberant sound in the enclosure. The ratio depends on the distance from the source, the polar distribution of the radiated sound power, and the period of reverberation of the enclosure.

acoustic reactance See **acoustic impedance.**

acoustic resonance Enhancement of response to an acoustic pressure of a frequency equal or close to the **eigenfrequency** of the responding system. When a system is at resonance, the imaginary part of its impedance is zero. Prominent in Helmholtz resonators, organ and other pipes, and vibrating strings.

acoustic saturation The aural effectiveness of a source of sound amid other sounds: it is low for a violin, but high for a triangle. The relative saturation of instruments indicates the number required in an auditorium of given acoustic properties.

acoustic scattering Irregular and multidirectional reflection and diffraction of sound waves produced by multiple reflecting surfaces the dimensions of which are small compared to the wavelength; or by certain discontinuities in the medium through which the wave is propagated.

acoustic spectrometer An instrument designed to analyse a complex sound signal into its wavelength components and measure their frequencies and relative intensities.

acoustic streaming Generation of constant flows by a strong sound wave. Acoustic streaming is a non-linear effect. It is responsible for the motion of the light particles (Lycopodium spores) in a **Kundt's tube.**

acoustic telescope Array of microphones. The signals of the microphones are added with certain phase-delays so as to generate desired directivities.

ACR Abbrev. for **Approach Control Radar.**

ACS Abbrev. for (1) **Active Control System**; (2) *Attitude Control System*; (3) *Air Conditioning System*.

ACT Abbrev. for *Active Control Technology.*

active array Antenna array in which the individual elements are separately excited by integrated circuit or transistor amplifiers.

active control Modern technique of noise or vibration control employing one or more sources that generate signals with the aim of making the resulting total signal smaller. Used e.g. for the control of low-frequency air-borne noise and vibration of machinery. See **antisound.**

active control system An advanced automatic flight control system designed to provide several special features, e.g. activation of flight control surfaces to minimize gust loads and bending stresses in the wing by detection and response to normal accelerations, provision of stability to a naturally unstable aircraft and implementation of pilot

manoeuvre demands. All these characteristics improve aircraft behaviour and performance, but the ACS demands extensive integration between aerodynamics, structure and electronic system design to achieve these advantages with reliability and safety.

active homing A guidance system where the missile contains the transmitter for illuminating the target and the receiver for the reflected energy.

active satellite Satellite equipped for sending out probing signals and receiving returned information. A *passive satellite* only receives information on the state of the target.

ACV Abbrev. for *Air Cushion Vehicle* (hovercraft).

adaptive array A radar antenna (either a *phased array* or an *active array*) whose gain, directivity and side lobes can be adjusted automatically to optimize the radar's performance under specific operating conditions.

ADD Abbrev. for *Airstream Direction Detector* (for stall protection).

ADF Abbrev. for **Automatic Direction Finding.**

adhesion Intermolecular forces which hold matter together, particularly closely contiguous surfaces of neighbouring media, e.g. liquid in contact with a solid. US *bond strength.*

adiabatic Without loss or gain of heat.

adiabatic change A change in the volume and pressure of the contents of an enclosure without exchange of heat between the enclosure and its surroundings.

adiabatic curve The curve obtained by plotting P against V in the adiabatic equation.

adiabatic lapse rate The rate of decrease of temperature which occurs when a parcel of air rises adiabatically through the atmosphere.

A-display Co-ordinate display on a CRT in which a level time base represents distance and vertical deflexions of beam indicate echoes.

adjustable-pitch propeller See **propeller.**

Adrastea A tiny natural satellite of **Jupiter**, discovered by the Voyager 2 mission in 1979.

advection The transference of any quantity by horizontal motion of the air.

advection fog Fog produced by the **advection** of warm moist air across cold ground.

AERO Abbrev. for *Air Education and Recreation Organisation* (UK).

aero-acoustics See panel on p. 4.

aerodynamic balance (1) A balance,

---

aero-acoustics

Aero-acoustics is concerned with sound generated aerodynamically, e.g. the noise of propellers, jet engines, turbo-machinery blading, combustion and jet efflux, wing vortices, boundary layer turbulence and shock waves from supersonic aircraft and missiles. The *sound pressure level* is measured in **decibels (dB)** which are defined as $20 \times \log_{10}$ of the *sound pressure at a point* over a *reference pressure*. The latter for air is usually taken as 2×10^{-4} bar. The *sound pressure spectrum level* refers to a specific frequency with a band width of 1 Hz. The *sound power ratio* is $10 \times \log_{10}$ of the *sound power per unit area* over the *reference power per unit area*.

A modification of the dB unit, called *dB(A)*, was introduced in the 1930s which weights the sound amplitudes at different frequencies to accord with a person's hearing sensitivity. *Composite Noise Rating* (CNR), *Perceived Noise Level* (PNL), *Noise Exposure Forecast* (NEF), *Noise and Number Index* (NNI) and *Psychological Assessment of Aircraft Noise Index* (PAANI) are all criteria taking account of community acceptance of noise, time of day etc.

See **boundary layer noise, buffeting, noise footprint, noise suppressor, sonic boom**.

---

usually but not necessarily in a wind tunnel, designed for measuring aerodynamic forces or moments. (2) Means for balancing air loads on flying control surfaces, so that the pilot need not exert excessive force, particularly as speed increases. The principle is to use aerodynamic forces, either directly on a portion of the control surface ahead of the hinge line, or indirectly through a small auxiliary surface with a powerful moment arm, to counterbalance the main airloads. Examples of the first are **horn balance**, *inset hinge* and *Frise balances*, and of the second the **balance tab**.

aerodynamic braking (1) The means of retarding a vehicle by creating high drag from operable brake flaps or parachutes. (2) Use of a planet's atmosphere to reduce the speed of space vehicles.

aerodynamic centre The point about which the pitching moment co-efficient is constant for a range of aerofoil incidence.

aerodynamic co-efficient A non-dimensional measure of aerodynamic force, pressure or moment that expresses the characteristics of a particular shape at a given incidence to the airflow. Typically lift co-efficient $C_L = \text{Lift}/\tfrac{1}{2}\,\rho V^2 S$, where ρ is air density, V is air speed and S is a typical area of the body (e.g. wing area). Similarly for drag co-efficient.

aerodynamic damping The suppression of oscillations by the inherent stability of an aircraft or of its control surfaces.

aerodynamic force That force created on a vehicle by the normal and shearing surface pressures by its passage through and disturb-

ance of the air. Conventionally, *lift* normal to the vehicle velocity and *drag* parallel to it.

aerodynamic heating Heating caused by shearing stress in the boundary layer and compression of the air by the vehicle. Heat is transferred to the surfaces and can be extremely high at hypersonic speeds esp. at body noses and wing leading edges.

aerodynamics See panel on p. 5.

aerodynamic sound See **flow noise**.

aerodyne Any form of aircraft deriving lift in flight principally from aerodynamic forces. Commonly, *heavier-than-air aircraft*, e.g. **aircraft, glider, kite, helicopter**.

aero-elastic divergence Aero-elastic instability which occurs when aerodynamic forces, or moments, increase more quickly than the elastic restoring forces or couples in the structure. Generally applied to wing weakness where the incidence at the tips increases under load, so tending to twist the wings off.

aero-elasticity The interaction of aerodynamic forces and the elastic reactions of the structure of an aircraft. Phenomena are most prevalent when manoeuvring at very high speed, e.g. *flutter*.

aero-embolism Release of nitrogen bubbles into the blood stream resulting from too rapid a reduction in ambient air pressure. Cf. **caisson disease**, the *bends*, encountered by undersea divers.

aero-engine See panel on p. 6.

aerofoil A body shaped so as to produce an aerodynamic reaction (lift) normal to its direction of motion, for a small resistance (drag) in that plane. A wing, plane, aileron,

aerodynamics

Aerodynamics is the science of air in motion, and covers the interactions between the air and solid surfaces where there is either relative motion, as in *aeronautics*, or where motion is caused by heating, as in *meteorology*. It also applies to spacecraft during exit and entry. Air molecules respond to external disturbances in many ways and there are therefore many branches of the subject.

Physics: air, at sea level, has a *density* (mass per unit volume) of 1.225 kg/m^3 at 15°C, decreasing with altitude; *heat transfer* results from viscous energy and involves specific heat, C_p at constant pressure and C_v at constant volume; *pressure* (the force per unit area) is 101.3 kPa at sea level; the *speed of sound* is 340.3 m/s at normal temperature and pressure, and equals Mach 1 and, generally, is 20.1 $\sqrt{T}$ m/s where T is the local absolute temperature.

Regions of airflow are considered under **boundary layer**. *Shock waves* occur at surfaces of discontinuity across which there is either an abrupt increase in pressure e.g. at the rear of a local supersonic region of flow around a surface at transonic speeds, or at an abrupt change in direction e.g. the bow shock wave ahead of a body moving at supersonic speeds or in any compression corner at supersonic or hypersonic speeds.

A number of types of fluid flow are defined: *ideal* flow is that of a mathematically perfect fluid without viscosity and hence without boundary layer or heating; *viscous* flow is real flow with viscosity present; *Newtonian* flow is that in a rarefied gas in which the molecular mean free path is long compared to that of a body moving in it; *free-molecule* flow is like Newtonian but the molecules impact and rebound from the body without influencing each other.

A number of *speed regimes* are defined for relative motion between a body and the fluid: *subsonic* with speeds less than that of sound; *sonic* at the speed of sound; *transonic* around the speed of sound, usually taken to be between Mach 0.7 and 1.3; *supersonic* faster than the speed of sound (Mach>1); *hypersonic* much faster than the speed of sound (Mach>5). At very high hypersonic speeds *magnetohydrodynamics* becomes important because heating will have dissociated the molecules and the resulting flow of free electrons behave as electric currents which can then create and be influenced by magnetic fields. *Hypervelocity* refers, in physics, to velocities approaching the speed of light and, in aerospace, to those exceeding the Earth satellite speed.

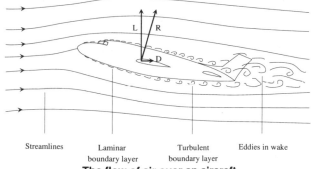

| Streamlines | Laminar boundary layer | Turbulent boundary layer | Eddies in wake |

The flow of air over an aircraft

L is the lift force, D is the drag force and R is the resultant of the aerodynamic pressure forces. In the diagram these forces are resolved along 'wind axes'. In guided flight 'body axes' are used with axial and normal forces resolved along and normal to the body.

See **aerodynamic-braking, -co-efficient, -damping, -force, -heating, aeroelasticity, -thermochemistry, -thermodynamics, -thermo-elasticity, boundary layer, computational fluid dynamics, wind tunnel.**

aero-engines

Aircraft propulsion units of a wide range of type depending on the power, speed and altitude at which the aircraft is to fly.

Reciprocating. Mainly *internal* combustion, *Otto cycle*, engines in which pistons oscillate in cylinders to rotate a crankshaft and hence, via gears, a propeller. Pistons can be arranged in-line, vee, radial and double-bank radial. In the *rotary* engine of the early 20th century the crankshaft was fixed to the airframe and the cylinder block rotated for cooling purposes. *Superchargers*, in which a compressor, usually centrifugal, is driven from the crankshaft or *turbochargers*, in which the compressor is driven by a turbine powered by the exhaust, are often used to increase performance especially at high altitude.

Gas turbine. An engine in which air is compressed by a rotary compressor (axial or centrifugal), then heated by combustion normally with kerosine with the exhaust gases passing through a radial or axial turbine system, providing power for the compressor. Types include: the *turbojet* (see A in diagram on next page) in which thrust is provided by the exhaust efflux of the gas turbine acting as a jet, and which may use *afterburners* (adding fuel to the exhaust gases after the turbine) to give a large increase in thrust for take-off or combat and *thrust reversers* (a mechanical means of deflecting exhaust gases forward and sideways) to shorten the landing run; *bypass turbojet*, in which part of the low pressure delivery air is bypassed round the combustion zone and turbine to deliver a cool, slow propulsive jet when mixed with the residual efflux from the turbine; *turbo-propeller*, a shaft turbine where the torque output is transmitted to a propeller through a reduction gearbox and which may take a single or double shaft or free-turbine form; *turbofan* or *ducted fan* (B in diagram) in which part of the power output of the gas turbine drives a fan mounted inside a duct; *unducted fan* (UDF) in which an advanced high-speed propeller or contra-rotating propellers are driven by a gas turbine; *variable cycle gas turbine* in which the flow of gases can be altered in different flight regimes by variable nozzles or deflectors for e.g. VTOL or operation over a wide speed range.

Ramjet. Also ATHODYD, abbrev. for *Aero THermO DYnamic Duct*. An airbreathing aircraft or missile engine (D in diagram), operating at supersonic speeds, in which the air compression, provided by a mechanical compressor in a gas turbine, is here derived from the increase in pressure of the air as it flows through and slows down in the intake duct. The combustion chamber and propelling nozzle are as in the jet engine. Includes the *turbo-ramjet*, which is a variable cycle jet engine operating at low speeds in a gas turbine mode and as a ramjet at high supersonic speeds.

Air-turbo-rocket. This is a composite engine (C in diagram) in which rocket propellant is used to activate a turbine driving a compressor whose air delivery then augments the rocket efflux to produce a propulsive jet.

Air-ram-rocket. *Ducted rocket*. This is a hybrid engine which is basically a rocket but incorporates an integral ramjet. The rocket is used at low and high speeds and the ramjet at intermediate supersonic speed.

Rocket. This can be solid or liquid fuelled and is used on aircraft to power *Rocket Assisted Take-Off Gear* (RATOG) or for fighter boost.

See **bleed, hi/lo stages, pressure ratio, propulsive lift, rating, turbine blade temperature, turbine entry temperature.**

See drawings on next page

aero-engines (contd)

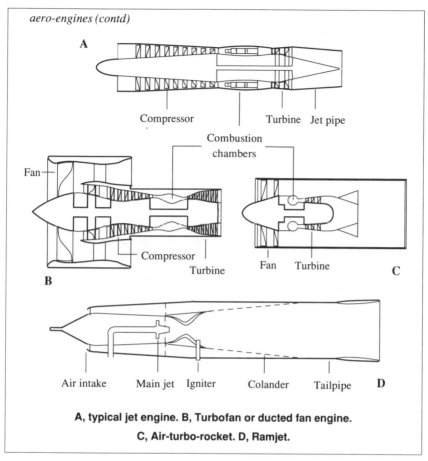

A, typical jet engine. B, Turbofan or ducted fan engine.

C, Air-turbo-rocket. D, Ramjet.

tailplane, rudder, elevator etc.

aerofoil section The cross-sectional shape or profile of an aerofoil.

aero-isoclinic wing A sweptback wing which has its torsional and flexural stiffness so adjusted that the angle of attack remains constant as the wing bends under flight loads, instead of decreasing with deflexion towards the tip, which is the normal geometric effect.

aerological diagram A thermodynamic diagram used for plotting the results of upper-air soundings usually containing, as reference lines, isobars, isotherms, **dry adiabatics**, **saturated adiabatics** and lines of constant **saturated humidity mixing ratio**.

aerology The study of the **free atmosphere**.

aeronautical engineering That branch of engineering concerned with the design, production and maintenance of aircraft structures, systems, and power units.

aeronautical fixed services A telecommunication service between fixed stations for the transmission of aeronautical information, particularly navigational safety, and flight planning messages. Abbrev. *AFS*.

aeronautics All activities concerned with aerial locomotion.

aeroplane See **aircraft**.

aerospaceplane Aircraft-like vehicle which can take off from and land on runways, manoeuvre in the atmosphere, operate in space and re-enter the atmosphere.

aerostat Any form of aircraft deriving support in the air principally from its buoyancy, e.g. a balloon or airship.

aerothermochemistry The chemical reactions which occur with airflow heating, e.g. a candle flame in air or the combustion of kerosene in a jet engine.

aerothermodynamics Particular branch of thermodynamics relating to the heating effects associated with the dynamics of a gas; in particular the physical effects produced in the air flowing over a vehicle during launch and re-entry, but also combustion of gases and by shock waves.

aerothermo-elasticity Aero-elasticity complicated by heating effects.

AFCS *Automatic Flight Control System.* A category of **automatic pilot** for the control of an aircraft while *en route*. It can be monitored by speed and altitude data signals, signals from *ILS* and *VOR*, has automatic approach capability, and is disengaged before landing. Cf. **autoflare, autoland, autothrottle**.

aft cg limit See **cg limits**.

afterbody Rear portion of a flying-boat hull, aft of the main step.

afterburner See **reheat**.

AGS Abbrev. for **Aircraft General Standard**.

AIAA Abbrev. for *American Institute of Aeronautics and Astronautics.*

aileron droop The rigging of ailerons so that under static conditions their trailing edges are below the wing trailing-edge line, pressure and suction causing them to rise in flight to the aerodynamically correct position.

ailerons Surfaces at the trailing edge of the wing, controlled by the pilot, which move differentially to give a rolling motion to the aircraft about its longitudinal axis.

air absorption Absorption of sound waves propagating in air. It is caused by molecular relaxation processes and viscosity.

air brake An extendable device, most commonly a hinged flap on wing or fuselage, controlled by the pilot, to increase the drag of an aircraft. Originally a means of slowing bombers to enable them to dive more steeply, it is an essential flight control on clean jet aircraft and sailplanes.

air conduction The passing of noise energy along an air path, as contrasted with structure-borne conduction of vibrational energy.

aircraft Any mechanically driven heavier-than-air flying machine with wings of fixed or variable sweep angle. Subdivisions: landplane, seaplane (float seaplane and flying-boat), amphibian.

aircraft design See panel on p. 9.

aircraft engine See **aero-engine**.

Aircraft General Standard Term referring to small parts or items such as bolts, nuts, rivets, fork joints etc. which are common to all types of aircraft. Abbrev. *AGS*.

aircraft noise Noise from propeller, engine, exhaust, and that generated aerodynamically over the surfaces; characterized by unstable low frequencies. See **aero-acoustics, jet noise**.

air data system A centralized unit into which are fed the essential physical measurements for flight, e.g. airspeed, **Mach number**, pitot and static pressure, barometric altitude, stagnation air temperature. From this central source data are transmitted to the cockpit dials, to flight and navigational instruments and to computers etc. Abbrev. *ADS*.

air drag Resistance to the motion of a body passing through the Earth's atmosphere, most serious in the lower regions, producing changes in the geometry of the orbit, even causing the body to re-enter. More generally the term *atmospheric drag* is used in reference to other planets.

airflow meter An instrument, mainly experimental, for measuring the airflow in ducts.

airframe The complete aircraft structure without power plant, systems, equipment, furnishings and other readily removable items.

air frost A screen temperature below 0°C. See **wind frost**.

airglow The faint permanent glow of the night sky, due to light-emission from atoms and molecules of sodium, oxygen and nitrogen, activated by sunlight during the day.

air intake Any opening introducing air into an aircraft; the opening for the main engine air is usually implied if unqualified.

air-intake guide vanes Radial, toroidal or volute vanes which guide the air into the compressor of a gas turbine, or the supercharger of a reciprocating engine.

air log An instrument for registering the distance travelled by an aircraft relative to the air, not to the ground.

air manometer A pressure gauge in which the changes in volume of a small quantity of air enclosed by mercury in a glass tube indicate changes in the pressure to which it is subjected.

airmanship Skill in piloting an aircraft.

air mass A part of the atmosphere where the horizontal temperature gradient at all levels within it is very small, perhaps of the order of 1°C/100 km.

air mass flow In a gas turbine power plant, the quantity of air which is ingested by the compressor, normally expressed in pounds or kilograms per second.

air-mileage unit An automatic instrument which derives the air distance flown and feeds it into other automatic navigational instruments.

air miles per gallon The number of miles

aircraft design

An aircraft results from the conscious process of design in which concepts are brought together in a creative way and in which compromises are struck between conflicting needs of performance, economics, safety etc. The resulting aircraft's essential characteristics are referred to as its *design* including its external shape, which is primarily an aerodynamic function, and its internal systems. The term *configuration* is of US origin and refers usually to the gross spatial arrangement of wings, bodies, engines and control surfaces, and how these are related to each other.

Throughout the history of aeronautics new concepts offer advantages of speed, economy or military power and, despite the risks of innovation, when they succeed, as in the change from biplane to monoplane or from piston engine to jet, then a fashion is set and many other examples follow.

The following show types of design or configuration, of which some are illustrated.

Biplane (see Wright biplane drawing), Sesquiplane, Monoplane.
All-wing or tail-less.
Delta wing.
Swept-back wing.
Swept-forward wing.
Variable sweep or variable geometry wing. See Tornado drawing which shows positions for subsonic and supersonic (dotted lines) flight.
Foreplane or canard (so-called because the wing is aft and it resembles a duck in flight),
Swept-back wing and podded engines on pylons. Boeing 707, B-52, 747 etc. See Tristar drawing.
Slender delta wing, as in Concorde.
Multistage as the Mayo composite flying boat.

Wright biplane

Tristar

Tornado

Concorde

Specific terms in design specification: *gross weight, landing weight, load* and *maximum weight* are all legal values in the specification and contract; *design points* are specific combinations of flight variables on which the design is based; *verification* is the proof of the viability of design by flight and ground tests; *wing area* has an agreed standard for calculation which is that enclosed by the wing's outline with all moving surfaces stowed, excluding fillets, but taken across the engine nacelles and fuselage to the centre-line.

flown through the air for each gallon of fuel burnt by the propulsion unit(s).

air pocket Colloq. for a localized region of rising or descending air current. Causes an abrupt vertical acceleration as an aircraft passes through it, severity increasing with

speed and also with low wing loading. Also *bump*. See **vertical gust**.

airport markers Particoloured boards defining areas on an airfield, e.g. *boundary markers* which indicate the limits of the landing area, *taxi-channel markers* for taxi tracks, *obstruction markers* for ground hazards, and *runway visual markers*, situated at equal distances, by which visibility is gauged in bad weather.

airport meteorological minima The minimum cloud base (vertical) and visibility (horizontal) in which landing or takeoff is permitted at a particular aerofield. *ICAO* standards: Cat. 1, 200 ft (60 m) height, 2600 ft (800 m) RVR; Cat. 2, 100 ft (30 m) height, 1300 ft (400 m) RVR; Cat. 3, zero height, (*a*) 700 ft (210 m), (*b*) 150 ft (45 m), (*c*) zero RVR. See **runway visual range** (RVR).

air position The geographical position which an aircraft would reach in a given time if flying in still air.

air-position indicator An automatic instrument which continually indicates air position, incorporating alterations of course and speed.

Air Registration Board The airworthiness authority of the UK until its functions were taken over in 1972 by the **Civil Aviation Authority**. Abbrev. *ARB*.

air route In organized flying, a defined route between two aerodromes; usually provided with direction-finding facilities, lighting, emergency-landing grounds, etc. See **airway**.

airscrew Defined in 1951 to be 'any type of screw designed to rotate in air'. Term now obsolete and replaced by **propeller**, a device for propelling aircraft and **fan**, a rotating bladed device for moving air in ducts or e.g. wind tunnels. See **rotor**.

airship Any power-driven **aerostat**. There are various types. *Non-rigid airship*, one with the envelope so designed that the internal pressure maintains its correct form without the aid of a built-in structure; small, and used for naval patrol work. *Rigid airship*, one having a rigid structure to maintain the designed shape of the hull, and to carry the loads; usually a number of ballonets or gas bags inside the frame; large, used for military purposes in World War I, and having limited commercial use until 1938. *Semi-rigid airship*, one having a partial structure, usually a keel only, to distribute the load and maintain the designed shape of the envelope or ballonets; intermediate size.

air space That part of the atmosphere which lies above a nation and which is therefore under the jurisdiction of that nation.

airspeed Speed measured relative to the air in which the aircraft or missile is moving, as distinct from ground speed. See **equivalent airspeed, indicated airspeed, true airspeed**.

airstrip A unidirectional landing area, usually of grass or of a makeshift nature.

air superiority fighter Combat aircraft intended to remove hostile aircraft from a volume of airspace and so establish control of the air.

air-traffic control The organized control, by visual and radio means, of the traffic on air routes, and into and out of aerodromes. ATC is divided into general *area control*, including defined *airways*; *control zones*, of specified area and altitude, round busy aerodromes; *approach control* for regulating aircraft landing and departing; and *aerodrome control* for directing aircraft movements on the ground and giving permission for takeoff. Air-traffic control operates under two systems: *visual flight rules* and, more severely, *instrument flight rules*. Since World War II great advances in radar technology have enabled *air-traffic controllers* to be given very complete 'pictures' of the position of aircraft, not only in flight, but also when manoeuvring on the ground. Abbrev. *ATC*.

air-traffic control centre An organization providing (1) air-traffic control in a control area and (2) **flight information** in a region.

air-traffic controller One who is licensed to give instructions to aircraft in a control zone.

Air Transport Association A US organization noted particularly for its specification which sets a standard to which manufacturers of aircraft and associated equipment are required to produce technical manuals for the aircraft operator's use. The specification is accepted by *IATA* as the basis for international standardization. Abbrev. *ATA*.

airway A specified 3-dimensional corridor (the lower as well as the upper boundary being defined) between **control zones** which may only be entered by aircraft in radio contact with **air-traffic control**.

airworthiness See panel on p. 11.

albedo A measure of the reflecting power of a non-luminous surface, expressed as the ratio of energy reflected in all directions to total incident energy.

algal corrosion Impairment of structure and systems by algae and other microorganisms.

Algol A star, β, in the constellation of Perseus which is the prototype of the eclipsing binary, where one component passes in front of the other at each revolution, causing an eclipse and a systematic fluctuation of magnitude.

aliasing Error in making real time spectra of short signals or of directivity in sound fields. Caused by insufficient number of data points.

airworthiness

The fitness of an aircraft, its engines and all internal systems, for flight operations in all possible environments and feasible circumstances for which the aircraft (or device) has been designed. Regulations exist to ensure that an aircraft meets government and international standards, and are enforced by the issue of the appropriate certificates.

The *Certificate of Airworthiness* is that issued or required by the Civil Aircraft Authority, *CAA*, confirming that a civil aircraft is *airworthy* in every respect to fly within the limitations of at least one of six categories. Abbrev. *C of A*. In US an Approved Type Certificate, *ATC*, is issued.

The *Certificate of Compliance* certifies that parts of an aircraft have been overhauled, repaired or inspected etc. to comply with airworthiness requirements. Abbrev. *C of C*.

The *Certificate of Maintenance* certifies that an aircraft has been inspected and maintained in accordance with its maintenance schedule. Abbrev. *C of M*.

If an unsafe condition is evident, the national airworthiness authority issues an Airworthiness Directive, *AD*. If the fault is critical for flight safety the AD requires immediate correction before flight. Otherwise there is a specific time limit for rectification.

A-licence Basic private pilot's licence.

all-burnt The moment at which the fuel of a missile or spacecraft is completely consumed.

allithium Aluminium-lithium alloys.

all-moving tail A one-piece **tailplane**, also controlled by the pilot as is the **elevator**. Also *flying tail, stabilizer*. See **T-tail**.

allowable deficiencies Aircraft systems or certain items of their equipment, tabulated in the flight or operating manual, which even if unserviceable will not prevent an aircraft from being flown or create a hazard in flight.

allowances In airline terminology, fuel reserves are frequently referred to as allowances, and are usually specified as time factors under certain conditions, as distance plus descent, or as a percentage (by weight or volume) of the cruising fuel for a given stage.

Almagest The Arabic form of the title of Claudius Ptolemy's great astronomical treatise, 'The Mathematical Syntaxis', written in Greek about AD 140.

almucantar A small circle of the celestial sphere parallel to the horizontal plane. The term is also applied to an instrument for measuring altitudes and azimuths.

Alpha Centauri The brightest star in the constellation Centaurus, actually three stars, the faintest of which, *Proxima Centauri*, is the nearest star to the Sun.

alpha particle The nucleus of a helium atom of mass number four, consisting of two neutrons and two protons and so doubly positively charged. Emitted from natural or radioactive isotopes. It is often written as α–particle.

alpha radiation Alpha particles emitted from radioactive isotopes.

alpha rays Stream of **alpha particles**.

altazimuth A type of telescope in which the principal axis can be moved independently in altitude (swinging on a horizontal axis) and azimuth (swinging on a vertical axis). Used in very large *optical telescopes* and *radio telescopes*.

alternate airfield One designated in a *flight plan* at which a pilot will land if prevented from alighting at his destination.

altimeter An aneroid barometer used for measuring altitude by the decrease in atmospheric pressure with height. The dial of the instrument is graduated to read the altitude directly in feet or metres, the zero being set to ground or aerodrome level. See **encoding-**, **radio-**, **recording-**.

altitude (1) The height in feet or metres above sea level. For precision in determining the performance of an aircraft, this must be corrected for the deviation of the meteorological conditions from that of the standardized atmosphere (*International Standard Atmosphere*). See **cabin altitude**, **pressure altitude**. (2) The angular distance of a heavenly body measured on that great circle which passes, perpendicular to the plane of the horizon, through the body and through the zenith. It is measured positively from the horizon to the zenith, from 0° to 90°.

altitudes by barometer　See **Babinet's formula for altitude**.

altitude switch　A switching device generally comprising electrical contacts, actuated by an aneroid capsule which in turn is deflected by change in atmospheric pressure. The contacts are adjusted to make or break a warning circuit at the pressure corresponding to a predetermined altitude.

altitude valve　A manually- or automatically-operated valve fitted to the carburettor of an aero-engine for correcting the mixture-strength as air density falls with altitude.

altocumulus　White and/or grey patch, *sheet* or *layer* of *cloud*, generally with shading, composed of laminae, rounded masses, rolls etc. which are partly fibrous or diffuse and which may or may not be merged; most of the irregularly arranged small elements usually have an apparent width of between 1° and 5°. Occurs between 3000–7500 m. Abbrev. *Ac*.

altostratus　Greyish or bluish *cloud sheet* or *layer* of striated, fibrous or uniform appearance, totally or partly covering the sky, and having parts thin enough to reveal the Sun at least vaguely, as through ground glass. Altostratus does not show halo phenomenon and occurs at 3000–7500 m. Abbrev. *As*.

Amalthea　The fifth natural satellite of **Jupiter**, discovered in 1892.

ambient noise　The noise existing in a room or any other environment, e.g. the ocean.

ambient noise level　Random uncontrollable and irreducible noise level at a location, or in a valve or circuit.

ampere　SI unit of electric current. Defined as that current which, if maintained in two parallel conductors of infinite length, of negligible cross-section, and placed 1 metre apart in vacuum, would produce between the conductors a force equal to 2×10^{-7} newtons per metre of length. One of the SI fundamental units.

amphibian　Aircraft capable of taking off and alighting on land or water, e.g. seaplane or flying boat with retractable landing gear or landplane with **hydroskis**.

anabatic wind　A local wind blowing up a slope heated by sunshine, and caused by the difference in density between the warm air in contact with the ground and the cooler air at corresponding heights in the **free atmosphere**.

anafront　A situation at a front, warm or cold, where the warm air is rising relative to the **frontal zone**.

analog(ue)　A previous *weather map* similar to the current map. The developments following the analogue aid forecasting.

Ananke　The twelfth natural satellite of **Jupiter**.

Andromeda nebula　The spiral galaxy M31 in Andromeda, visible to the naked eye. Of comparable size with our own Galaxy, it is about 2 million light-years distant.

anechoic room　One in which internal sound reflections are reduced to an ineffective value by extremely high sound absorption, e.g. glass-fibre wedges. Also *dead room*.

anemograph　See **anemometer**.

anemometer　An instrument for measuring the speed of the wind. A common type consists of four hemispherical cups carried at the ends of four radial arms pivoted so as to be capable of rotation in a horizontal plane, the speed of rotation being indicated on a dial calibrated to read wind speed directly. An *anemograph* records the speed and sometimes the direction.

aneroid barometer　One having a vacuum chamber or syphon bellows of thin corrugated metal, one end diaphragm of which is fixed, the other being connected by a train of levers to a scale pointer which records the movements of the diaphragm under changing atmospheric pressure.

angels　Radar echoes from an invisible and sometimes undefined origin. High-flying birds, insect swarms and certain atmospheric conditions can be responsible.

angled deck　The flight deck of an aircraft carrier prolonged diagonally from one side of the ship, so that aircraft may fly off and land on without interference to or from aircraft parked at the bows. US *canted deck*.

angle of approach light　A light indicating an approach path in a vertical plane to a definite position in the landing area.

angle of attack　The angle between the *chord line* of an aerofoil and the relative airflow, normally the immediate flight path of the aircraft. Also, erroneously, *angle of incidence*.

angle-of-attack indicator　An instrument which senses the *true* angle of incidence to the relative airflow, and presents it to the pilot on a graduated dial or by means of an indicating light.

angle of bank　See **angle of roll**.

angle of incidence　Angular setting of any aerofoil to a reference axis. See **angle of attack**.

angle of roll　The angle through which an aircraft must be turned about its longitudinal axis to bring the lateral axis horizontal. Also horizontal *angle of bank*.

angle of stall The angle of attack which corresponds with the maximum lift coefficient.

angular acceleration The acceleration of a spacecraft around an axis, resulting in pitch, roll or yaw.

angular diameter Observed diameter of any celestial object expressed as the angle subtended by its diameter as perceived by the observer.

angular distance of stars Observed angular separation of two stars as perceived by the observer.

angular momentum The moment of the linear momentum of a particle about an axis. Any rotating body has an angular momentum about its centre of mass, its *spin angular momentum*. The angular momentum of the centre of mass of a body relative to an external axis is its *orbital angular momentum*. In atomic physics, the orbital angular momentum of an electron is *quantized* and can only have values which are exact multiples of the **Dirac constant**. In particle physics, the angular momentum of particles which appear to have spin energy is quantized to values that are multiples of half the Dirac constant.

angular velocity The rate of change of angular displacement, usually expressed in radians per second.

anharmonic Said of any oscillation system in which the restoring force is non-linear with displacement, so that the motion is not simple harmonic.

anhedral See **dihedral angle**.

ANIK Canadian remote sensing satellite system.

annihilation Spontaneous conversion of a particle and its antiparticle into radiation, e.g. positron and electron yielding two γ-ray photons each of energy 0.511 MeV.

annihilation radiation The radiation produced by the annihilation of an elementary particle with its corresponding antiparticle.

annoyance The psychological effect arising from excessive noise. There is no absolute measure, but the annoyance caused by specified classes of noise can be correlated.

annual equation One of four terms describing the orbit of the Moon, which arises from the eccentricity of the Earth's orbit round the Sun. Its period is one year.

annual parallax The motion of the Earth round the Sun causes minute changes in the apparent positions of the stars. The regular annual displacement is the *annual parallax*. It is largest, at 0.71 seconds of arc, for the star **Proxima Centauri**.

annular combustion chamber A gas turbine combustion chamber in which the perforated **flame tube** forms a continuous annulus within a cylindrical outer casing.

annular eclipse See **eclipse**.

anomalistic month The interval (amounting to 27.554 55 days) between two successive passages of the Moon in its orbit through perigee.

anomalistic year The interval (equal to 365.259 64 mean solar days) between two successive passages of the Sun, in its apparent motion, through perigee. See **time**.

anomaly The angle between the radius vector of an orbiting body and the major axis of the orbit, measured from perihelion (periastron etc.) in the direction of motion.

antapex See **solar antapex**.

anthelion A mock sun appearing at a point in the sky opposite to and at the same altitude as the Sun. The phenomenon is caused by the refraction of sunlight by ice crystals.

antibaryon Antiparticle of a baryon, i.c. a hadron with a baryon number of −1. The term **baryon** is often used generically to include both.

anti-clutter Refers to a circuit or part of a radar system designed to eliminate unwanted echoes, *clutter*, and permit the display of signals which might otherwise be obscured. Often takes the form of a gain control which automatically reduces gain immediately after the transmitted pulse and gradually restores it during the interval leading up to the anticipated return echo.

anti-collision beacon A flashing red or blue light which is mounted above and below an aircraft to make it conspicuous when flying in **control zones** or other busy areas.

anticyclone A distribution of atmospheric pressure in which the pressure increases towards the centre. Winds in such a system circulate in a clockwise direction in the northern hemisphere and in a counterclockwise direction in the southern hemisphere. Anticyclones give rise to fine, calm weather conditions, although in winter fog is likely to develop.

anti-*g*-suit A close-fitting garment covering the legs and abdomen, which is inflated, either automatically or at will by the wearer, so that counter-pressure is applied when blood is displaced away from the head and heart during high-speed manoeuvres. Colloq. *g-suit*.

anti-*g*-valve (1) A spring-loaded mass type of air valve which automatically regulates the inflation of an *anti-g-suit* according to the acceleration (*g*) loads being imposed. (2) A valve incorporated in some aircraft fuel systems to prevent engines being starved of fuel under specific *g* loads.

anti-icing Protection of aircraft against icing by preventing ice formation on e.g. wind-shield panels, leading edges of wings, tail units and turbine engine air intakes. The most common methods are to apply continuous heating by hot air tapped from an engine, by electrical heating elements or periodically inflating rubber bags. Cf. **de-icing**.

antilepton An antiparticle of a **lepton**. Positron, positive muon, antineutrinos and the tau-plus particle are antileptons.

antimuon Antiparticle of a **muon**.

antineutrino Antiparticle to the **neutrino**. As there are four types of neutrino there are also four types of antineutrino.

antineutron Antiparticle with spin and magnetic moment oppositely orientated to those of a neutron.

antiparticle A particle that has the same mass as another particle but has opposite values for its other properties such as charge, baryon number, strangeness etc. The antiparticle to a fundamental particle is also fundamental, e.g. the electron and positron are particle and antiparticle. Interaction between such a pair means simultaneous **annihilation**, with the production of energy in the form of radiation.

antiproton Short-lived particle, half-life 0.05 μsec, identical with proton, but with negative charge; annihilating with normal proton, it yields mesons.

antiquark The antiparticle of a **quark**.

antisound Sound signal with same amplitude but opposite phase of some unwanted sound signal so that both signals cancel each other when superimposed. Used in **active control**.

antispin parachute A small parachute, normally in a canister, which may be fixed to the tail (occasionally to the wing tips) of an aircraft or glider for release in emergency to lower the nose into a dive and so assist recovery from a spin. It is jettisoned after use. Colloq. *spin chute*.

antisurge valve A valve for bleeding off surplus compressor air to suppress the unstable airflow due to **surge** in a gas turbine engine.

antitrades Winds, at a height of 900 m or more, which sometimes occur in regions where tradewinds are prevalent, their direction being opposite to that of the tradewinds.

anti-transmit-receive tube A gas discharge tube which isolates a pulsed radar transmitter from the antenna so that echoes can be received. Abbrev. *ATR tube*. Cf.

transmit-receive tube.

anvil cloud A common feature of a thundercloud, consisting of a wedge-shaped projection of cloud suggesting the point of an anvil.

aperture The effective area over which an aerial extracts power from an incident plane wave. The aperture (A) and gain (G) are related by the equation $G=4\pi A/\lambda^2$.

aperture synthesis Two or more radio telescope antennas are connected as pairs of **interferometers** in this technique. The amplitude and phase of the interference pattern is continuously recorded. The interferometer baseline is normally variable, and the rotation of the Earth changes the position angle with respect to a distant radio source. The *Fourier transform* (see **Fourier analysis**) of the amplitude and phase patterns are then used to compute a map of the radio source. By means of long baselines, achieved by linking telescopes on different continents, it is possible to achieve a *resolving power* of 0.001 arc seconds or so.

aphelion The farthest point from the Sun on a planet's, comet's or spacecraft's orbit. Pl. *aphelia*.

Apjohn's formula A formula which may be used for determining the pressure of water vapour in the air from readings of the wet and dry bulb hygrometer. The formula is: $p_t = p_w$ −0.000 75$H(t-t_w)$ [1−0.008($t-t_w$)], where p_w is the saturated vapour pressure at the temperature (t_w) of the wet bulb, H is the barometric height, and t is the temperature of the dry bulb.

aplanatic Said of an optical system which produces an image free from spherical aberration.

apochromatic lens A lens so designed that it is corrected for chromatic aberration for three wavelengths thus reducing the secondary spectrum.

apogee The point in the orbit of the Moon or an artificial satellite which is furthest from the Earth. Also the highest altitude attained by a missile.

apogee motor Engine fired at the apogee of an elliptical orbit to establish a circular orbit whose altitude is that of the apogee of the original orbit. Similarly a *perigee motor* for transforming a circular orbit into an eccentric one.

Apollo The NASA programme for putting a man on the Moon and returning him to Earth. The Apollo spacecraft consisted of a *Command Module* (CM), a *Service Module* (SM) and a *Lunar Excursion Module* (LEM). The CM-SM combination remained in lunar orbit (with one astronaut) while the lunar landing (with two astronauts) was performed by the LEM. Earth re-entry was effected by the CM only, the crew and CM being recovered from the sea. The ensemble was launched by a Saturn V rocket. Using this system, Neil Armstrong and Buzz Aldrin became the first men to step on the Moon (July 1969).

Apollo asteroid An **asteroid** whose orbit brings it within 1 AU of the Sun.

apparent magnitude See **magnitude**.

apparent solar day The interval, not constant owing to the Earth's elliptical orbit, between two successive transits of the true Sun over the meridian.

apparent solar time Time as measured by the apparent position of the Sun in the sky, e.g. by a sundial.

Appleton layer Same as **F-layer**.

approach control radar A surveillance radar which shows on a CRT display the positions of aircraft in an aerodrome's traffic control area. Abbrev. *ACR*.

approach lights Lights indicating the desired approach to a runway, usually of sodium or high-intensity type and laid in a precise pattern of a lead-in line with cross-bars at set distances from the **runway threshold**.

approach speed The indicated air speed at which an aircraft approaches for landing.

appulse Seemingly close approach of two celestial objects as perceived by an observer, particularly the close approach of a planet or asteroid to a star without the occurrence of an eclipse.

apron A firm surface of concrete or 'tarmac' laid down adjacent to aerodrome buildings to facilitate the movement, loading and unloading of aircraft.

apse line The diameter of an elliptic orbit which passes through both foci and joins the points of greatest and least distance of the revolving body from the centre of attraction. Also *line of apsides*.

APU See **auxiliary power unit**.

ARABSAT Communication satellite system set up by Arab States.

Arago point The bright spot found along the axis in the shadow of a disk illuminated normally.

ARC Abbrev. for (1) *Aeronautical Research Council*, UK, (2) *Ames Research Centre*, US.

areal velocity The rate, constant in elliptic motion, at which the radius vector sweeps out unit area.

area rule An aerodynamic method of reducing drag at transonic speeds by maintaining a smooth cross-sectional variation throughout the length of an aircraft. Because of the effect of the wing, this often results in a 'wasp-

waist' on the fuselage or the addition of bulges to the wing or fuselage.

ARINC *Aeronautical Radio INCorporated*, an organization (US) whose membership includes airlines, aircraft constructors and *avionics* component manufacturers. It publishes technical papers and agreed standards, and finances research.

ARM Abbrev. for *Anti-Radiation Missile*.

armillary sphere Celestial globe, first used by the Greek astronomers, in which the sky is represented by a skeleton framework of intersecting circles, the Earth being at the centre. In antiquity, of major importance for measuring star positions.

arrester gear (1) A device on aircraft carriers and some military aerodromes, usually consisting of a number of individual transverse cables held by hydraulic shock-absorbers, which stop an aircraft when its **arrester hook** catches a cable. (2) A barrier net, usually of nylon or webbing, attached to heavy drag weights, which stops fast aircraft from over-running the end of the runway in an emergency.

arrester hook A hook extended from an aircraft to engage the cable of an arrester gear, mainly on aircraft carriers.

articulated blade A rotorcraft blade which is mounted on one or more hinges to permit flapping and movement about the **drag axis**.

artificial ear Device for testing earphones which presents an acoustic impedance similar to the human ear and includes facilities for measuring the sound pressure produced at the ear.

artificial feel In an aircraft flying control system, esp. with *automatic control of flying surfaces*, in which the pilot's control actions are modified to provide forces moving the flying controls, a natural feel, opposing the pilot's actions, is fed back from the controls. Since these forces vary mostly with dynamic air pressure $(q = \frac{1}{2}ev^2)$; artificial feel is sometimes known as *q-feel*.

artificial horizon See **gyro horizon**.

artificial satellite Man-made space vehicle whose velocity is sufficient to maintain it in orbit about another body. Earth satellites are used for the purposes of observation of the surface, the atmosphere, the Sun and deep space, as communication links, in weather forecasting and for the performance of microgravity and other technological experiments.

artificial stability An automatic flight control system which provides positive stability to an otherwise unstable or neutrally stable aircraft.

artificial voice Loudspeaker and baffle for simulating speech in testing of microphones.

ascending node For Earth, the point at which a satellite crosses the equatorial plane travelling from south to north.

asdic *Allied Submarine Detection Investigation Committee.* Underwater acoustic detecting system which transmits a pulse and receives a reflection from underwater objects, particularly submarines, at a distance. Also used by trawlers to detect shoals of fish. Equivalent to US *sonar.*

ashen light A faint glow sometimes seen in that part of the disk of Venus that is not directly illuminated by the Sun.

aspect See **attitude.**

aspect ratio The ratio of span/mean chord line of an aerofoil (usually in wing); defined as (span)^2/area. Important for **induced drag** and range/speed characteristics. Normal figure between 6 and 9, lesser values than 6 being *low aspect ratios,* greater than 9 *high aspect ratios.*

aspheric surface A lens surface which departs to a greater or lesser degree from a sphere, e.g. one having a parabolic or elliptical section.

aspirated psychrometer A psychrometer which uses a forced draught of at least 12 km/h (8 mi/h) over the wet bulb.

assisted take-off Supplementing the full power of the normal engines by auxiliary means, which may or may not be jettisonable. Small turbojet or rocket motor units, powder, or liquid rockets may be used. See **JATO, RATOG.**

asterism A conspicuous or memorable group of stars, smaller in area than a constellation, e.g. the *Plough.*

asteroid One of thousands of rocky objects normally found between the orbits of Mars and Jupiter, ranging in size from 1 to 1000 km. A few, e.g. Eros, pass close to the Earth.

astigmatism A defect in an optical system on account of which, instead of a point image being formed of a point object, 2 short line images (focal lines) are produced at slightly different distances from the system and at right angles to each other. Astigmatism is always present when light is incident obliquely on a simple lens or spherical mirror.

astrocompass A non-magnetic instrument that indicates true north relative to a celestial body.

astrodome A transparent dome, fitted to some aircraft usually on the top of the fuselage, with calibrated optical characteristics, for astronomical observations.

astrolabe Ancient instrument (ca 200 BC) for showing the positions of the Sun and

bright stars at any time and date. If fitted with sights, also used for measuring the altitude above the horizon of celestial objects, and, in this mode, a 15th c. forerunner of the **sextant.**

astrometry The precise measurement of position in astronomy, generally deduced from the co-ordinates of images on photographic plates.

astronaut A man or woman who flies in space; early astronauts had to be very fit, whereas nowadays more benign conditions lead to requirements of normal fitness and high technical qualifications. Also *cosmonaut, spationaut.*

astronautics The science of space flight.

Astronomer Royal Formerly the title of the Director of the Royal Greenwich Observatory. Since 1972, a purely honorary title awarded to a distinguished British astronomer.

astronomical clock An elaborate clock showing astronomical phenomena such as the phases of the Moon and principally found in medieval cathedrals, e.g. Wells in Somerset and Strasbourg. In modern observatories the term is applied to any clock displaying **sidereal time.**

Astronomical Ephemeris Annual handbook (*ephemeris*) published a few years in advance by Her Majesty's Nautical Almanac Office, essentially identical to *The American Ephemeris* and issued in an abridged form as the *Nautical Almanac.*

astronomical telescope See panel on p. 17.

astronomical triangle Triangle on the celestial sphere formed by a heavenly body S, the zenith Z, and the pole P. The 3 angles are the hour-angle at P, the azimuth at Z, and the parallactic angle at S.

astronomical twilight The interval of time during which the Sun is between 12° and 18° below the horizon, morning and evening. See **civil twilight, nautical twilight.**

astronomical unit Mean distance of the Earth from the Sun, 1.496×10^8 km or about 93 million miles. Abbreviated AU and commonly used as a unit of distance within the solar system. There are 63 240 AU in one light-year.

astrophysics That branch of astronomy which applies the laws of physics to the study of interstellar matter and the stars, their constitution, evolution, luminosity etc.

asymmetric flight The condition of flying with asymmetrically balanced thrust, weight, drag or lift forces, as could occur e.g. with one external weapon mounted under one wing or in a twin-engined aircraft with one engine inoperative.

ATA Abbrev. for **Air Transport Association.**

astronomical telescope

Any instrument that is specifically designed to collect, detect and record radiation from any cosmic source may be termed an astronomical telescope. The collector may be a mirror, often dish-shaped, a lens or an array of dipoles (for a radio telescope). The collector concentrates the radiation at a focus, where it is detected by eye, in a photographic emulsion or by an electronic device. Finally this detected radiation is recorded, either photographically or digitally. Almost all telescopes therefore have three essential elements: collector, detector and recorder. The great range of telescope designs is a consequence both of observing across the entire electromagnetic spectrum, radio waves to gamma rays, and of the variety of objects (Sun, planets, stars and galaxies) to be studied.

Galileo first applied the optical **refracting telescope** to astronomy in 1610 and even with this crude instrument the power of the technique was immediately apparent. He resolved the *Milky Way* into stars and discovered four natural satellites of **Jupiter**. The first lenses suffered badly from **chromatic aberration**, and to avoid this fault Newton designed the **reflecting telescope** which used a paraboloid metallic mirror to form an image (see diagram). Further

development of this instrument by W. Herschel and Lord Rosse led to the construction of giant reflectors in the nineteenth century with which the distant galaxies and nebulas could be catalogued and studied for the first time. About the turn of the century, astronomers in California discovered the advantages of siting telescopes on mountains. The Mount Wilson Observatory in Pasadena led the way with a 2.5 metre telescope (1917), followed by a 5-metre Hale Telescope at Mount Palomar (1949). These instruments showed the immense size of the cosmos and the diversity of objects in the universe.

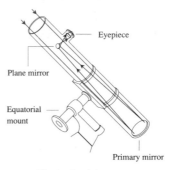

Newtonian telescope

A modern optical telescope functions as a multi-purpose instrument. For surveying the sky a **Schmidt telescope** (see diagram) is generally the best; it uses a thin corrector plate to produce an undistorted field over an area of about 6° square. The prime focus (see diagram on following page) may be used for direct photography. The *Cassegrain focus*, behind the main mirror is favoured for most investigations as it is simpler to mount instruments such as **spectrographs** or **CCD** detectors there.

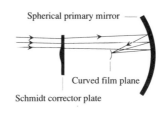

Schmidt telescope

For long focal lengths, favoured for high resolution spectroscopy, the coudé focus is available. In the 1970s several 4-metre telescopes were constructed, but design changes indicate that 8-metres will be the preferred size in the 1990s. Lightweight mirrors and the use of computer controlled optics have significantly reduced the costs of large ground based instruments.

continued on next page

astronomical telescope (contd)

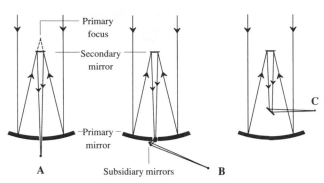

Focal positions of an astronomical telescope

A is the Cassegrain focus, B the coudé and C the Nasmyth. The secondary mirror is removed when the primary focus is used.

There are two basic ways of mounting any telescope. It must be able to swing on two axes at right angles in order to reach all parts of the sky. In the *equatorial mount* one axis is parallel to the Earth's rotation axis; the telescope is driven around this axis only in order to follow the motion of the sky. Until recently this arrangement was favoured for all types of telescope. From about 1970 computer developments offered astronomers the possibility of the *altazimuth mount*, which has a vertical axis. This is much cheaper to engineer but both axes must be driven when observing, so computer drives are essential.

Radio telescopes gather radio waves from the Sun, hot gas clouds and active galaxies. They need a far larger aperture than an optical telescope in order to reach a useful resolving power. In practice this is achieved by linking single dishes into an **interferometer**. Baselines up to 10,000 km are available by linking observatories on different continents in very long baseline interferometry.

Progress in satellite technology has enabled astronomers to design X-ray and gamma ray telescopes which must operate from a space platform as these radiations are absorbed in the atmosphere. In this part of the spectrum extremely hot interacting **binary stars**, **neutron stars** and possibly black holes are detected. The Hubble Space Telescope operating from a space platform in the optical and ultraviolet gives unrivalled images of very faint objects that cannot be seen at ground level.

ATC Abbrev. for **air-traffic control**.
ATCRBS Abbrev. for *Air-Traffic Control Radar Beacon System*. A direct development of the World War II **IFF** system. Operating at about 1 GHz, it gives air-traffic controllers three-dimensional positional information and full identification of aircraft.
Atlas The fifteenth natural satellite of **Saturn**, discovered in 1980.
atm Abbrev. for **standard atmosphere** (1). See **atmospheric pressure**.
atmospheric absorption (1) The absorption of the light of the stars by the Earth's atmosphere; it is practically negligible above

45° altitude, but the extinction amounts to about half a magnitude at 20°, 1 magnitude at 10°, and 2 magnitudes at 4° altitude. (2) Diminution of intensity of a sound wave in passing through the air, apart from normal inverse square relation, and arising from transfer of sound energy into heat.
atmospheric acoustics That concerned with the propagation of sound in the atmosphere, of importance in sound ranging and aircraft noise.
atmospheric boundary layer See panel on p. 19.
atmospheric electricity That causing in-

atmospheric boundary layer

Abbrev. *ABL*. That region of the Earth's atmosphere in which direct interaction with the Earth's surface is predominant. The airflow is largely turbulent with eddies on scales roughly equal to their height above the surface and with turbulent shearing stresses comparable in magnitude with pressure gradients, buoyancy and **Coriolis forces**; there are significant vertical flux divergence of momentum, heat and moisture due directly to surface influence. The top of the boundary layer is often marked by a shallow stable region which normal turbulent motions cannot penetrate (although strongly convective elements may). This top is usually at a height of 1–2 km. Much nearer the ground, at a height of some tens of metres, there is frequently a sublayer called the *surface-flux layer* (sometimes *constant-flux layer*) where fluxes may be assumed to be almost constant with height; in this layer the Coriolis force may be neglected. See **boundary layer** for applications to aircraft etc.

Theoretical and mathematical treatment of transfer processes in the ABL is extremely difficult owing to the turbulent nature of the flow. The methods employed usually consist of similarity theory and dimensional analysis combined with empirical results to give formulae which should hold in a reasonably wide range of natural conditions; a considerable degree of statistical scatter is nevertheless inevitable. An important parameter is the *surface roughness length* which is characteristic of the surface and ought to be independent of the flow. However, it is possible for roughness lengths to be different for transfers of momentum, heat and water vapour, and their estimation is in any case difficult; consider, for example, bare soil, growing crops, forests and built-up areas. Another important parameter is the non-dimensional **Richardson number** which is derived from a consideration of the relative magnitudes of the rate of production of mechanical turbulence, depending on the vertical shear of the mean wind, and its enhancement or suppression, depending on thermal stability. Turbulent transfers of heat or momentum may also be usefully considered, at least in special cases, as analogues of transfers by molecular diffusion and simple equations may be derived employing constant 'exchange co-efficients' which are analogues of viscosity or thermal conductivity. A demonstration of the variation of horizontal wind through the ABL and its approach to the 'free stream' value in the **Ekman spiral** is easily obtained by such methods.

Virtually all life on Earth exists within the ABL; almost all the Sun's energy, which drives the **general circulation of the atmosphere** and creates the weather, enters the atmosphere by means of it.

creasing potential with height, about 100 V/m, in calm conditions, altered considerably by thunder-clouds. See **lightning**.

atmospheric pollution See panel on p. 20.

atmospheric pressure The pressure exerted by the atmosphere at the surface of the Earth is due to the weight of the air. Its standard value is $1.013\ 25\times10^5$ N/m^2, or 14.7 lbf/in^2. Variations in atmospheric pressure are measured with a barometer. See **barometric pressure, standard atmosphere**.

atmospheric tides The changes of atmospheric pressure arising directly from changes in temperature due to the Earth's rotation. See **diurnal range**.

atomic bomb Bomb in which the explosive power, measured in terms of equivalent TNT, is provided by nuclear fissionable material such as ^{235}U or ^{239}Pu. The bombs dropped on Hiroshima and Nagasaki (1945) were of this type. Also *A-bomb*, *atom bomb*, *fission bomb*. See **hydrogen bomb**.

atomic clock A clock whose frequency of operation is controlled by the frequency of an atomic or molecular process. The inversion of the ammonia molecule with a frequency of 23 870 Hz provides the basic oscillations of the *ammonia clock*. The difference in energy between two states of a caesium atom in a magnetic field giving a frequency of 9 192 631 770 Hz is the basis of the *caesium clock* which has an accuracy

atmospheric pollution

The presence in the atmosphere of gases and aerosols that are in one way or another harmful to plant and animal life, or cause damage to buildings, and are due either entirely to human activity or occur in much greater concentrations than would be natural without such activity. For example, carbon dioxide occurs naturally, but quantities have considerably increased since the industrial revolution, owing to the burning of fossil fuels; similarly, ammonia and oxides of nitrogen and sulphur are produced by lightning flashes and volcanoes, but their concentrations have steadily increased owing to intensive farming practices and general industrial activity. Chlorofluorocarbons (*CFC*s), on the other hand, are new and have been made artificially during the last few decades for use in refrigerators and aerosol sprays; they have recently been discovered to play a part in destroying atmospheric ozone and enhancing the **greenhouse effect**. Photochemical smog is also new, being produced by the action of strong sunlight on high concentrations of exhaust fumes from internal combustion engines; it is particularly common in sunny, populous and affluent areas where winds are light, e.g. Los Angeles and its neighbourhood.

Noxious compounds, after emission from chemical plants, power stations etc., are diffused widely by turbulent motions in the **atmospheric boundary layer**, and can subsequently spread through the **troposphere** and even penetrate into the **stratosphere**. They may also reach the ground either by *dry deposition* or by being dissolved in raindrops or fog droplets, i.e. *wet deposition*. Dry deposition occurs as a combined effect of gravitational settling, impaction of particulates on vegetation and absorption of chemically reactive gases; it is most important near the original source of pollution. The popular term 'acid rain' is a rough synonym for *wet deposition*. 'Natural' pre-industrial rain was itself acidic with a pH of between 5 and 6; however, since about 1970 the pH of much rain in industrial areas has fallen to about 4 and on exceptional occasions to as low as 3, i.e. about as acid as vinegar. Wet deposition is much more efficient at depositing pollution than dry deposition and can rapidly bring down high concentrations of pollutants from the upper atmosphere at great distances from the original source, e.g. when radioactive material from the Chernobyl disaster was deposited on high ground in North Wales, the Lake District and the West of Scotland in the UK.

The effects of atmospheric pollution on animal and vegetable life are complex and difficult to unravel and there is still controversy over their precise nature.

---

of better than 1 in 10^{13}.

atomic disintegration Natural decay of radioactive atoms, by emission of radiation, into chemically different atomic products.

atomic frequency A natural vibration frequency in an atom used in the atomic clock.

atomic scattering That of radiation, usually electrons or X-rays, by the individual atoms in the medium through which it passes. The scattering is by the electronic structure of the atom in contrast to *nuclear scattering* which is by the nucleus.

ATR tube See **anti-transmit-receive tube**.

attenuation General term for reduction in magnitude, amplitude or intensity of a physical quantity, arising from absorption, scattering or geometrical dispersion. The latter, arising from diminution by the inverse-square law, is not generally considered as attenuation proper. See **absorption spectrum**.

attitude The *attitude* or *aspect* of an aircraft in flight, or on the ground, is defined by the angles made by its axes with the relative airflow, or with the ground, respectively.

attitude control The provision of a desired orientation to satisfy mission requirements; it is usually effected by a low thrust system in conjunction with a measuring instrument, such as a star sensor, and maintained by a stabilizing device, such as a gyroscope. Attitude control can also be maintained by spinning the spacecraft about one of its axes.

attitude indicator A *gyro horizon* which indicates the true attitude of the aircraft in pitch and roll throughout 360° about these axes. See **heading indicator**.

AU Abbrev. for **astronomical unit**.

audibility Ability to be heard; said of faint sounds in the presence of noise. The extreme range of audibility is 20 to 20 000 Hz in frequency, depending on the applied intensity; and from $2 \times 10^{-5} N/m^2$ (rms) at 1000 Hz (the zero of the phon scale, selected as the average for good ears) to 120 dB.

audiofrequency Frequency which, in an acoustic wave, makes it audible. In general, any wave motion including frequencies in the range of e.g. 30 Hz to 20 kHz.

audiogram Standard graph or chart which indicates the hearing loss (in **bels**) of an individual ear in terms of frequency. See **objective noise meter**, **sound-level meter**.

audiometer Instrument for measurement of acuity of hearing. Specifically to measure the minimum intensities of sounds perceivable by an ear for specified frequencies. See **noise audiometer**.

auditory canal Duct connecting the ear drum with the external ear (*pinna*), by which sound waves are transmitted from outer to inner ear.

augmentor Means of increasing forces: (1) afterburning in a gas turbine; (2) by induced airflow in a rocket; (3) in a wing of **STOL** aircraft by ducting compressed airflow from a gas-turbine into circulation-increasing slots and flaps to create high lift co-efficients and thereby giving slow landing speeds.

aureole (1) The reddish ring round the Sun or Moon, forming the inner part of a corona. (2) The bright indefinite ring round the Sun in the absence of clouds.

aurora Luminous curtains or streamers of light seen in the night sky at high latitudes, caused when electrically charged particles from the Sun are guided by the Earth's magnetic field to the polar regions, there colliding with atoms in the upper atmosphere. In the northern hemisphere known as *aurora borealis* and in the southern as *aurora australis*.

AUSAT Australian satellite communications system.

auto coarse pitch The setting of the blades of a propeller to the minimum drag position if there is a loss of engine power during take-off.

autoflare An automatic landing system which operates on the **flare out** part of the landing, using an accurate radio-altimeter.

autoland A landing in which the descent, forward speed, **flare-out**, alignment with the runway and touchdown are all automatically controlled. See **autoflare**, **autothrottle**.

automatic direction finding Airborne *navaid* tuned to radio source of known position. Using rotatable loop aerial mounted above an aircraft to detect the direction of the radio source by rotating until the signal is zero. Abbrev. *ADF*.

automatic mixture control A device for adjusting the fuel delivery to a reciprocating engine in proportion to air density.

automatic observer An apparatus for recording, photographically or electronically, the indications of a large number of measuring instruments on experimental research aircraft.

automatic parachute A parachute for personnel which is extracted from its pack by a static line attached to the aircraft.

automatic pilot *Autopilot*. A device for guiding and controlling an aircraft on a given path. It may be set by the pilot or externally by radio control. Colloq. *George*.

automatic stabilizer A form of automatic pilot, operating about one or more axes, adjusted to counteract dynamic instability. Colloq. *autostabilizer*. See **damper**.

automatic tracking Servo control of radar system operated by a received signal, to keep antenna aligned on target.

automatic weather station A transistorized and packaged apparatus which measures and transmits weather data for electronic computation.

autonomous vehicle Generally an unmanned aircraft operating without external assistance.

autopilot See **automatic pilot**.

autorotation (1) The spin; continuous rotation of a symmetrical body in a uniform air-stream due entirely to aerodynamic moments. (2) Unpowered rotorcraft flight, i.e. a helicopter with engine stopped, in which the symmetrical aerofoil rotates at high incidence parallel with the airflow.

autostabilizer See **automatic stabilizer**.

autothrottle A device for controlling the power of an aero-engine to keep the approach path angle and speed constant during an automatic blind landing.

autumnal equinox See **equinox**.

auxiliary air intake (1) An air intake for accessories, cooling, cockpit air etc. (2) Additional intake for turbojet engines when running at full power on the ground, usually spring-loaded so that it will open only at a predetermined suction value.

auxiliary power unit An independent airborne engine to provide power for ancillary equipment, electrical services, starting etc. May be a small reciprocating or turbine. Abbrev. *APU*.

auxiliary rotor A small rotor mounted at the

aviation fuels

Liquid hydrocarbons are the ideal fuel for aviation because of their high heat of combustion per unit fuel mass (specific energy) and volume (energy density). Other favourable characteristics include: ease of combustion, moderate volatility and viscosity and good thermal stability and capacity. Gasoline and kerosine (also *kerosene*) are the two major fuel types, as described below.

Aviation gasoline, ***AVGAS***. Blends of hydrocarbon liquids, almost all petroleum products boiling between 32° to 220°C, used as PE(4) fuel. Four grades currently available, covering the range of antiknock rating from 80 octane number to 145 performance number. Comparatively small quantities are now used.

Aviation kerosine for gas-turbine engines. Typically boils over the range 144 to 252°C. There are many variants designated currently as follows:

Jet A, with relatively high freezing point, for internal use in the US.

Jet A-1, ***AVTUR***, NATO F-35. (Designated JP-8, NATO F-34, when Fuel System Icing Inhibitor added for military use). The international jet fuel.

Jet B, ***AVTAG*** (Designated JP-5, NATO F-44 when FS11 added). Wide-cut fuel obtained by blending kerosine with naphtha (gasoline components) as a postwar measure to ensure availability of jet fuel. Now being phased out except for operation in cold climates.

AVCAT, NATO F-43 (Designated JP-5, NATO F-44, when FS11 added). High-flash for naval use to avoid vapour build-up in unvented sections of aircraft carriers.

JPTS, fuel of high thermal stability and low volatility for high-flying U2 type aircraft.

JP-7, blend of special aromatics-reduced stocks of high thermal stability and specific energy for high-altitude high-speed (Mach 3+) 'Blackbird' aircraft.

JP-10, high density hydrocarbon (exo-tetrahydro-dicyclopentadiene) for use in turbine and ramjet powered missiles.

AVPIN, aviation isopropyl nitrate.

AVPOL, aviation petrol, oil and lubricant.

AVGARD, TN for an additive to JP-1 and other jet fuels to produce anti-misting kerosine.

Methane in liquid form is of interest since it is rich in hydrogen and thus of a high specific energy (50 MJ/kg net compared to 43.4 MJ/kg net for AVTUR). It is lower in CO_2 emissions, but cryogenic.

Hydrogen in liquid form is also of particular interest because of its exceptionally high specific energy (120.24 MJ/kg net) but it has a very low energy density (8.42 MJ/l compared to 34.7 MJ/l for AVTUR) which makes it unsuitable for small aircraft. Cryogenic, it can only be used in bulky, insulated tanks within the fuselage rather than within mainplanes. It has been used experimentally in the US and recently in the USSR.

Boron was also used experimentally as a basis for a high energy liquid fuel for military application. Pentaborane (B_5H_9) is about 50% better than kerosine by weight and 20% by volume but costs a hundred times more.

tail of a helicopter, usually in a perpendicular plane, which counteracts the torque of the main rotor; used to give directional and rotary control to the aircraft.

auxiliary tanks See **fuel tanks**.

available potential energy That part of the total potential energy of the atmosphere available for conversion into kinetic energy by adiabatic redistribution of its mass so that the density stratification becomes horizontal everywhere.

aviation fuels See panel on this page.

aviation kerosine, kerosene See panel.

aviatrix Female aviator.

avionics

The collective term for aircraft-, missile- and spacecraft-subsystem elements which involve electronic principles. It is derived from *AVIation electronICS* and refers specifically to air and space-borne equipment, and is not strictly applicable to their ground-based equivalents. They include computing, sensing and data transmission circuits and systems. The circuits can be analog or digital, or a hybrid combination with electrons as the basic means of operating. *Photonics* is a possible system of the future, employing photons rather than electrons. As these are smaller, more data can be handled at a faster rate than with electronics. Already optical digital data are transmitted in optical fibres, and **lasers** are used for ranging and sensing with much current research on the difficult processes of computing with photons.

There are three major subdivisions of avionic systems. (1) *Communications* by radio systems which can be in the form of voice or data (analog or digital). Such communications can be between: ground and aircraft, aircraft and aircraft, spacecraft and aircraft, ground and sea. In *direct voice input* (DVI) a pilot can command his aircraft's controls by talking to them. (2) *Navigation*, the means whereby an air- or space-craft's position is determined, is achieved by instruments in the vehicle, on the ground or in space. There are also *collision avoidance systems* which establish the distance, velocity and orientation between an aircraft and its neighbours, and warn of potential collisions with a recommended avoidance procedure. (3) *Flight control* by instruments and/or autopilot. Systems in common use include: electronic flight controls, all-weather landing, electronic engine control, auto-navigation, automatic control systems, management of mechanical systems, fault diagnosis and reporting, fire warning and energy management systems.

Military avionics embodies bomb-sighting systems, weapon control and launch equipment, electronic countermeasures and radar warning devices.

Missile avionics includes sensors for navigation and homing, autoflight control, fusing and arming circuits, and warhead initiation.

See **control**, **navigation sytems**, **radar**, **radio communication**, **stability and control**.

avionics See panel on this page.

avpin, avpol, avtag, avtur See panel on aviation fuels.

AWACS Abbrev. for *Airborne Warning And Control System*.

axial engine Turbine engine with an axial-flow compressor.

axial-flow compressor A compressor in which alternate rows of radially-mounted rotating and fixed aerofoil blades pass the air through an annular passage of decreasing area in an axial direction.

axial-flow turbine Characteristic aero-engine turbine, usually of 1 to 3 rotating stages, in which the gas flow is substantially axial.

axial response The response of a microphone or loudspeaker, measured with the sound-measuring device on the axis of the apparatus being tested.

axis The 3 axes of an aircraft are the straight lines through the centre of gravity about which change of attitude occurs: *longitudinal* or *drag* axis in the plane of symmetry (roll); *normal* or *lift* axis vertically in the plane of symmetry (yaw) and the *lateral* or *pitch* axis transversely (pitch). See **wind axes**.

azimuth The azimuth of a line or celestial body is the angle between the vertical plane containing the line or celestial body and the plane of the meridian, conventionally measured from north through east in astronomical computations, and from south through west in triangulation and precise traverse work.

azimuth marker Line on radar display made to pass through target so that the bearing may be determined.

azimuth stabilized PPI Form of plan position indicator display which is stabilized by a gyrocompass, so that the top of the screen always corresponds to north.

B

[B] A Fraunhofer line in the red of the solar spectrum, due to absorption by the Earth's atmosphere. [B] is actually a close group of lines having a head at a wavelength 686.7457 nm.

Babinet's formula for altitude The altitude in metres is given by

$$\frac{32(500 + t_1 + t_2)(B_1 - B_2)}{B_1 + B_2}$$

t_1 and t_2 are the respective temperatures in degrees Celsius, and B_1 and B_2 the barometric heights at sea level and at the station whose altitude is required.

background A general problem in physical measurements, which limits the ability to detect or accurately measure any given phenomenon. Background consists of extraneous signals arising from any cause which might be confused with the required measurements, e.g. in electrical measurements of nuclear phenomena and of radioactivity, it would include counts emanating from amplifier noise, cosmic rays, insulator leakage etc. Cf. **signal/noise ratio.**

background noise *Ground noise.* Extraneous noise contaminating sound measurements and which cannot be separated from wanted signals. Residual output from microphones, pickups, lines etc., giving a **signal/noise ratio.**

backing The changing of a wind in a counterclockwise direction. Cf. **veering.**

baffle Any device to impede or divide a fluid flow (a) in a tank, to reduce sloshing of liquid propellants, (b) plates fitted between cylinders of air-cooled engines to assist cooling.

baffle Extended surface surrounding a diaphragm of a sound source (loudspeaker) so that an acoustic short-circuit is prevented.

Baily's beads A phenomenon, first observed by Baily in 1836, in which, during the last seconds before a solar eclipse becomes total, the advancing dark limb of the Moon appears to break up into a series of bright points.

balance equation An equation used in meteorology to express the balance between the non-divergent part of the horizontal wind-field and the corresponding field of *geopotential* on a constant pressure surface. It is

$$f\nabla^2\psi + \nabla\psi\cdot\nabla f + 2\left\{\frac{\partial^2\psi}{\partial x^2}\frac{\partial^2\psi}{\partial y^2} - \left(\frac{\partial^2\psi}{\partial x\partial y}\right)^2\right\} = \nabla^2\phi$$

Where Ψ is the stream function, Φ the geopotential, and f the **Coriolis parameter.** Winds derived from the balance equation, which requires numerical solution on a computer, are closer to their actual values than those derived from the **geostrophic approximation,** esp. in regions where the isobars are markedly curved.

balance tab A *tab* whose movement depends upon that of the main control surface. It helps to balance the aerodynamic loads and reduces the *stick forces.* Cf. **servo tab, spring tab, trimming tab.**

ballistic missile See missile.

ball lightning A slowly-moving luminous ball, which is occasionally seen at ground level during a thunderstorn. It appears to measure about 0.5 m in diameter.

ballonnet An air compartment in the envelope of an aerostat, used to adjust changes of volume in the filler gas.

balloon A general term for aircraft supported by buoyancy and not driven mechanically.

balloon barrage An anti-aircraft device consisting of suitably disposed tethered balloons.

Balmer series A group of lines in the hydrogen spectrum named after the discoverer and given by the formula:

$$v = R_H\left(\frac{1}{2^2} - \frac{1}{n^2}\right),$$

where n has various integral values, v is the wave number and R_H is the hydrogen Rydberg number ($=1.096\ 775\ 8\times10^7$ m^{-1}).

BALPA Abbrev. for *British Airline Pilots Association.*

band spectrum Molecular optical spectrum consisting of numerous very closely spaced lines which are spread through a limited band of frequencies.

banking Angular displacement of the wings of an aircraft about the longitudinal axis, to assist turning.

bar Unit of pressure or stress, 1 bar $= 10^5$ N/m^2 or pascals $= 750.07$ mm of mercury at 0°C and lat. 45°. The *millibar* (1 mbar $= 100$ N/m^2 or 10^3 dyn/cm^2) is used for barometric purposes. (NB Std. atmos. pressure $= 1.013\ 25$ bar.) The *hectobar* (1 hbar $= 10^7$ N/m^2, approx. 0.6475 tonf/in^2) is used for some engineering purposes.

Barlow lens A plano-convex lens between the objective and eyepiece of a telescope to increase the magnification by increasing the effective focal length.

Barnard's star A red dwarf star in Ophiuchus, found in 1916 to have the largest proper motion yet measured, amounting to 10 seconds of arc per annum.

baroclinic atmosphere An atmosphere which is not **barotropic.**

barograph A recording barometer, usually of the aneroid type, in which variations of atmospheric pressure cause movement of a pen which traces a line on a clockwork-driven revolving drum.

barometer An instrument used for the measurement of atmospheric pressure. The **mercury barometer** is preferable if the highest accuracy of readings is important, but where compactness has to be considered, the **aneroid barometer** is often used. For *altitudes by barometer*, see **Babinet's formula for altitude.**

barometric corrections Necessary corrections to the readings of a mercury barometer for index error, temperature, latitude and height.

barometric pressure The pressure of the atmosphere as read by a barometer. Expressed in *millibars* (see **bar**), the height of a column of mercury, or (SI) in hectopascals.

barometric tendency The rate of change of atmospheric pressure with time. The change of pressure during the previous three hours.

barostat A device which maintains constant atmospheric pressure in a closed volume, e.g. the input and output pressure of fuel metering device of a gas turbine to compensate for atmospheric pressure variation with altitude.

barotropic atmosphere An atmosphere with zero horizontal temperature gradient at all levels so that the **isopleths** of density and pressure coincide and the **thickness chart** has no pattern.

barrage balloon A small captive kite balloon, the cable of which is intended to destroy low-flying aircraft.

barrier penetration The passage of a sound wave, at an angle for which **Snell's law** predicts zero transmission, through a very thin layer.

baryon A *hadron* with a baryon number of +1. Baryons are involved in strong interactions. Baryons include neutrons, protons and hyperons.

baryon number An intrinsic property of an elementary particle. The baryon number of a baryon is +1, of an antibaryon −1. The baryon number of mesons, leptons, and gauge bosons is zero. Baryon number is conserved in all types of interaction between particles.

Quarks have a baryon number of + 1/3 and antiquarks of − 1/3.

basic six The group of instruments essential for the flight handling of an aircraft and consisting of the *airspeed indicator, vertical speed indicator, altimeter, heading indicator, gyro horizon* and *turn and bank indicator.*

basic T A layout of flight instruments standardized for aircraft instrument panels in which four of the essential instruments are arranged in the form of a T. The pitch and roll attitude display is located at the junction of the T flanked by airspeed on the left and attitude on the right. The vertical bar portion of the T is taken up by directional information.

bass frequency One towards the lower limit of frequency in an audiofrequency signal or a channel for such, e.g. below 250 Hz.

Baumé hydrometer scale The continental Baumé hydrometer has the rational scale proposed by Lunge, in which 0° is the point to which it sinks in water and 10° the point to which it sinks in a 10% solution of sodium chloride, both liquids being at 12.5°C.

Bayard and Alpert gauge One for measuring very low gas pressure by collecting ions on a fine wire inside a helical grid.

B-display Rectangular radar display with target bearing indicated by horizontal coordinate and target distance by the vertical co-ordinate, the targets appearing as bright spots.

beaching gear Floatable, detachable, temporary trolleys which enable a seaplane to be run on and off the shore or slipway.

beacon (1) System of visual lights which indicates fixed features, e.g. masts, reefs. (2) *Radio beacons,* which can be of any frequency but are usually VHF, and can be omni-directional or of directional beam type. *Vertical fan marker* radio beam type beacons are used to identify particular spots in control zones and on approach patterns. A *non-directional beacon* (abbrev. *NDB*) is a transmitter, the bearing of which can only be determined by an aircraft equipped for direction finding. See **instrument landing system.**

beats The subjective difference tone when two sound waves of nearly equal frequencies are simultaneously applied to one ear. It appears as a regular increase and decrease of the combined intensity.

Beaufort notation A code of letters used for indicating the state of the weather; e.g. *b* stands for *blue sky, o* for *overcast, r* for *rain.*

Beaufort scale A numerical scale of wind force, ranging from 0 for winds less than 1 knot to 12 for winds within the limits 110 to 118 knots. Where V is the mean wind speed in miles/hour, and B is the Beaufort wind force, then $V = 1.87\sqrt{AB^3} = (1.52B)^{3/2}$.

becquerel SI unit of radioactivity; 1 becquerel is the activity of a quantity of radioactive material in which 1 nucleus decays per second. Abbrev. *Bq*. Replaces the **curie**. $1\ Bq = 2.7 \times 10^{-11}$ Ci.

belly tank See **ventral tank**.

bending wave Wave observed on thin plates and bars. The motion is perpendicular to the direction of propagation. Important for sound radiation from walls and enclosures.

Bergeron-Findeisen theory That the initiation of precipitation in a cloud consisting mainly of supercooled water droplets is due to the presence of ice crystals which grow at the expense of the droplets because the saturation vapour pressure, with respect to ice, is lower than that with respect to liquid water at the same temperature.

Bergstrom's method A method of assessing the stresses in concrete pavements with particular reference to aerodrome runways and taxiing tracks.

Bernoulli's law For a non-viscous, incompressible fluid in steady flow, the sum of the pressure, potential and kinetic energies per unit volume is constant at any point. It is a fundamental law of fluid mechanics.

beta decay *Beta disintegration*. Radioactive disintegration with the emission of an electron or positron accompanied by an uncharged antineutrino or neutrino. The mass number of the nucleus remains unchanged but the atomic number is increased by one or decreased by one depending on whether an electron or positron is emitted. See **electron capture**.

beta disintegration energy For electron, β^-, emission it is the sum of the energies of the particles, the neutrino and the recoil atom. For positron, β^+, emission there is in addition the energy of the rest masses of two electrons.

beta particle An electron or positron emitted in beta decay from a radioactive isotope. Beta rays are streams of β-*particles*.

Bethe cycle See **carbon cycle**. (H.A. Bethe, American astrophysicist.)

BGA Abbrev. for *British Gliding Association*.

b-group A close group of Fraunhofer lines in the green of the solar spectrum, due to magnesium.

bifilar micrometer An instrument attached to the eyepiece end of a telescope to enable the angular separation and orientation of a visual double star to be measured.

Big Bang See panel on p. 27.

Big Dipper The asterism (chiefly US) in the constellation Ursa Major, known by a great variety of popular names.

binary See **binary star**.

binary star A double star in which the two components revolve about their common centre of mass under the influence of their gravitational attraction. See **eclipsing binary**, **spectroscopic binary**. Also *binary*.

binoculars A pair of telescopes for use with both eyes simultaneously. Essential components are an objective, an eyepiece and some system of prisms to invert and reverse the image.

binomial array A linear array in which the current amplitudes are proportional to the co-efficients of a binomial expansion. Such an array has no side lobes.

bipropellant Rocket propellant made up of two liquids, one being the fuel and the other the oxidizer, which are kept separate prior to combustion.

birefringent filter One based on the polarization of light which enables a narrow spectral band of .1 nm to be isolated, i.e. effectively a **monochromator**; used for photographing solar flares etc.

Bjerknes circulation theorem For relative motion with respect to the Earth, the rate of change $\dfrac{dC}{dt}$ of the circulation C along a closed curve, always consisting of the same fluid particles, is equal to the number N of isobaric−isosteric (i.e. pressure−density) solenoids enclosed by the curve, minus the product of twice the angular velocity of the Earth and the rate of change $\dfrac{dA}{dt}$ of the area A defined by the projection of the curve on the equatorial plane. If $\mathbf{V}$ is the velocity of the fluid, $\mathbf{dl}$ an element of the curve, p the pressure, and ρ the density, then C and N are defined by

$$C = \oint \mathbf{V} \cdot \mathbf{dl}$$

and

$$N = -\oint \frac{dp}{\rho}.$$

black body A body which completely absorbs any heat or light radiation falling upon it. A *black body* maintained at a steady temperature is a full radiator at that temperature, since any black body remains in equilibrium with the radiation reaching and leaving it.

black-body radiation Radiation that would be radiated from an ideal black body. The energy distribution is dependent only on the temperature and is described by **Planck's radiation law**. See **Stefan-Boltzmann law**,

---

Big Bang

The standard Big Bang model for the origin and evolution of the Universe is widely accepted as the most plausible account of the history of the cosmos. Like all modern theories of **cosmology** it is firmly grounded in the **general theory of relativity**, with its account of **gravitation** and the four-dimensional structure of spacetime.

Two observations underpin the Big Bang theory. The first, dating from the 1920s, is the **Hubble law** which showed that galaxies are receding from us with velocities that increase with distance. The second is the **microwave background**, a uniform radio emission with thermal temperature of 2.7 K that covers the whole sky. Both observations are consistent with an explosive and intensely hot origin for the Universe. The galaxies continue to rush away into space, from the explosion, and the background radiation is a fossil relic of the intense heat.

The standard model is extemely detailed and has been investigated thoroughly through the laws of physics. It is much more than a picturesque description. The key feature is that as we extrapolate our understanding of the universe back in time, we encounter an instant in the remote past when all matter and radiation were concentrated in one location. Properly termed a *singularity,* this is popularly known as the *primeval atom*. At the singularity, the density was almost infinite. The expansion event was about 10–20 billion years ago, and the Universe has expanded ever since.

In the earliest phase of the Big Bang, after the first microsecond, most of the energy was in *leptons* (electrons, neutrinos and photons). As the Universe expands and cools, it becomes energetically favourable for electrons to annihilate, and this happens at about 1 second, flooding the Universe with photons. For the next million years, or *radiation era*, the Universe is dominated by photons. In the first few minutes of the radiation era, interactions between protons and neutrons led to the synthesis of helium, which accounts for 25% of the matter in the universe, with 75% remaining as hydrogen. (All other elements were formed much later, in exploding stars.) When 1 million years had elapsed, the temperature had fallen sufficiently to allow electrons to attach themselves to nuclei and form stable atoms. At this *recombination era* the Universe became transparent to radiation for the first time. The background radiation comes from this horizon.

The standard model leaves unanswered questions. It cannot explain why the Universe has the density we observe, does not tell us how galaxies form, and cannot explain the extraordinary homogeneity on the large scale. These difficulties are circumvented in the **inflationary universe** picture developed since about 1980. This remarkable synthesis of elementary particle physics and cosmology pushes the laws of physics from a microsecond back to 10^{-35} seconds and describes a phase change, during which the Universe expanded by a factor of 10^{75} or so. This very successful model implies that we can only ever see a tiny fraction of the physical Universe.

Wien's laws.

black-body temperature The temperature at which a **black body** would emit the same radiation as is emitted by a given radiator at a given temperature. The *black-body temperature* of carbon-arc crater is about 3500°C, whereas its true temperature is about 4000°C.

black box See **flight recorder**.

black frost An air frost with no deposit of

hoar frost.

black hole A region of *spacetime* from which matter and energy cannot escape. A black hole could be a star or galactic nucleus which has collapsed in on itself to the point where its **escape velocity** exceeds the speed of light. Some **binary stars** which strongly emit X-rays may have black hole companions. Black holes are also invoked as the energy sources of **quasars**.

blade activity factor The capacity of a propeller blade for absorbing power, expressed as a non-dimensional function of the surface and expressed by the formula

$$AF = \left[\frac{5}{R}\right]^5 \int_{0 \cdot 2R}^{R} cr^3 dr,$$

where R equals diameter, and c equals blade chord at any radius r.

blade angle The angle between blade chord and plane of rotation at any radius. It is not constant because of the higher airspeed toward the tip, the incidence being progressively reduced to maintain optimum thrust. Change of blade angle from root to tip is called *blade twist*.

blade loading The thrust of a helicopter rotor divided by the total area of the blades.

blade twist See **blade angle**.

blazar A type of extremely luminous extragalactic object, similar to a **quasar** except that the optical spectrum is almost featureless.

bleed The air drawn from a gas turbine compressor (1) to prevent **surging** or (2) to operate some other equipment, e.g. *blown flaps*, reaction controls for VTOL or a cabin conditioning or de-icing system.

BLEU The *Blind Landing Experimental Unit* operated by the Royal Aircraft Establishment which developed a fully automatic blind landing system. See **autothrottle, flight director**.

B-licence Commercial pilot's licence.

blimp Colloq. for *non-rigid airship*.

blind flying, blind landing The flying and landing of an aircraft by a pilot who, because of darkness or poor visibility, must rely on the indication of instruments. See **instrument landing system, ground-controlled approach**.

blind flying instruments A group of instruments, often on an individual central panel, essential for **blind flying**. Commonly airspeed indicator, altimeter, vertical speed, turn-and-slip, artificial horizon, directional gyro. See **basic six, basic T**.

blink comparator An instrument in which two photographic plates of the same region are viewed simultaneously, one with each eye, any difference being detected by a device which alternately conceals each plate in rapid succession.

blinking Modification of a loran transmission, so that a fluctuation in display indicates incorrect operation. See **LORAN**.

blip Spot on CRT screen indicating radar function.

blivet Flexible bag for transporting fuel, often slung beneath a helicopter.

blocking action The effect of a well established, extensive high-pressure area in blocking the passage of a **depression**.

block time The time elapsed from the moment an aircraft starts to leave its loading point to the moment when it comes to rest. It is an important factor in airline organization and scheduling. Also *chock-to-chock, buoy-to-buoy* (sea planes) and *flight time*.

blown flap A **flap**, the efficiency of which is improved by blowing air or other gas over its upper surface to maintain attached airflow even at high angles of deflection.

blue of the sky Sunlight is 'scattered' by molecules of the gases in the atmosphere and by dust particles. Since this scattering is greater for short waves than for long waves, there is a predominance of the shorter waves of visible light (i.e. blue and violet) in the scattered light which we see as the blue of the sky.

BMEWS Abbrev. for *Ballistic Missile Early Warning System*. An *over the horizon* radar system for the detection of intercontinental ballistic missiles, with linked sites in the UK, Alaska and Greenland.

Bode's law A numerical relationship linking the distances of planets from the Sun, discovered by J. Titius (1766) and published by J. Bode (1772). The basis of this relationship is the series 0,3,6,12,...,384, in which successive numbers are obtained by doubling the previous one. If 4 is added to create the new series 4,7,10,16,...388, the resulting numbers correspond reasonably with the planetary distances on a scale with the Earth's distance equal to 10 units. It is believed to be a chance coincidence.

bogie landing gear A main landing gear carrying a pair or pairs of wheels in tandem and pivoted at the end of the shock strut or *oleo*. This arrangement helps to spread the weight of an aircraft over a larger area and also allows the wheel size to be minimized for easier stowage after retraction.

boiling point The temperature at which a liquid boils when exposed to the atmosphere. Since, at the boiling point, the saturated vapour pressure of a liquid equals the pressure of the atmosphere, the boiling point varies with pressure; it is usual, therefore, to state its value at the standard pressure of 101.325 kN/m². Abbrev. *bp*.

Bok globule Small dark nebula (10^3–10^5 AU in diameter) in the Milky Way, thought to be regions of star formation.

bolide A brilliant **meteor**, generally one that explodes: a fireball.

bonding (1) The electrical interconnection

of metallic parts of an aircraft normally at earth potential for the safe distribution of electrical charges and currents. Protects against charges due to precipitation, static, and electrostatic induction due to lightning strikes. Reduces interference and provides a low resistance electrical return path for current in earth-return systems. (2) Joining structural parts by adhesive. May be performed at high temperature and pressure.

boom Enhanced reverberation or resonance in an enclosed space at low frequencies, due to reduced acoustic absorption of the surfaces for low frequencies.

boost control A device regulating reciprocating-engine manifold pressure so that supercharged engines are not over-stressed at low altitude.

boost control over-ride In a supercharged piston aero-engine fitted with **boost control**, a device (sometimes lightly wire-locked so that its emergency use can be detected), which allows the normal maximum manifold pressure to be exceeded. Also *boost control cut-out.*

booster A rocket engine, or cluster of engines, part of a launch system, either the first stage or auxiliary stage, used to provide an initial thrust greater than the total lift-off weight.

booster coil A battery-energized induction coil which provides a starting spark for aero-engines.

booster pump A pump which maintains positive pressure between the fuel tank and the engine, thus intensifying the flow. Any pump to increase the pressure of the liquid in some part of a pipe circuit.

booster rocket See **booster, take-off rocket.**

boost gauge An instrument for measuring the manifold pressure of a supercharged aero-engine in relation to ambient atmosphere or in absolute terms. Also used for racing and other car sports.

bootstrap cold-air unit A unit of the compressor-turbine type in which the air charging an aircraft cabin passes through the compressor and, via an **intercooler**, the turbine.

borsic Boron fibre coated with silicon carbide.

Bouguer law of absorption The intensity *p* of a parallel beam of monochromatic radiation entering an absorbing medium is decreased at a constant rate by each infinitesimally thin layer *db*,

$$\frac{-dp}{p} = k\,db$$

where *k* is a constant that depends on the nature of the medium and on the wavelength.

boundary layer See panel on p. 30.

boundary layer control See panel on p. 30.

boundary layer noise The noise occurring at high speeds due to the oscillations in the turbulent boundary layer at many frequencies and heard in cockpit and cabin.

boundary lights Lights defining the boundary of the landing area.

boundary lubrication A state of partial lubrication which may exist between two surfaces in the absence of a fluid oil film, due to the existence of adsorbed mono-molecular layers of lubricant on the surfaces.

boundary markers See **airport markers.**

Boussinesq approximation An approximation to the equations of motion in which variations of density from the mean state are ignored except when they are multiplied by the acceleration of gravity.

Bowen ratio The ratio of the amount of sensible heat (enthalpy) to latent heat lost by a surface to the atmosphere, by conduction and turbulence.

box baffle Box, with or without apertures and damping, one side fitted with an open diaphragm loudspeaker unit, generally coil-driven.

BPA Abbrev. for *British Parachute Association.*

bracing wires The wires used to brace the wings of biplanes and the earlier monoplanes. See **drag wires, landing wires.**

Brackett series A group of spectral lines of atomic hydrogen in the infrared given by the formula

$$\nu = R_\mathrm{H}\left(\frac{1}{n_1^2} - \frac{1}{n_2^2}\right)$$

in which ν = the wave number; $R_\mathrm{H} = 1.096\ 775\ 8 \times 10^7\ \mathrm{m}^{-1}$; $n_1 = 4$; $n_2 =$ various integral values.

brake parachute One attached to the tail of some high-performance aircraft and streamed as a brake for landing. Sometimes a ribbon canopy is used for greater strength and on large aircraft a cluster of two or three is required to give sufficient area with convenient stowage. Also *landing parachute, parabrake.*

BRASILSAT Brasilian communications satellite system.

brightness See **luminance.** As a quantitative term, *brightness* is deprecated.

Brockenspectre See **spectre of the Brocken.**

Brunt-Väisälä frequency The frequency $N/2\pi$ of small vertical oscillations of a parcel

boundary layer

The thin layer of fluid (air) adjacent to the surface of a body moving through it in which viscous forces exert a major influence on the motion of the fluid. Air particles near a surface encounter molecular forces which cause them to adhere to the surface so that the airspeed is reduced to zero. This creates a shearing stress responsible for skin friction drag; a number of characteristic airflow patterns result, depending on the conditions. See diagram.

Airflow over a flat plate

The two graphs plot air velocity against distance from the surface of the plate.

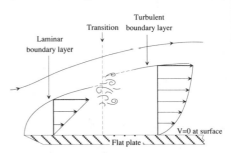

The *laminar boundary layer* is that in which airspeed increases linearly with distance outward from the surface; there are no large eddies present and skin friction is low. It may occur towards the front of an aerofoil if the surface is smooth and the pressure gradient favourable. Such a laminar flow is difficult to maintain and eventually breaks down at the *transition point* with the formation of eddies of many sizes. Downstream from this point the *turbulent boundary layer* occurs in which the skin friction increases by between 3 and 10 times that of the laminar layer. At extremely high altitude (e.g. for re-entering spacecraft) the reduced air viscosity cannot bring the moving air to rest on the surface causing a residual surface velocity and a low skin friction.

There are a number of other associated terms. A *boundary layer bleed* is a means of preventing low-velocity boundary layer air entering a jet engine intake and thus reducing its efficiency. This may take the form of a streamwise plate standing off the surface, the boundary layer then passing away or through a boundary layer duct to a region of favourable pressure. Alternatively a suction slot may be used. *Boundary layer control* is the modification of the airflow in the boundary layer to increase lift and/or decrease drag. Means include: removal of the boundary layer by sucking it through slots or porous surfaces; the use of *vortex generators* to re-energize sluggish surface flow; ejection of high-speed air through slits; blowing, by propulsive efflux, over wing surfaces. A vortex generator or *boundary layer energizer* is a plate mounted at right angles to a surface, inclined at an angle to the flow which creates a vortex in the main flow generally parallel to the surface, so bringing higher velocity air from above the boundary layer towards the surface and thus reducing separation. *Boundary layer heating* occurs at high airspeeds when viscous forces heat the air and the surface. This requires high temperature-resistant metals at supersonic speed and possibly also cooling at hypersonic speeds. *Boundary layer separation* occurs when the turbulent boundary layer thickens as it flows downstream, reducing in speed and increasing in eddy size until it can no longer adhere to the surface and breaks away. This causes drastic loss of lift and/or control.

There are many non-aerospace examples such as the boundary layer over the surface of the Earth which extends tens of metres above it, the increase in wind speed above the sea's surface which must be allowed for in sailing vessels, and that around motor vehicles which causes noise and vibration.

of air about its equilibrium position in a stable atmosphere. N is given by

$$N^2 = \frac{g}{\theta}\frac{\partial \theta}{\partial z}$$

where g is the acceleration due to gravity and $\dfrac{\partial \theta}{\partial z}$ the vertical gradient of the potential temperature.

buffet boundary The limiting values of **Mach number** and altitude at which an aircraft can be flown without experiencing buffet in unaccelerated flight.

buffeting (1) An irregular oscillation of any part of an aircraft, caused and maintained by an eddying wake from some other part; commonly, tail buffeting in the downwash of the main plane, which gives warning of the approach of the **stall**. (2) A very low frequency vibration in the air accompanying a main flow breakaway over the surface of an aircraft.

bulkhead (1) In fuselages, a major structural transverse dividing wall providing access between several internal sections, or a strengthened and sealed wall at the front and rear designed to withstand the differential pressure required for pressurization. (2) In power plant nacelles, a structure serving as a firewall.

bump See **air pocket**.

bunt A manoeuvre in which an aircraft performs half an inverted loop, i.e. the pilot is on the outside where he experiences **negative g**.

buoyancy The vertical thrust on an aircraft due to its immersion, either wholly or partly, in a fluid. Equal to the weight of air displaced by the gas-bags in the case of airship; equal to the weight of water displaced by the immersed portions of the floats of a sea-plane, or the body of a flying-boat. See **reserve buoyancy**.

Buran The name of the Soviet **space shuttle**, which made its first unmanned flight in 1989.

burn Controlled firing of rocket engine for adjusting course, re-entry initiation etc.

burn-out velocity The maximum velocity achieved by a rocket when all the propellant has been consumed.

burst Unusually large pulse arising in an ionization chamber caused by a cosmic-ray shower.

bus Physical means (usually a good electrical conductor) of distributing and/or collecting electrical energy or signals. Sometimes, a universal platform for diverse space experiments and applications.

butterfly diagram A graphical presentation of the occurrence of **sunspots** in the 11-year sunspot cycle.

butterfly tail See **vee-tail**.

Buys Ballot's law If an observer stands with his back to the wind, the lower atmospheric pressure is on his left in the northern hemisphere, on his right in the southern hemisphere.

buzz (1) Severe vibration of a control surface in transonic or supersonic flight caused by separation of the airflow due to compressibility effects. (2) To interfere with an aircraft in flight by flying very close to it.

by-pass ratio The ratio of the by-passed airflow to the combustion airflow in a *dual-flow turbojet* having a single air intake.

by-pass turbojet A turbojet in which part of the compressor delivery is by-passed round the combustion zone and turbine to provide a cool, slow propulsive jet when mixed with the residual efflux from the turbine. See **turbofan**.

C

c The symbol used for the velocity of electromagnetic radiation *in vacuo*. Its value, according to the most accurate recent measurements, is $2.997\,924\,56 \times 10^8$ m s^{-1}.

[C] One of the Fraunhofer lines in the red of the solar spectrum. Its wavelength is 656.3045 nm; it is due to hydrogen.

CAA Abbrev. for **Civil Aviation Authority**, UK.

CAB Abbrev. for *Civil Aeronautics Board*, US.

cabin altitude The nominal pressure altitude maintained in the cabin of a pressurized aircraft.

cabin blower An engine-driven pump, usually of displacement type, for maintaining an aircraft cockpit or cabin above atmospheric pressure. Also *cabin supercharger*.

cabin differential pressure The pressure in excess of that of the surrounding atmosphere which is needed to maintain comfortable conditions at high altitude. For an aircraft flying at 9000 m this differential would be about 60 kN/m^2.

cabin supercharger See **cabin blower**.

cable-angle indicator An indicator showing the vertical angle between the longitudinal axis of a glider and its towing cable, also its yaw and roll attitude relative to the towing aircraft.

cadmium red line Spectrum line formerly chosen as a reproducible standard of length. Wavelength = 643.8496 nm.

calibrated airspeed Indicated airspeed corrected for *position error* and instrument error only. Not to be confused with **equivalent airspeed** or **true airspeed**. Abbrev. *CAS*. Also *rectified airspeed*.

Callisto The fourth natural satellite of **Jupiter**, 4800 km in diameter, discovered by Galileo.

calorie The unit of quantity of heat in the CGS system. The 15°C calorie is the quantity of heat required to raise the temperature of 1 g of pure water by 1°C at 15°C; this equals 4.1855 J. By agreement, the International Table calorie (cal$_{IT}$) equals 4.186 J exactly and the thermochemical calorie equals 4.184 J exactly. There are other designations, e.g. gram calorie, mean calorie, and large or kilocalorie (= 1000 cal, used particularly in nutritional work). The calorie has now been largely replaced by the SI unit of the joule (J).

calorimetry The measurement of thermal constants, such as specific heat, latent heat, or calorific value. Such measurements usually necessitate the determination of a quantity of heat, by observing the rise of temperature it produces in a known quantity of water or other liquid.

Calypso The fourteenth natural satellite of **Saturn**, discovered in 1980 and associated with **Telesto** and **Tethys**.

camber The curvature of an aerofoil, relative to the chord line. Colloq. the curved surface of an aerofoil.

camber flap See **plain flap**.

camouflage Treatment of objects so that there is ineffective reflection of radar waves.

Campbell-Stokes recorder See **sunshine recorder**.

canard See **tail-first aircraft**.

candela Fundamental SI unit of luminous intensity. If, in a given direction, a source emits monochromatic radiation of frequency 540×10^{12} Hz, and the radiant intensity in that direction is 1/683 watt per steradian, then the luminous intensity of the source is 1 candela. Abbrev. *cd*.

cannular combustion chamber A gas turbine combustion system with individual flame tubes inside an annular casing.

canopy (1) The transparent cover of a cockpit. (2) The fabric (nylon, silk or cotton) body of a parachute, which provides high air drag. Usually hemispherical, but may be lobed or rectangular in shape. See also **ribbon parachute**.

canted deck See **angled deck**.

cap (1) The covering of cloud which congregates at the top of a mountain. (2) The transient top of detached clouds above an increasing cumulus. Also *pileus*.

captive balloon A balloon anchored or towed by a line. Usually the term refers to spherical balloons only. Special shapes for stability etc. are called *kite balloons*.

car In an airship, the part intended for the carrying of the load (crew, passengers, goods, engines etc.). It may be suspended below, or may be inside the hull or envelope.

carbon cycle A chain of nuclear fusion reactions, believed to take place in stars more massive than the Sun, the net effect of which is to transmute protons in helium nuclei, the carbon atoms effectively acting as a catalyst. The reaction is strongly dependent on temperature and for this reason is believed to be the main source of energy in hot massive stars. Also *Bethe cycle*.

carbon dating *Radiocarbon dating*. Atmospheric carbon dioxide contains a constant proportion of radioactive ^{14}C, formed by cosmic radiation. Living organisms absorb this isotope in the same proportion. After death it decays with a half-life 5.57×10^3 years. The proportion of ^{12}C to the residual ^{14}C indicates the period elapsed since death.

carbon-nitrogen cycle See **carbon cycle**.

carbon star Rare giant star showing molecular bands of carbon compounds in its spectrum.

cardinal points Name given to the four principal points of the horizon, north, south, east and west, corresponding to **azimuths** 0°, 180°, 90° and 270°. See **quadrantal points**.

cardioid directivity Special shape of a directivity. It is produced by superimposing the fields of a monopole and a dipole, and has the shape of a cardioid.

Carme The eleventh natural satellite of **Jupiter**.

carrier-controlled approach S y s t e m used for landings on aircraft carriers.

carrier noise Noise which has been introduced into the carrier of a transmitter before modulation.

cartridge starter A device for starting aero-engines in which a slow-burning cartridge is used to operate a piston or turbine unit which is geared to the engine shaft.

CAS Abbrev. for *Collision Avoidance System*.

cascades Fixed aerofoil blades which turn the airflow round a bend in a duct, e.g. in wind tunnels or engine intakes.

Cassegrain telescope A form of reflecting telescope in which the rays after reflection at the main mirror fall on a small convex mirror placed inside the prime focus. The rays are thus reflected back through a hole in the main mirror, forming an image beyond it. It is similar to the **Gregorian telescope**.

Cassini's division A dark ring concentric with the ring of Saturn and dividing it into two parts; first observed by G.D.Cassini in 1675.

Cassiopeia A Strongest radio source in the sky, after the Sun, located in the constellation Cassiopeia. About 3 **kiloparsec** away, it is the remnant of a **supernova** seen to explode about 1667.

catadioptric An optical system using a combination of refracting and reflecting surfaces designed to reduce aberrations in a telescope.

catapult An accelerating device for launching an aircraft in a short distance. It may be fixed or rotatable to face the wind. It is usually used on ships which have no landing deck, having been superseded on aircraft

carriers by the **accelerator**. During World War II, fighters were carried on *CAMS* (Catapult Armed Merchant Ships) for defence against long-range bombers. Land catapults have been tried but have been superseded by **RATOG** and **STOL** aircraft.

cat E Category E damage to an aircraft; equivalent to a total loss or 'write-off'.

catoptric element A component of an optical system that uses reflection, not refraction, in the formation of an image.

Cavendish experiment An experiment, carried out by Henry Cavendish in 1798, to determine the constant of *gravitation*. A form of torsion balance was used to measure the very small forces of attraction between lead spheres.

cavitation Generation of cavities (e.g. bubbles) in liquids by rapid pressure changes like those induced by ultrasound. When cavity bubbles implode, they produce shock waves in the liquid. Components can be damaged by cavitation if it is induced by turbulent flow.

CCD See **charge-coupled device**.

CCD array An array of many thousands of photodiodes, whose response to an image focused on the surface of the array can be converted into a video signal by employing CCD electronic circuits. An alternative to vacuum tubes in television cameras.

CCV Abbrev. for **Control-Configured Vehicle**.

C-display Display in which bright spot represents the target, with horizontal and vertical displacements representing the bearing and elevation, respectively.

celestial equator The great circle in which the plane of the Earth's equator cuts the celestial sphere; the primary circle to which the co-ordinates, right ascension and declination are referred.

celestial mechanics The study of the motions of celestial objects in gravitational fields. This subject, founded by Newton, classically deals with satellite and planetary motion within the solar system, using Newtonian gravitational theory, which is much simpler than the **general theory of relativity**.

celestial poles The two points in which the Earth's axis, produced indefinitely, cuts the celestial sphere.

celestial sphere An imaginary sphere, of indeterminate radius, of which the observer is the centre. On the surface of all stars, independent of their real distance, are points specified by two co-ordinates, referred to some chosen great circle of the sphere.

Celsius scale The SI name for **Centigrade**

scale. The original Celsius scale of 1742 was marked zero at the boiling point of water and 100 at the freezing point, the scale being inverted by Strömer in 1750. Temperatures on the International Practical Scale of Temperature are expressed in degrees Celsius. See **Kelvin thermodynamic scale of temperature.**

Centigrade scale The most widely used method of graduating a thermometer. The fundamental interval of temperature between the freezing and boiling points of pure water at normal pressure is divided into 100 equal parts, each of which is a *Centigrade degree*, and the freezing point is made the zero of the scale. To convert a temperature on this scale to the Fahrenheit scale, multiply by 1.8 and add 32; for the Kelvin equivalent add 273.15. See **Celsius scale.**

centre of action A position occupied, more or less permanently, by an anticyclone or a depression, which largely determines the weather conditions over a wide area. The climate of Europe is dependent on the Siberian anticyclone and the Icelandic depression.

centre of mass Centroid of all distributed masses of a flight vehicle. Important design feature for determination of size of wings, layout, stability and payload. Also operationally in loading aircraft to safe limits for flight.

centre of pressure The point at which the resultant of the aerodynamic forces (lift and drag) intersects the chord line of the aerofoil. Its distance behind the leading edge is usually given as a percentage of the chord length. Abbrev. *cp.*

centre section The central portion of a wing, to which the main planes are attached; in large aircraft it is often built into the fuselage and may incorporate the engine nacelles and the main landing gear.

centrifugal-flow compressor A compressor in which a vaned rotor inspires air near the axis and throws it towards a peripheral diffuser, the pressure rising mainly through centrifugal forces. The maximum pressure ratio for a single stage is about four. Centrifugal compressors are universal for aero-engine superchargers and were widely used in the earlier turbojets.

Cepheid parallax See **period-luminosity law.**

Cepheid variable A class of variable star with a period of 1 to 50 days, characterized by precise regularity. There is an exact correlation between the period and **luminosity**: the longer the period, the more luminous the star. This relationship means that the ob-

served period of a Cepheid is an indicator of its distance, and for this reason they have been of immense importance for the determination of the distance scale of the universe. The prototype is beta Cephei.

Cerenkov radiation Radiation emitted when a charged particle travels through a medium at a speed greater than the velocity of light in the medium. This occurs when the refractive index of the medium is high, i.e. much greater than unity, as for water.

cermet Ceramic articles bonded with metal. Composite materials combining the hardness and high temperature characteristics of ceramics with the mechanical properties of metal, e.g. cemented carbides.

Certificates of Airworthiness, Compliance and Maintainance See **airworthiness.**

CFC Abbrev. for **chlorofluorocarbon.**

cg limits The forward and aft positions within which the resultant centre of gravity of an aircraft must lie if balance and control are to be maintained. Cf. **loading and cg diagram.**

chaff Radar reflective strip or particles dispensed from aircraft, missiles or guns to confuse radar. First used in World War II and known as 'Window', some kinds can be cut to desired length in flight to cope with different radar frequencies.

Chandrasekhar limit The upper limit, 1.44 solar masses, for a **white dwarf** star.

characteristic spectrum Ordered arrangement in terms of frequency (or wavelength) of radiation (optical or X-ray) related to the atomic structure of the material giving rise to them.

characteristic velocity The change of velocity and the sum of changes of velocity (accelerations and decelerations) required to execute a space manoeuvre, e.g. transfer between orbits.

charge-coupled device A semiconductor device which relies on the short-term storage of minority carriers in spatially defined depletion zones on its surface. The charges thus stored can be moved about by the application of control voltages via metallic conductors to the storage points, in the manner of a shift register. Increasingly used for the conversion of an optical image into electrical signals as in the videocamera. Abbrev. *CCD.*

Charon Pluto's only known satellite, of uncertain diameter, discovered photographically in 1978.

chart comparison unit Type of display superimposed upon navigational chart. Also *map comparison unit.*

chassis Rigid base on which electronic

units or other electrical components are mounted.

chemical hygrometer See **absorption hygrometer**.

chemosphere The atmospheric layers between 20 and 200 km.

Cherenkov. See **Cerenkov.**

chine The extreme outside longitudinal member of the planing bottom of a flying-boat hull, or of a seaplane float; runs approximately parallel with the keel. Also used on sharp fuselage edge of supersonic aircraft, e.g. SR 71.

chip detector Magnetic device fitted in the lubrication system of aero-engines to collect chips of steel from worn or broken parts.

chirp radar A radar system using linear frequency-modulated pulses.

Chladni figure The visual pattern produced by a fine powder on a vibrating plate. The powder particles accumulate at the nodes where the plate does not move.

chlorofluorocarbons Abbrev. *CFCs*. Compounds consisting of ethane or methane with some or all of the hydrogen replaced by fluorine and chlorine. Used as refrigerants, in fire-extinguishers, as aerosol fluids for insecticides etc. because of their low bp and chemical inertness, and for insulating atmospheres in electrical apparatus because of their high breakdown strengths. Their use is now deprecated because they destroy atmospheric ozone and thus contribute to the **greenhouse effect**. See **atmospheric pollution**.

chock-to-chock See **block time**.

choke In a waveguide, a groove or discontinuity of such a shape and size as to prevent the passage of guided waves within a limited frequency range.

choke flange Type of waveguide coupling which obviates the need for metallic contact between the mating flanges, and yet offers no obstruction to the guided waves. One of the waveguide flanges has a slot formed in it of dimensions which prevent energy leakage within the desired frequency range.

chord line A straight line joining the centres of curvature of the leading and trailing edges of any aerofoil section. The chord (distance) is that between leading and trailing edges measured along this line.

chromatic aberration An enlargement of the focal spot caused (1) in a cathode ray tube, by the differences in the electron velocity distribution through the beam and (2) in an optical lens system using white light, by the refractive index of the glass varying with the wavelength of the light, resulting in coloured fringes surrounding the image.

chromosphere Part of the outer gaseous layers of the Sun, visible as a thin crescent of pinkish light in the few seconds of a total **eclipse** of the Sun. Located above the **photosphere** and below the **corona**, it has a temperature of 4 500 K.

circular velocity The horizontal velocity of a body, required to keep it in a circular orbit, at a given altitude, about a planet. For a near Earth orbit, assuming no air drag, the circular velocity (V_c) is given by:

$$V_c = \sqrt{Rg} = 7.91 \text{ km/sec}$$

where R is the radius of the Earth and g the acceleration due to gravity. Also *orbital velocity*.

circulation Used to describe the lift-producing airflow round an aerofoil, but strictly defined as the integral of the component of the fluid velocity along any closed path with respect to the distance round the path. See **super-circulation**.

circulator Device having a number of terminals, usually three, with internal circuits which ensure that energy entering one terminal flange flows out through the next in a particular direction. They may appear as waveguide, co-axial or stripline components.

circumpolar stars Those stars which, for a given locality on the Earth, do not rise and set but revolve about the elevated celestial pole, always above the horizon. To be circumpolar, a star's declination must exceed the co-latitude of the place in question.

cirrocumulus Thin white patch, *sheet* or *layer* of *cloud* without shading, composed of very small elements in the form of grains, ripples etc., merged or separate, and more or less regularly arranged; most of the elements have an apparent width of less than one degree. Abbrev. *Cc*.

cirrostratus Transparent whitish *cloud veil* of fibrous (hair-like) or smooth appearance, totally or partly covering the sky, and generally producing halo phenomena. Abbrev. *Cs*.

cirrus Detached *clouds* in the form of white, delicate filaments of white or mostly white patches or narrow bands. These clouds have a fibrous (hair-like) appearance, or silky sheen, or both. Abbrev. *Ci*.

Civil Aviation Authority An independent body which controls the technical, economic and safety regulations of British civil aviation. In 1972 it took over the relevant functions of the Department of Trade and Industry, and those of the Air Registration Board and the Air Transport Licensing Board. It is responsible for the civil side of the Joint National Air Traffic Control

Services. Abbrev. *CAA*.

civil twilight The interval of time during which the Sun is between the horizon and 6° below the horizon, morning and evening.

clamshell (1) Cockpit canopy hinged at front and rear. (2) Hinged part of thrust reverser in gas turbine. (3) Hinged door of cargo aircraft.

clangbox Deflector fitted to jet engine to divert gas flow for e.g. V/STOL operation.

classical flutter See **flutter**.

classification of clouds See panel on p. 37.

clean room A special facility for handling material, destined for space activities, in a sterile and dust-free environment.

clear air turbulence Turbulence in the free atmosphere that is not associated with cumulus or cumulonimbus clouds. It occurs mainly in the upper troposphere or lower stratosphere, esp. in the vicinity of jet streams.

clearing manoeuvre Alteration of aircraft attitude to give better view of other air traffic.

clevis joint Fork and tongue joint secured by metal pin as used in joining solid rocket motor cases.

climate The statistical ensemble of atmospheric conditions characteristic of a particular locality over a suitably long period (e.g. 30 years) including relevant parameters such as mean and extreme values, measures of variability, and descriptions of systematic seasonal variations. Aspects considered include temperature, humidity, rainfall, solar radiation, cloud, wind and atmospheric pressure.

climate modelling See panel on p. 39.

climatic change See panel on p. 40.

climatic zones The Earth may be divided into zones, approximating to zones of latitude, such that each zone possesses a distinct type of climate. Eight principal zones may be distinguished: a zone of tropical wet climate near the equator; 2 subtropical zones of steppe and desert climate; 2 zones of temperate rain climate; 1 incomplete zone of boreal climate with a great range of temperature in the northern hemisphere; and 2 polar caps of arctic snow climate.

climatology The study of climate and its causes.

C-line Fraunhofer line in spectrum of Sun at 656.28 nm, arising from ionized hydrogen in its atmosphere.

closed-jet wind tunnel Any wind tunnel in which the working section is enclosed by rigid walls.

cloud A mass of water droplets or ice particles remaining more or less at constant altitude. Cloud is usually formed by condensation brought about by warm moist air which has risen by convection into cooler regions and has been cooled thereby, and by expansion, below its dew point.

cloud and collision warning system A primary radar system with forward scanning which gives a VDU display of dangerous clouds and high ground at ranges sufficient to allow course to be altered for their avoidance. A second sytem is usually necessary to give short-range warning of the presence of other aircraft.

cloudiness The amount of sky covered by cloud irrespective of type. Estimated in eighths (*oktas*). Overcast, 8; cloudless, 0.

cluster variables Short-period variable stars first observed in globular clusters; they are typical members of Population II. The periods are less than one day. See **RR Lyrae variable**.

clutter Unwanted echoes on a radar display, usually due to terrain in the immediate vicinity of the antenna, or to rain or sea.

CNES Abbrev. for *Centre National d'Etudes Spatiales*, the French national space agency.

Coal Sack A large obscuring dust cloud visible to the naked eye as a dark nebula in the Milky Way near the Southern Cross.

co-altitude See **zenith distance**.

Coanda effect Named after its discoverer, it is the tendency of a fluid jet to attach itself to a downstream surface roughly parallel to the jet axis. If this surface curves away from the jet the attached flow will follow it, deflecting from the original direction.

coarse scanning A rapid radar scan carried out to determine approximate location of any target.

coated lens The amount of light reflected from a glass surface can be considerably reduced by coating it with a thin film of transparent material (blooming). The film has a thickness of a quarter of a wavelength and its refractive index is the geometric mean of that of air and glass. Used to increase the light transmission through an optical system by reducing internal reflections, thus also reducing **flare**.

co-axial propellers Two propellers mounted on concentric shafts having independent drives and rotating in opposite directions.

cobalt bomb Theoretical nuclear weapon loaded with ^{59}Co. The long-life radioactive ^{60}Co, formed on explosion, would make the surrounding area uninhabitable. Also radioactive source comprising ^{60}Co in lead shield with shutter.

classification of clouds

International cloud classification is still largely based on the system proposed by Luke Howard, a London pharmacist, in 1803. He divided clouds into three fundamental forms: sheet or layer clouds (*stratus*), heaped-up clouds (*cumulus*) and fibrous or tufty clouds (*cirrus*). Howard also added a form *nimbus*, a rain cloud, but this word is only used now in composite forms, e.g. *nimbostratus* (a raining cloud sheet) or *cumulonimbus* (a cumulus cloud producing a shower of rain or hail). The prefix *alto-* is used to indicate clouds with bases in the middle troposphere, well away from the immediate influence of the ground. It is now known that cirrus clouds are composed of ice particles which accounts for their fibrous appearance, very different from that of clouds formed of water droplets whose edges (if visible) appear hard and sharp. Cirrus normally occurs in the high troposphere.

Clouds can also be classified into *low* (with bases up to 2 km or 7000 ft), *medium* (with bases between 2 km and about 6 km and composed of water droplets) and *high* (base above about 6 km and composed of ice particles). Various Latin words and prefixes may be used to describe details of their structure and appearance, e.g. *lenticularis* (lens-shaped), *cumulogenitus* (arising from the spreading of cumulus), *castellanus* (with a turreted structure) and so on. *Mamma* are udder-shaped protuberances found on the underside of decaying thunder clouds (cumulonimbus).

We thus have the following main types of cloud allowed for in the International Classification, each of which is conventionally given a 'C code' digit:

High clouds	Cirrus (*Ci*), C=0	Feathery clouds in tufts or hooks.
	Cirrocumulus (*Cc*), C=1	Very small elements arranged in rippling lines or as a network, 'mackerel sky'.
	Cirrostratus (*Cs*),C=2	A thin veil of continuous ice cloud, often giving haloes or other optical effects.
Medium clouds	Altocumulus (*Ac*), C=3	White or grey cloudlets, often arranged in rows or lens-shaped particles; often in distinct thin layers.
	Altostratus (*As*), C=4	An extensive, often thick, continuous layer, light or dark grey.
	Nimbostratus, (*Ns*), C=5	A continuous dark grey layer thick enough to hide the Sun completely, often with rain or snow falling from it.
Low clouds	Stratocumulus (*Sc*), C=6	An extensive layer with a gross dappled or undulating pattern, often with small clear patches.
	Stratus (*St*), C=7	A grey featureless layer, usually shallow.
	Cumulus (*Cu*), C=8	Separate heaped-up clouds, sometimes shallow and sometimes towering to a considerable height.
	Cumulonimbus (*Cb*), C=9	Very tall and massive shower or thunderstorm clouds whose tops become glaciated or anvil shaped.

continued on next page

classification of clouds (contd)

All clouds are produced by the cooling of air until it becomes saturated with water vapour which then condenses to form cloud particles: tiny water droplets or ice crystals. This cooling has four main causes:

1. Large-scale ascent of air ahead of and around deepening depressions (cyclones) producing layer clouds at all levels (St, As, Cs, Ns).

2. Local ascent in convection currents over warm surfaces producing heap clouds (Cu, Cb).

3. Transfer of heat from moist warm airstreams to cool surfaces producing widespread low stratus (St).

4. Local ascent caused by flow over irregular terrain producing wave clouds (*stratocumulus lenticularis*, see below) and low stratus (St).

Additionally, internal convection due to radiation effects produces the dappling or banding characteristic of most shallow clouds (Sc, Ac).

Wave clouds are produced by vertical oscillations, often in the mid or upper troposphere, in airstreams flowing over mountain ranges; such *lee waves* can persist over 100 km downwind, with condensation in the wavecrests and evaporation as the air sinks again and warms. Wave clouds do not move with the wind but remain quasi-stationary relative to the ground. Rather similar small clouds can form over a rapidly developing cumulonimbus because of the powerful upward convection current; such clouds are known as *pileus* or cap clouds.

For transmission of weather information over national and international communication channels, low, medium and high cloud formations observed at official stations are each given a code number in the range 1 to 9. For example, 'low cloud, type 8' indicates 'cumulus and stratocumulus present together with bases at different levels', 'medium cloud, type 2' is 'opaque altostratus or nimbostratus' and 'high cloud, type 3' is 'dense cirrus originating from cumulonimbus present'. Full details are given in official publications.

International cloud classification is designed to be used by observers with limited scientific knowledge in order to provide information useful for general forecasting and aircraft safety. It is questionable, however, how far the fully developed listing shown in the International Cloud Atlas with its pseudo-Linnean arrangement of genera, species, varieties and 'complementary features' is of practical or scientific value. A bad feature of the official coding is that wave clouds, although an important independent class comparable to cumulus and stratus, are all lumped togther within 'medium cloud, type 4' which also includes 'altocumulus continually changing shape', the opposite behaviour to that of waveclouds. Although for official reporting it is essential to classify clouds by objective appearance only, for most laymen and many meteorologists the interesting thing is to understand how clouds develop, how they illustrate otherwise invisible physical processes and dramatic small-scale vertical and horizontal motions and how they may give warning of imminent violent phenomena like tornadoes, line squalls and dangerous falls of hail.

cockpit The compartment in which the pilot or pilots of an aircraft are seated. It is so called even where it forms a prolongation of the cabin. Also *flight deck*.

coding Process of subdividing a relatively long pulse of transmitted power into a pre-determined pattern of shorter pulses. The receiver uses **autocorrelation** or **matched filter** techniques to respond only to echoes bearing the transmitted code. This increases the radar's range, resolution and immunity to interference or jamming.

co-efficient of absorption See **absorption co-efficient**.

---

climate modelling

The construction of physical and mathematical models of the Earth's atmosphere, including the influence of seas and oceans, and all relevant aspects of the surface of the land, so that extended **numerical weather predictions** (NWP) may be made over periods of simulated time long enough to accumulate sufficient statistical data to define a *model climate*. Such data cover not only values of temperature, pressure, wind and rainfall at the surface but at all upper levels of the model, together with estimates of mean fluxes of heat, energy and water vapour which may be compared with similar estimates for the real atmosphere. Models used for climate modelling need to take much more accurate account of certain physical processes than do those adequate for predictions only a few days ahead; these processes are mainly concerned with energy transformations due to radiation and the influence on these of cloud amount and type, heat transfer between atmosphere and ocean, and the forming and melting sea-ice and snow-cover over land. Computational methods for solving the equations of the model by numerical approximation must be devised so as to conserve energy and angular momentum very accurately, and to minimize spurious computational smoothing of gradients. Many of the physical processes which need to be considered, e.g. turbulent transfers of heat and moisture at the Earth's surface, or absorption, reflection and transformation of solar and long-wave radiation by clouds, must perforce be *parametrized* as in ordinary NWP, but more extensively and accurately.

The solution of any set of simultaneous hyperbolic partial differential equations, such as a NWP model, depends on both initial and boundary conditions; in *climate modelling*, initial conditions soon become irrelevant and the statistics of interest will depend only on the boundary conditions. A hierarchy of models may be conceived in the simplest of which only the atmosphere is modelled with fixed values of solar constant, sea-surface temperature, soil moisture, surface roughness and orography. In more complex models, interactions at the various interfaces will themselves be modelled so that sea-surface temperature and ice-cover, surface albedo (depending on snow-cover, land use and growth of vegetation), soil moisture etc. will all be derived from the calculations as they proceed. For the longest time-scales, in which climatic changes such as the onset of the next ice-age will be involved, oceanic circulations and changes in continental ice-sheets will have to be included.

---

co-efficient of expansion The fractional expansion (i.e., the expansion of the unit length, area, or volume) per degree rise in temperature. Calling the coefficients of linear, superficial and cubical expansion of a substance α, β, and γ respectively, β is approximately twice, and γ three times, α.

co-efficient of viscosity The value of the tangential force per unit area which is necessary to maintain unit relative velocity between two parallel planes unit distance apart in a fluid; symbol η. That is, if F is the tangential force on the area A and (dv/dz) is the velocity gradient perpendicular to the direction of flow, then $F = \eta A(\delta\varpi/\delta\zeta)$. For normal ranges of temperature, η for a liquid decreases with increase in temperature and is independent of the pressure. Unit of measurement is the *poise*, 10^{-1} $Nm^{-2}s$ in SI units and 1 dyne $cm^{-2}s$ in CGS units.

coelostat An instrument consisting of a mirror (driven by clockwork) rotating about an axis in its own plane, and pointing to the pole of the heavens. It serves to reflect, continuously, the same region of the sky into the field of view of a fixed telescope.

C of A Abbrev. for *Certificate of Airworthiness*.

C of C Abbrev. for *Certificate of compliance*.

C of M Abbrev. for *Certificate of Maintenance*.

coherence If two light waves are superposed so as to produce *interference* effects and there is a constant *phase* relation maintained between them, the waves are said to be coherent. Sources producing coherent light are necessary to produce observable interference effects and such sources can be formed by dividing the wave from one

climatic change

Change of climate on time-scales longer than that on which the usual working definition of 'climate' is based, i.e. 30–50 years. Mean values of atmospheric variables such as temperature and rainfall taken over N years fluctuate less and less as N increases from 1 to 30 or so, and a period of 30 years is usually adequate in most regions to provide rough estimates of extremes likely to be encountered over much longer periods. However, it has become increasingly apparent that 30-year means can fluctuate to a degree greater than is to be expected by pure chance, and that even in historical times some periods of 200–300 years may be significantly colder or warmer than others, the best known of these being the 'Little Ice Age' from about 1550 to 1850 which affected much of Europe and the North Atlantic region.

Before the development of thermometers and rain gauges, evidence for weather and climate can be obtained from historical records of harvests, floods, frosts, snowfalls etc. and for prehistoric times, significant conclusions can often be drawn from *proxy data* including tree-rings, oxygen isotope ratios ($^{18}O/^{16}O$) in layers from borings in polar ice-caps and ocean-bed deposits, analysis of pollen and remains of climate-sensitive Coleoptera (beetles) in deep layers of peat and undisturbed soil, varves (year-layers in lake beds) etc. Dating deposits by radiocarbon methods are useful back to about 40 000 years BP (before present), uranium analysis to 300 000 years BP, while for still earlier periods, records of geomagnetic field reversals provide clues to dating.

On the longest time-scale, there is good geological evidence that during the last 10^9 years ice-age epochs, each several million years long, have occurred at intervals of 200 to 300 million years. The last of these epochs began to affect the northern hemisphere about 3 or 4 million years ago (although the Antarctic has been continuously glaciated for 15 million years), possibly as a result of major changes to the oceanic circulation due to continental drift. Since then, the northern hemisphere seems to have experienced glacial or near-glacial conditions for about 90% of the time but with much warmer interglacials, each lasting about 10 000 years and occurring at roughly 100 000 year intervals; there have also been *interstadials*, which are periods of a a few thousand years when conditions were warmer than those typical of full glaciation but not as consistently warm as in an interglacial.

In Britain, the last ice-age (the Devensian) ended about 10 000 years BP, the previous (Ipswichian) interglacial having ended about 110 000 years BP, with interstadials at 60 000 and 45 000 years BP. (There was a further brief warm spell about 12 000 BP). Between 8000 and 4000 BP the climate in Europe was probably warmer than at present by about 2° and 3°C, while much of what is now the Sahara desert received generous rainfall and was forested, with animals such as hippopotamus and elephant being depicted in cave paintings; these conditions were most likely caused by the persistence of deep ice-sheets over much of North America for much longer than over northern Europe which caused a persistent deep trough of low pressure over the western Atlantic. For the last 3000 years, variations have not been so extreme but have nevertheless been of considerable social and economic significance. The Little Ice Age, already referred to, had very severe effects on areas with marginal climates such as Greenland, Iceland, Scandinavia and Scotland. It also produced, owing to the strengthening of the thermal gradients in the latitudes of the British Isles, many violent storms over England and the Low Countries with massive flooding as well as the notable winters portrayed in much Flemish painting.

continued on next page

---

climatic change (contd)

As regards causes, there is increasing evidence that the sequence of glacial, interstadial and interglacial periods, on a a time scale of 10^4 to 10^5 years, is due in some way not yet fully explained to variations in the parameters of the Earth's orbit (Milankovitch theory), but very possibly involving modifications of the **greenhouse effect** by variations in the absorption of atmospheric CO_2 by phytoplankton in the oceans. Suggested causes for variation on the historical scale are still more speculative, including veiling of the Sun's radiation by volcanic dust, and complex feed-back processes involving atmospheric and oceanic circulations.

There is good evidence derived from pollen analysis and assemblages of Coleoptera that major climatic changes, e.g. the end of the Ipswichian intergla-cial, can take place within two or three centuries, though the development and decay of major continental ice-sheets require several millenia. There is increas-ing speculation at the present time whether human activity associated with rapidly increasing population, industrialization, deforestation and intensive agriculture may affect the climate on a global scale. See **atmospheric pollution.**

---

source into two parts. Coherence can be thought of in terms of both time and space and *lasers* are capable of producing light of great time and spatial coherence.

coherent oscillator One which is stabil-ized by being phase-locked to the transmitter of a radar for beating with the echo, and used with radar-following circuits.

coincidence phenomenon Equality of wavelength of two sound-carrying spheres, e.g. the special form of interaction between the bending wave on a plate and sound waves in the surrounding medium, causing increased sound transmission. See **limiting frequency** (1).

col The region between two centres of high pressure or anticyclones.

co-latitude The complement of the latitude, terrestrial or celestial. On the celestial sphere it is, therefore, the angular distance between the celestial pole and the observer's zenith, and also the meridian altitude of the celestial equator above the observer's horizon.

cold front The leading edge of an advancing mass of cold air, often attended by line-squalls and heavy showers.

cold pool See **thickness chart.**

coleopter Aircraft having an annular wing, the fuselage and engine lying on the centre line. Some French designs had VTOL capa-bility.

collective pitch control A helicopter con-trol by which an equal variation is made in the blades of the rotor(s), independently of their azimuthal position, to give climb and descent.

collimation The process of aligning the various parts of an optical system. (The word

is falsely derived from the L. *collineare,* to bring together in a straight line).

colour excess The amount by which the colour index of a star exceeds the accepted value for its spectral class; used as a measure of absorption of starlight.

colour index The difference between the photographic and the visual magnitudes of a star, from which may be deduced the effec-tive temperature.

colour-luminosity array A variant of the **Hertzsprung-Russell diagram** in which the absolute magnitudes of stars are plotted as a function of their colours.

Columbus The name given the European Space Agency's programme to ensure a permanent presence in low-Earth orbit. It includes the Columbus Attached Labora-tory which is part of the space station, Freedom, ensemble, the Columbus Free-Flying Laboratory and Columbus Polar Platform. The two former are manned sys-tems whereas the latter is unmanned.

colures The great circles passing through (1) the poles of the celestial equator and ecliptic, and through both solstitial points and (2) the poles of the celestial equator and both equinoctial points, these two great cir-cles being the solstitial and equinoctial colures respectively.

coma The visible head of a comet.

combat rating See **power rating.**

combustion chamber The chamber in which combustion occurs: (1) the cylinder of a reciprocating engine; (2) the individual chambers or single annular chamber of a gas turbine; (3) the combustion zone of a ramjet duct; (4) the chamber, with a single venturi

communications satellite

An artificial satellite to aid global communications by the relay of data, voice and television. The satellite can be purely reflective, but usually fulfills a repeater role using an on-board **transponder**. The latter amplifies the signal and changes the frequency to avoid interference between the incoming and outgoing electromagnetic signals. The satellite may orbit the Earth at any altitude but a **geosynchronous** (in this case **geostationary**) orbit is normally used to give continuous coverage over a large area.

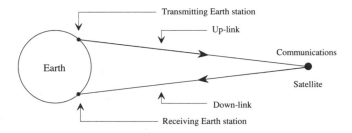

The communications system (see diagram) is made up of a space *segment* consisting of the satellite and its associated on-board equipment and the Earth segment which comprises the ground stations suitably equipped to handle transmission and reception of the signals. The microwave frequencies used are normally in the gigaHertz range (1 gigaHertz = 10^9 cycles/sec.). Large diameter antennas are used to concentrate the signals. Three satellites, suitably placed, in geostationary orbits (about 36 000 km altitude) provide adequate world coverage and almost instantaneous and interference-free communication is possible between any two points on Earth. A direct broadcast satellite (DBS), operating at high power, can provide direct services, such as TV programmes, to geographically remote homes. The domestic reception is usually via a roof-top mounted dish of about 1 m diameter.

The organization INTELSAT provides international public and private communications services by satellite links throughout the world. EUTELSAT fulfills a similar role in Europe.

outlet, of a rocket.

combustion noise Noise caused by combustion. It can be particularly loud if combustion takes place in an acoustic resonator (e.g. industrial burner, jet aircraft) where a feedback between the released heat and the sound waves can lead to **instabilities**.

comet A member of the solar system, of small mass, becoming visible as it approaches the Sun, partly by reflected sunlight, partly by fluorescence excited by the solar radiation. A bright nucleus is often seen, and sometimes a tail. This points away from the Sun, its gases and fine dust being repelled by radiation pressure and the solar wind.

command guidance The guidance of missiles or aircraft by electronic, optical or wire-borne signals from an external source controlled by human operator or automatically.

communications satellite See panel on this page.

comparison spectrum A spectrum formed alongside the spectrum under investigation, for the purpose of measuring the wavelengths of unknown lines. It is desirable that the comparison spectrum should contain many standard lines of known wavelength. The spectrum of the iron arc is often used for this purpose.

compass safe distance The minimum distance at which equipment may safely be positioned from a direct-reading magnetic compass, or detector unit of a remote-indicating compass, without exceeding the values of maximum compass deviation change.

compensated scale barometer See **Kew-pattern barometer**.

complex amplitude Complex number with amplitude and phase information of a harmonic signal, e.g. sound pressure.

compressed-air wind tunnel See **variable-density wind tunnel**.

compressibility drag The sharp increase of drag as airspeed approaches the speed of sound and flow characteristics change from those of a viscous to those of a compressible fluid, causing the generation of shock waves.

compression rib See **rib**.

compressor Compresses the air supply to a **gas turbine**.

compressor drum A cylinder composed of a series of rings or, more usually, disks wherein the blades of an axial compressor are mounted.

Compton wavelength Wavelength associated with the mass of any particle, such that $\lambda = h/mc$ where λ=wavelength, h= Planck's constant, m=rest mass and c=velocity of light.

computational fluid dynamics The calculation of the flow around a surface or in a passage by the solution of mathematical equations over a suitable computing grid and hence, the calculation of surface pressures, temperatures, overall forces and moments.

COMSAT Abbrev. for *Communications Satellite Corporation*, an organization which provides satellite services for international transmission of data.

concave grating A diffraction grating ruled on the surface of a concave spherical mirror, made usually of speculum metal or glass. Such a grating needs no lenses for collimating or focusing the light. Largely on this account it is the most useful means of producing spectra for precise measurement.

concave lens A divergent lens.

concave mirror A curved surface, usually a portion of a sphere, the inner surface of which is polished. It is capable of forming real and virtual images, their positions being given by the equation $2/r = 1/l' + 1/l$, where r is the radius of curvature of the surface, l the distance of the object and l' the distance of the image from the mirror. (Cartesian **convention of signs**.)

condensation The process of forming a liquid from its vapour. When moist air is cooled below its dewpoint, water vapour condenses if there are extended surfaces or nuclei present. These nuclei may be dust particles or ions. Mist, fog and cloud are formed by nuclear condensation.

condensation trails Artificial clouds caused by the passage of an aircraft due either to condensation following the reduction in pressure above the wing surfaces, or to condensation of water vapour contained in the engine exhaust gases. Also *contrails*.

con-di nozzle See **convergent-divergent nozzle**.

conditional instability The condition of the atmosphere when the **temperature lapse rate** lies between the *dry* and the **saturated adiabatic lapse rate** i.e. the atmosphere is stable for unsaturated air but unstable for saturated.

conditional instability of the second kind (CISK) A process whereby low-level **convergence** in the wind field produces convection and cumulus formation thereby releasing latent heat which enhances the convergence and increases convection; this 'positive feed-back loop' may lead to the formation of a large-scale disturbance.

configuration The mode of arrangement of the major elements of an aircraft or spacecraft. See **aircraft design**.

conical camber An expression applied to the adjustment of the **camber** of a wing across the span to meet the variation of the upflow of the air from the fuselage side to the tip. Used on high-speed aircraft, the 'twist' is applied mainly to the leading edge. The name originates from the conic lofting process used in deriving the aerofoil sections.

conical scanning Similar to **lobe switching**, but circular. The direction of maximum response generates a conical surface; used in missile guidance.

coning angle Angle between longitudinal axis of blade of lifting rotor and the tip-path plane in helicopters.

conjunction Term signifying that two heavenly bodies have the same apparent geocentric longitude. Applied to Venus and Mercury it is subdivided into *inferior conjunction* and *superior conjunction*, according as the planet is between the Earth and Sun, or the Sun between the Earth and planet, respectively.

conservation laws In the interaction between particles certain quantities remain the same before and after the event. (1) Dynamical quantities such as mass-energy and momentum are conserved. (2) Intrinsic properties of charge and *baryon number* are conserved in nuclear interactions. In addition, in *strong* and *electromagnetic* interactions between elementary particles, the intrinsic properties of strangeness, charm, topness and bottomness are also conserved but not in *weak* interactions.

conservation of energy The total energy

of an isolated system is constant. Energy may be converted from one form to another, but is not created or destroyed. If the system has only conservative forces, then the total mechanical energy (kinetic and potential) is constant. See **mass-energy equation**.

conservation of momentum The sum of the momenta in a closed system (i.e. one in which no influences act upon it from outside) is constant and is not affected by processes occurring within the system.

conservation of movement of the centre of gravity The state of rest or of motion of the centre of gravity of a system can never be altered by the action of internal forces within the system.

constant-level chart An upper air chart showing isobars for a particular level. Cf. **constant-pressure chart**.

constant-pressure chart An upper air chart showing contours of the geopotential height above sea level at which a particular pressure occurs. Cf. **constant-level chart**. See **thickness chart**.

constant-speed propeller See **propeller**.

constellation A group of stars, not necessarily connected physically, to which have been given a pictorial configuration and a name (generally of Greek mythological origin) which persist in common use although of no scientific significance.

contact flight Navigation of an aircraft by the pilot observing the ground only.

continental climate A type of climate found in continental areas not subject to maritime influences, and characterized by more pronounced extremes between summer and winter; the winters become colder to a greater degree than the summers become hotter; also relatively small rainfall and low humidities.

continuous creation See **steady-state cosmology**.

continuous spectrum One which shows continuous non-discrete changes of intensity with wavelengths or particle energy.

contour fringes Interference fringes formed by the reflection of light from the top and bottom surfaces of a thin film or wedge. The fringes correspond to optical thickness. Also called *Fizeau fringes*. See **Newton's rings**.

contraction ratio The ratio of the maximum cross-sectional area of a wind tunnel to that of the working section.

contrails See **condensation trails**.

control See panel on p. 45.

control column The lever supporting a hand-wheel or hand-grip by which the ailerons and elevator of an aircraft are operated. It may be a simple 'joystick', pivoted at the foot and rocking fore-and-aft and laterally. On military aircraft, usually fighters, it is often hinged halfway up for lateral movement; on transports it is usually either 'spectacle' or 'ram's horn' shape.

control-configured vehicle One designed with artificial stability giving e.g. reduced wing size and control surfaces, enhanced manoeuvrability, reduced gust response and flutter suppression.

controllable-pitch propeller See **propeller**.

controlled air space Areas and lanes clearly defined in 3 dimensions wherein no aircraft may fly unless it is under radio instructions from **air-traffic control**.

control reversal See **reversal of control**.

control zone A volume of controlled air space, precisely defined in plan and altitude, including airports, in which flight rules additional to those in a control area pertain. ICAO defines a specific upper limit.

convective transfer The transfer of energy from one part of a star to another by convection.

convergence Negative divergence.

convergent-divergent nozzle A venturi type of nozzle in which the cross-section first decreases to a throat and then increases to the exit, such a form being necessary for efficient expansion of steam, as in the steam-turbine, or of other vapours or gases, as in the supersonic aero-engine, ramjet or rocket. Abbrev. *con-di nozzle*.

convertiplane A **VTOL** aircraft which can take off and land like a helicopter, but cruises like an aircraft; by swivelling the rotor(s) and/or wing to act as propellers, or by putting the rotor(s) into **autorotation** and using other means for propulsion.

cooling drag That proportion of the total drag due to the flow of cooling air through and round the engine(s).

Copernican System The heliocentric theory of planetary motion; called after Copernicus, who introduced it in 1543. It superseded the geocentric, or Ptolemaic System.

core Gas generator portion of a gas-turbine engine which may be developed as a basis of several engines used on different types of aircraft.

Coriolis acceleration See panel on p. 46.

Coriolis effect See panel on p. 46.

Coriolis force See panel on p. 46.

Coriolis parameter See panel on p. 46.

corner reflector Metal structure of three mutually perpendicular sheets, for returning signals.

---

control

The term *control* is used in several ways in aeronautics and space as described here and referenced to other entries. There are four main uses:

Air-traffic control (*ATC*) is the organized control, by visual and radio means, of the traffic on air routes and into or out of airfields. ATC is divided into general *area* control, including defined airways; *control zones* of specified area and altitude around busy airfields; *approach control* for regulating landing and departing; and *airfield control* for directing aircraft movements on the ground and giving permission for take-off. Air-traffic control operates under two systems: *visual flight rules* and, more severely, *instrument flight rules*. Since World War II, great advances in radar technology have enabled *air-traffic controllers* to be given very complete 'pictures' of the position of aircraft, not only in flight, but also when manoeuvring on the ground. See **air-traffic controller, air-traffic control centre.**

Primary flight controls are described under **stability and control,** and for *guided missiles* in **guided atmospheric flight** where the term *guidance and control* is commonly used. See **autopilot, control column, terrain-avoidance** and **terrain-clearance systems.**

Photogrammetric control points are the fixed points on the surface used in aerial surveying.

For *space systems*, in which aerodynamic forces cannot be used, *attitude control* is effected by spinning the vehicle, exploiting the gradient of gravity and/or firing reaction-control thrusters. With the latter **gyros** and **momentum wheels** can be used between firings. Space vehicle controllers operate intermittently when sensors detect that the orientation has drifted away from a predetermined value. Guidance procedures to adjust a spacecraft's course can be effected by the vehicle's rocket engines. When the desired attitude is attained engine firing is initiated by command.

---

corona A system of coloured rings seen round the Sun or Moon when viewed through very thin cloud. They are caused by diffraction by water droplets. The diameter of the corona is inversely proportional to the size of the droplets.

coronagraph A type of telescope designed by Lyot in 1930 for observing and photographing the solar corona, prominences etc. at any time.

corpuscular radiation A stream of atomic or subatomic particles, which may be charged positively, e.g. α-particles, negatively, e.g. β-particles, or not at all, e.g. neutrons.

corpuscular theory of light The view, held by Newton, that the emission of light consisted of the emission of material particles at very high velocity. Although this theory was discredited by observations of interference and diffraction phenomena, which could only be explained on the wave theory, there has been, to some extent, a return to the corpuscular idea in the conception of the photon.

corridor (1) Safe track for intruding aircraft. (2) Path through atmosphere of re-entering aerospace craft above which there is insufficient air density for lift control, and below which kinetic heating is excessive.

cosmic abundance The relative proportion of each atomic element found in the universe, determined from studies of the solar spectrum and the composition of the Earth, Moon and **meteorites.**

cosmic background radiation See **microwave background.**

cosmic rays See panel on p. 47.

cosmic string Some **Grand Unified Theories** of matter predict the existence of thread-like defects in the structure of space-time. These strings have thickness of 10^{-31} m and mass about 10^7 per light year. They may have played a part in the formation of clusters of galaxies during the very early universe.

cosmogenic Said of an isotope capable of being produced by the interaction of cosmic radiation with the atmosphere or the surface of the Earth.

cosmogony The science of the origins of stars, planets and satellites. It deals with the

---

Coriolis effect and acceleration

In a rotating reference frame, Newton's second law of motion is not valid, but it can be made to apply if, in addition to the real forces acting on a body, a (fictitious) *Coriolis force* and a *centrifugal force* are introduced. The effect of the Coriolis force is to deviate a moving body perpendicular to its velocity. So a body freely falling towards the Earth is slightly deviated from a straight line and will fall to a point east of the point directly below its initial position.

The magnitude of the **Coriolis acceleration** for a particle moving horizontally on the surface of the Earth is $2\Omega V \sin\phi$ where Ω is the angular velocity of rotation of the Earth, V is the speed of the particle relative to the Earth's surface, and ϕ is the latitude. The acceleration is perpendicular to the direction of V and is directed to the right in the northern hemisphere. Coriolis forces explain the directions of the trade winds in equatorial regions and would affect astronauts in an artificial g-environment produced by rotation.

For general three-dimensional motion, the Coriolis acceleration has some other small terms because the Earth's axis does not, normally, lie parallel to the local vertical; for meteorological purposes these additional terms are negligible.

The **Coriolis force** is that which, acting on a given mass, produces the *Coriolis acceleration*.

The **Coriolis parameter,** f, is defined by $f = 2\Omega\sin\phi$ where Ω is the angular velocity of rotation of the Earth, and ϕ is the latitude.

---

genesis of the Galaxy and the solar system.

cosmological principle The postulate, adopted generally in **cosmology**, that the universe is uniform, homogeneous and isotropic, i.e. that it has the same appearance for all observers everywhere in the universe, and there is no preferred position.

cosmological redshift A **redshift** in the spectrum of a **galaxy** that is caused by the recession velocity associated with the expansion of the universe.

cosmology The study of the universe on the largest scales of length and time, particularly the propounding theories concerning its origin, nature, structure and evolution. A cosmology is any model said to represent the observed universe. Western cosmology is entirely scientific in its approach, and has produced two famous models: the **Big Bang** and **steady-state cosmology**.

COSMOS General term applied to Russian satellites used for a variety of missions, e.g. surveillance, atmospheric research, communications, solar wind studies, testing propulsion units and military purposes. COSMOS-1 was launched in March 1962 and now over 2000 objects bearing the name have been put in orbit. The series continues.

coudé telescope An arrangement by which the image in an equatorial telescope is formed, after an extra reflection, at a point on the polar axis. It is then viewed by a fixed eyepiece looking either down or up the polar axis. This type of mounting is much used for high dispersion spectroscopy with modern large telescopes. Also *coudé mounting*. See **astronomical telescope**.

coulomb SI unit of electric charge, defined as that charge which is transported when a current of one ampere flows for one second. Symbol C.

count down A sequence of events in the preparation for the firing of a launch system, denoted by counting time backwards towards zero where zero represents lift-off. The count starts some hours before launch and is finally reckoned in minutes and seconds.

counterglow See **zodiacal light**.

couple A system of two equal but oppositely directed parallel forces. The perpendicular distance between the two forces is called its *arm* and any line perpendicular to the plane of the two forces its *axis*. The *moment* of a couple is the product of the magnitude of one of its two forces and its arm. A couple can be regarded as a single statical element (analogous to force); it is then uniquely specified by a vector along its axis having a magnitude equal to its moment. Couples so specified combine in accordance with the parallelogram law of addition of vectors.

coupled flutter See **flutter**.

course correction The firing or burning of a rocket motor, during a coasting flight, in a

cosmic rays

Cosmic rays were discovered by V.F. Hess in 1912 as natural ionizing radiation detected during a balloon flight. Their astrophysical significance did not become apparent until the 1960s when cosmic ray particle detectors could be flown on orbiting satellites. They are extemely energetic particles moving through the universe at practically the speed of light. Relativistic effects such as **time dilation** are therefore important in the study of cosmic rays. The energies are from 10^8 - 10^{20} electron volts (eV). (10^9 eV is the rest mass energy of the proton; 10^{11} eV is the largest energy attained in particle accelerators.)

Cosmic rays are atomic nuclei accelerated to very high energies. Their chemical composition mirrors the **cosmic abundance** as found in stars like the Sun, although there are small departures at the very highest energies. These observations are very important because cosmic rays are the only particles we can detect that have traversed the **Galaxy**. Indeed, the ones of the very highest energy may well be coming from **quasars** and active galactic nuclei, where they were created by unknown processes, probably explosive in nature, that pose a real challenge to modern astrophysics.

Lower energy cosmic rays (up to 10^{18} eV) are generated by sources within our Galaxy. Almost certainly they originate in **supernova** explosions, remnants like the **Crab Nebula**, and **pulsars**. The energy spectrum of cosmic rays is similar to that of the relativistic electrons that produce **synchroton radiation** emitted in these objects. These cosmic rays are trapped by the magnetic field of the Galaxy, probably for tens of millions of years. As a consequence, the direction of an incoming cosmic ray tells us nothing directly about its origin. **Solar flares** are a source of the lowest energy cosmic rays, which increase in intensity at times of *solar maximum*.

The particles above the atmosphere are known collectively as *primary* cosmic rays. The initial collision produces **pions** which decay rapidly to form gamma rays. These in turn produce electrons and positrons via **pair production**. A single incoming cosmic ray can generate a million secondary particles.

Primary cosmic radiation is studied with **scintillation counters** flown on balloons or spacecraft. The secondary radiation, studied through extensive arrays at ground level, provides most of the knowledge we have about the highest energy cosmic rays.

controlled direction and for a controlled duration to correct an error in course.

cowl flaps See **gills**.

cowling The whole or part of the streamlining covering of any aero-engine; in air-cooled engines, designed to assist cooling airflow.

cp See **centre of pressure**.

CPL Abbrev. for *Commercial Pilot's Licence*.

Crab Nebula An expanding nebulosity in Taurus which represents the remains of the supernova of 1054. It is a powerful source of radio waves and of X-rays. The nebulosity arises from a faint star at the centre, which is a rapid pulsar with a period of 0.033 sec. Both the X-ray and optical radiation show the same pulse, the period of which is slowly increasing.

crack stopper In structural design, a means

of reducing the progression of potential cracks by placing adjacent components across the likely direction of the crack.

crash recorder See **flight recorder**.

crater Circular depression on the surface of a planetary body. Those on Mercury, the Moon, and most of the natural satellites of planets, have been formed by impacts with **meteorites** in the remote past. The Moon, Mars and Io (one of Jupiter's satellites) have volcanic craters also.

crepuscular rays The radiating and coloured rays from the Sun below the horizon, broken up and made apparent by clouds or mountains; also the apparently diverging rays from the Sun passing through irregular spaces between clouds.

crescent wing A sweptback wing in which the angle of sweep and **thickness-chord ratio** are progressively reduced from root to tip so as to maintain an approximately constant **critical Mach number**.

critical Mach number, M$_{CRIT}$ This is the **Mach number** at which the airflow over the aircraft first becomes locally supersonic. It can be as low as M= 0.3 in leading edge slat gaps during high incidence climbing but more usually between M= 0.75 and M= 0.95 for wings of decreasing thickness. It is generally the Mach number above which compressibility effects noticeably affect handling characteristics.

critical speed (1) The speed during take-off at which it has to be abandoned if the aircraft is to stop in the available space. Cf. **accelerate-stop distance**. Also *decision speed*. Abbrev. V_1. (2) Rotational speed at which resonance or whirling may occur.

cross range The distance either side of a nominal re-entry track which may be achieved by using the lifting properties of a re-entering space vehicle.

cruise missile Missile launched from a mobile platform, following a low altitude course and guided by an inertial guidance system which takes account of minute gravitational anomalies over the terrain on the way to the target.

Cuban 8 Aerobatic manoeuvre in a vertical plane consisting of ¾ loop, ½ roll, ¾ loop, and ½ roll.

culmination The highest or lowest altitude attained by a heavenly body as it crosses the meridian. *Upper culmination* indicates its meridian transit above the horizon, *lower culmination* its meridian transit below the horizon or, in the case of a circumpolar, below the elevated pole.

cumulonimbus Heavy and dense *cloud*, with a considerable vertical extent, in the form of a mountain or huge towers. At least part of its upper portion is usually smooth, or fibrous or striated, and nearly always flattened; this part often spreads out in the shape of an anvil or vast plume. Under the base of this cloud, which is often very dark, there are frequently low, ragged clouds either merged with it or not, and precipitation sometimes in the form of **virga**. Abbrev. *Cb*.

cumulus Detached *clouds*, generally dense and with sharp outlines, developing vertically in the form of rising mounds, domes or towers, of which the bulging upper part often resembles a cauliflower. The sunlit part of these clouds are mostly brilliant white; their base are relatively dark and nearly horizontal. Sometimes ragged. Abcumulus are ragged. Abbrev. *Cu*.

curie Unit of radioactivity: 1 curie is defined as 3.700×10^{10} decays per second, roughly equal to the activity of 1 g of ^{226}Ra. Abbrev. *Ci*. Now replaced by the becquerel (*Bq*): 1 Bq = 2.7×10^{-11} Ci.

curve of pursuit That path followed by a combat aircraft steering towards the present position of an adversary.

cusps The horns of the Moon or of an inferior planet in the crescent phase.

CW radar *Continuous Wave radar*. One in which the transmitter emits a continuous radio-frequency signal; the receiving antenna is arranged so that a very small amount of the transmitted power enters it, along with the signal reflected from the target. Movement of the target causes Doppler frequency shift in the reflected signal and this difference can be detected at the output of a mixer. CW radar uses less bandwidth than conventional pulsed radar.

cyclic pitch control Helicopter rotor control in which the blade angle is varied sinusoidally with the blade azimuth position, thereby giving a tilting effect and horizontal translation in any desired direction.

cyclogyro An aircraft lifted and propelled by pivoted blades rotating parallel to roughly horizontal transverse axes.

cyclone (1) Same as **depression**. (2) A **tropical revolving storm** in the Arabian Sea, Bay of Bengal and South Indian Ocean.

cyclostrophic wind The theoretical wind which, when blowing round circular isobars, represents a balance between the pressure gradient and the centrifugal force, the **Coriolis force** being neglected; it is a useful approximation only at low latitudes e.g. in tropical cyclones.

Cygnus A Strongest radio source in Cygnus, identified with a distant peculiar galaxy which is also an X-ray source.

D

[d] A line in the blue of the solar spectrum, having a wavelength of 437.8720 nm due to iron.

[D] A group of 3 Fraunhofer lines in the yellow of the solar spectrum. [D₁] and [D₂], wavelengths 589.6357 and 589.0186 nm, are due to sodium, and [D₃], wavelength 587.5618 nm, to helium.

DALR See **dry adiabatic lapse rate**.

damper Widely used term applied to devices for the suppression of unfavourable characteristics or behaviour; e.g. *blade damper*, to prevent the hunting of a helicopter rotor; *flame damper*, to prevent visual detection at night of the exhaust of a military aircraft; *shimmy damper*, for the suppression of *shimmy*; *yaw damper* suppresses directional oscillations in high-speed aircraft, while a *roll damper* does likewise laterally, in both cases the frequency of the disturbances being too high for the pilot to anticipate and correct manually.

damping The capability of an aircraft of suppressing or resisting harmonic excitation and/or flutter. *Internal damping* is intrinsic to the materials, while *structural damping* is the total effect of the built-up structure. See **resonance test**.

damping Transfer of sound energy into heat. There are different mechanisms; structure-borne sound is damped e.g. by molecular displacement processes, and air-borne sound by friction on interfaces. See **Stokes layer**.

dangerous semicircle The right-hand half of the storm field in the northern hemisphere, the left-hand half in the southern hemisphere, when looking along the path in the direction a **tropical revolving storm** is travelling. Cf. **navigable semicircle**.

dark nebulae Obscuring clouds of dust and gases, common throughout the Milky Way, and also observed in other galaxies. See **Coal Sack**.

DARPA Abbrev. for *Defense Advanced Research Projects Agency*, US government agency. Originally *ARPA*.

data handling The management of the flow of data to-and-from a space vehicle; the on-board subsystem might include data buses, commutators, computers, recorders, multiplexers etc. whereas the ground segment uses equipment like de-multiplexers and display units to interpret the transmitted signal which is sent either directly or via a data relay satellite.

datum *Datum level* or *rigging datum*, is the horizontal plane of reference, in flying attitude, from which all vertical measurements of an aircraft are taken; *cg datum* is the point from which all mass moment arms are measured horizontally when establishing the centre of gravity and loading of an aircraft.

day See **apparent solar-, mean solar-, sidereal-**.

dB Abbrev. for **decibel**.

dBA, dBB, dBC Result of a **sound pressure level** measurement when the signal has been weighted with a frequency response of the A, B or C curve. The dBA curve approximates the human ear and is therefore used most in noise control regulations.

DBS Abbrev. for **Direct Broadcast Satellite**.

dead An enclosure which has a period of reverberation much smaller than usual for its size and audition requirements. Applied to sets in motion-picture production.

dead rise At any cross-section of a flying-boat hull or seaplane float, the vertical distance between the keel and the chine.

dead room See **anechoic room**.

debacle, débâcle The breaking up of the surface ice of great rivers in spring.

Decca Navigator TN for a navigation system of the radio position fixing type using continuous waves. The **fix**, given by the intersection of two hyperbolic position lines, is indicated on meters and can be plotted on a Decca chart or on a flight log which gives a continuous pictorial presentation of position. See **Dectra**.

deceleration Negative acceleration. The rate of diminution of velocity with time. Measured in metres (or feet) per second squared.

decibel 10 times the logarithm to base 10 of an energy ratio, e.g. the sound pressure level is measured in decibels and defined as 10 log p^2/p_0^2, where p is the r.m.s sound pressure and p_0 is a reference sound pressure. For air-borne sound, p_0 is the *threshold of hearing*.

decibel meter Meter which has a scale calibrated in logarithmic steps and labelled with decibel units.

decision speed See **critical speed**.

declination The angular distance of a heavenly body from the celestial equator measured positively northwards along the hour circle passing through the body.

declination circle A graduated circle on the declination axis of an equatorial telescope which enables the telescope to be set to a given declination or the declination of a given star to be read.

Dectra A radio position fixing system, based largely on **Decca Navigator** principles, designed to cover specific air route segments and transoceanic crossings. In addition to fix, location along a track and range information are given: hence *Decca, Track and Range*.

deferent See **epicycle**.

degenerate gas (1) That which is so concentrated, e.g. electrons in the crystal lattice of a conductor, that the Maxwell-Boltzmann law is inapplicable. (2) Gas at very high temperature in which most of the electrons are stripped from the atoms. (3) An electron gas which is far below its Fermi temperature so that a large fraction of the electrons completely fills the lower energy levels and has to be excited out of these levels in order to take part in any physical processes.

de-icing Method of protecting aircraft against icing by removing built-up ice before it assumes dangerous proportions. It may be based on pulsating pneumatic overshoes, chemical applications or intermittent operation of electrical heating elements. Cf. **anti-icing**.

Deimos One of two natural satellites of **Mars**.

delay Time shift which can be introduced into the transmission of a signal, by recording it magnetically on tape, disk or wire, and reproducing it slightly later. Used in public-address systems, to give illusion of distance and to coalesce contributions from original source and reproducers.

delayed drop A live parachute descent in which the parachutist deliberately delays pulling the ripcord until after a descent of several thousand metres.

delayed opening Delaying the opening of a parachute by an automatic device. In any flight above 40 000 ft (12 500 m), low temperature and pressure require that air-crew must reach lower altitude for survival as rapidly as possible, and it is usual to have a barostatic device to delay opening to a predetermined height. This is usually 15 000 ft (4500 m).

delta V The velocity change required to transform a particular space trajectory into another.

delta wing A sweptback wing of substantially triangular planform, the trailing edge forming the base. It is longitudinally stable and does not require an auxiliary balancing aerofoil, although tail or nose planes are sometimes fitted to increase pitch control and trim so that landing flaps can be fitted.

demonstrator A new aircraft, engine or system constructed to prove its novel features prior to embarking on full development.

density The mass of unit volume of a substance, expressed in such units as kg/m^3, g/cm^3 or lb/ft^3.

density of gases According to the **gas laws**, the density of a gas is directly proportional to the pressure and the rel. mol. mass and inversely proportional to the absolute temperature. At standard temperature and pressure the densities of gases range from $0.0899 \ g/dm^3$ for hydrogen to $9.96 \ g/dm^3$ for radon.

departure time The exact time at which an aircraft becomes airborne is an important factor in air-traffic control; estimated time of departure (abbrev. *ETD*).

depression A cyclone. That distribution of atmospheric pressure in which the pressure decreases to a minimum at the centre. In the northern hemisphere, the winds circulate in a counterclockwise direction in such a system; in the southern hemisphere, in a clockwise direction. A depression usually brings stormy unsettled weather.

depression of freezing point A solution freezes at a lower temperature than the pure solvent, the amount of the depression of the freezing point being proportional to the concentration of the solution, provided this is not too great. The depression produced by a 1% solution is called the *specific depression* and is inversely proportional to the molecular weight of the solute. Hence the depression is proportional to the number of moles dissolved in unit weight of the solvent and is independent of the particular solute used.

descending node For Earth, the point at which a satellite crosses the equatorial plane travelling north to south.

de-spun antenna One that turns in the opposite sense to the spinning space vehicle to which it is attached, thereby continuing to point in the required direction, usually an antenna on Earth.

devil A small whirlwind due to strong convection, which, in the tropics, raises dust or sand in a column.

dew The deposit of moisture on exposed surfaces which accumulates during calm, clear nights. The surfaces become cooled by radiation to a temperature below the dewpoint, thus causing condensation from the moist air in contact with them.

DEW line Line of radar missile warning stations along 70th parallel of latitude from Alaska to Greenland. (*Distant Early-Warning line*).

dewpoint The temperature at which a given

sample of moist air will become saturated and deposit dew, if in contact with the ground. Above ground, condensation into water droplets takes place.

dewpoint hygrometer A type of hygrometer for determining the dewpoint, i.e. the temperature of air when completely saturated. The relative humidity of the air can be ascertained by reference to vapour pressure tables.

dew pond A water-tight hollow, usually on elevated land, where the combined effects of rainfall and fog drip exceed that of evaporation. The effect of dew is negligible and the name is the result of an ancient misunderstanding.

DFVLR See **DLR**.

diaphragm A vibrating membrane as in a loudspeaker, telephone and similar sound sources; also in receivers, e.g. the human eardrum.

dichotomy The half-illuminated phase of a planet, as the Moon at the quarters, and Mercury and Venus at greatest elongation.

diffluence The spreading apart of streamlines.

diffraction The phenomenon, observed when waves are obstructed by obstacles or apertures, of the disturbance spreading beyond the limits of the geometrical shadow of the object. The effect is marked when the size of the object is of the same order as the wavelength of the waves, and accounts for the alternately light and dark bands, diffraction fringes, seen at the edge of the shadow when a point source of light is used. It is one factor that determines the propagation of radio waves over the curved surface of the Earth and it also accounts for the audibility of sound around corners.

diffraction angle That between the direction of an incident beam of light, sound or electrons, and any resulting diffracted beam.

diffraction grating One of the most useful optical devices for producing spectra. In one of its forms, the diffraction grating consists of a flat glass plate in the surface of which have been ruled, with a diamond, equidistant parallel straight lines, which may be as close as 1000 per millimetre. If a narrow source of light is viewed through a grating it is seen to be accompanied on each side by one or more spectra. These are produced by diffraction effects from the lines acting as a very large number of equally spaced parallel slits.

diffraction pattern That formed by equal intensity contours as a result of diffraction effects, e.g. in optics or radio transmission.

diffuser A means for converting the kinetic energy of a fluid into pressure energy;

usually it takes the form of a duct which widens gradually in the direction of flow; also fixed vanes forming expanding passages in a compressor delivery to increase the pressure.

diffuser Irregular structure, pyramids and cylinders, to break up sound waves in rooms. See **scatterer**.

diffuse sound Sound which is reflected in all directions inside a volume.

dihedral angle Acute angle at which an aerofoil is inclined to the transverse plane of reference. A downward inclination is *negative dihedral*, sometimes *anhedral*.

Dione The fourth natural satellite of **Saturn**, 1100 km in diameter. It is heavily cratered.

Dione B The twelfth natural satellite of **Saturn**, a tiny object just 20 km in diameter. It is orbitally associated with the much larger **Dione**.

dip The angle measured in a vertical plane between the Earth's magnetic field at any point and the horizontal. Also *inclination*.

dip circle An instrument consisting of a magnetic needle pivoted on a horizontal axis; accurate measurements of magnetic dip can be obtained with it.

dip needle Dipping needle, inclinometer. Magnetic needle on horizontal pivot, which swings in vertical plane. When set in magnetic north-south plane, its inclination shows the *angle of dip*.

dipole Radiator producing a sound field of two adjacent **monopoles** in antiphase. A localized fluctuating force is the prototype of a dipole. The directivity of a dipole has the shape of an eight.

Dirac's constant *Planck's constant* (h) divided by 2π. Usually termed *h-bar*, and written . Unit in which *electron spin* is constant. See **Planck's law**.

direct broadcast satellite *DBS*. A communications satellite, operating at high power, that permits the broadcast of direct TV services to any home possessing a suitable receiver dish and decoder. Olympus satellites can be used for this purpose.

direct injection The injection of metered fuel (for a spark-ignition engine) into the super-charger eye of the cylinders, which eliminates the freezing and poor high-altitude behaviour of carburettors.

direct-injection pump A fuel-metering pump for injecting fuel direct to the individual cylinders.

directional antenna One in which the transmitting and/or receiving properties are concentrated along certain directions, used in space over very long distances.

direct sound The sound intensity arising

from a source to a listener, as contrasted with the reverberant sound which has experienced reflections between the source and the listener.

direct voice input A means by which a pilot can command an aircraft to respond to his spoken instructions for such functions as change of radio frequency, flight performance, and possibly weapon aiming and delivery.

dirigible A navigable balloon or airship.

dish Colloq. for *parabolic reflector*, which may be made of sheet metal or mesh; a form of microwave antenna used for point-to-point and satellite communication and for radio astronomy and satellite broadcast reception.

disk area The area of the circle described by the tips of the blades of a rotorcraft; similarly applied to propellers.

disk brakes Type in which two or more pads close by caliper action on to a disk which is connected rigidly to the landing or car wheel-hub; more efficient than drum type, owing to greater heat dissipation.

disk loading The lift, or upward thrust, of a rotor divided by the disk area.

dispersion The dependence of wave velocity on the frequency of wave motion; a property of the medium in which the wave is propagated. In the visible region of the electromagnetic spectrum, dispersion manifests itself as the variation of refractive index of a substance with wavelength (or colour) of the light. Dispersion enables a prism to form a spectrum.

dispersive power The ratio of the difference in the refractive indices of a medium for the red and violet to the mean refractive index diminished by unity. This may be written

$$v = \frac{n_V - n_R}{n - 1}.$$

displacement The mass of the air displaced by the volume of gas in any lighter-than-air craft, or water by a seaplane hull or float.

disposable load Maximum ramp weight minus *operating weight empty (OWE)*; includes crew, fuel, oil and payload (civil) or armament (military).

dissipation trails Lanes of clear atmosphere formed by the passage of an aircraft through a cloud. Also *distrails*.

distance mark Mark on the screen of a CRT to denote distance of target.

distance-measuring equipment Air borne secondary-radar which indicates distance from a ground transponder beacon. Abbrev. *DME*

distrails See **dissipation trails**.

ditching Emergency alighting of a landplane on water.

diurnal During a day. The term is used in astronomy and meteorology to indicate the variations of an element during an average day.

diurnal libration The name given to the phenomenon by which, owing to the finite dimensions of both the Earth and the Moon, an observer can see rather more than half the Moon's surface when his observations at different times or from different places on the Earth are combined. The effect is one of **parallax**, the term *libration* being a misnomer in this case.

diurnal parallax The change in the apparent position of a celestial object which results from the change in the observer's position caused by the Earth's daily rotation. Geometrically it is the angle subtended at the object by the Earth's radius. It is significant only for members of our solar system.

diurnal range The extent of the changes which occur during a day in a meteorological element such as atmospheric pressure or temperature.

diurnal variation A variation of the Earth's magnetic field as observed at a fixed station, which has a period of approximately 24 hours.

dive A steep descent with or without power. Also *nosedive*.

dive-recovery flap An **air brake** in the form of a flap to reduce the **limiting velocity** of an aircraft.

divergence (1) In *aircraft stability*, a disturbance that increases without oscillation; *lateral divergence* leads to a spin or an accelerating spiral descent; *longitudinal divergence* causes a nosedive or a stall. (2) If, in *meteorology*, the components of the vector wind are u,v,w the divergence is defined as

$$\frac{\partial u}{\partial x} + \frac{\partial v}{\partial y} + \frac{\partial w}{\partial z}.$$

The horizontal divergence is defined as $\frac{\partial u}{\partial x} + \frac{\partial v}{\partial y}$ which is usually almost exactly compensated by $\frac{\partial w}{\partial z}$. Furthermore the integrated divergence throughout a column of the atmosphere is almost zero, i.e. is a small residual of larger positive and negative values; this is known as the *Dines compensation*. Hence strong negative values near the surface are matched by strong positive values at high levels..

divergence speed The lowest equivalent **airspeed** at which **aero-elastic divergence** can occur.

DLR Formerly *DFVLR*. Abbrev. for *Deutsche Forschung und Versuchanstalt für Luft und Raumfahrt*, the German centre for aerospace research.

DME Abbrev. for **Distance-Measuring Equipment**.

Dobson spectrophotometer An instrument used in the routine measurement of atmospheric ozone. It compares, using a photomultiplier and an optical wedge, the intensities of two wavelengths in the solar spectrum in the region of partial ozone absorption (0.30 to 0.33 μm), and from the result the total amount of ozone in a vertical column can be calculated. The instrument may be used to obtain the vertical distribution of ozone from the **umkehr effect**.

docking The physical attaching of one space vehicle to another.

doldrums Regions of calm in equatorial oceans. Towards the solstices, these regions move about 5° from their mean positions, towards the north in June and towards the south in December.

domain wall Some **grand unified theories** of matter predict the existence of wall-like defects in the structure of spacetime. These are major obstacles in certain theories of the expanding universe but the difficulties can be circumvented in the theory of the **inflationary universe**.

doping A chemical treatment with nitrocellulose or cellulose acetate dissolved in thinners, which is applied to fabric coverings, for the purposes of tautening, strengthening, weatherproofing etc.

Doppler broadening Frequency spread of radiation in single spectral lines, because of Maxwell distribution of velocities in the molecular radiators. This also broadens the resonance absorption curve for atoms or molecules excited by incident radiation.

Doppler effect The apparent change of frequency (or wavelength) because of the relative motion of the source of radiation and the observer. For example, the change in frequency of sound heard when a train or aircraft is moving towards or away from an observer. For sound, the observed frequency is related to the true frequency by

$$f_o = \frac{V - V_o(+W)}{V - V_s(+W)} \cdot f_s$$

where V_s, V_o are velocities of source and observer, V is the phase velocity of the wave and W is the velocity of the wind. For electromagnetic waves, the *Lorentz transformation* is used to give

$$f_o = \sqrt{\frac{1 - v/c}{1 + v/c}} \cdot f_s$$

where V is the *relative* velocity of the source and observer and C is the velocity of light. In astronomy the measurement of the frequency shift of light received from distant galaxies, the **redshift**, enables their recession velocities to be found.

Doppler navigator Automated dead reckoning by a device which measures true ground speed by the Doppler frequency shift of radio beam echoes from the ground and computes these with compass readings to give the aircraft's true track and position at any time. Entirely contained in the aircraft, this system cannot be affected by radio interference or hostile jamming.

Doppler radar Any means of detection by reflection of electromagnetic waves, which depends on measuremcent of change of frequency of a signal after reflection by a target having relative motion. See **CW radar**, **pulsed Doppler radar**.

doran Doppler ranging system for tracking missiles (*Doppler range*).

dorsal fins Forward extensions along the top of the fuselage to increase effectiveness of the main fin in sideslip, esp. in **asymmetric flight**.

double-entry compressor A centrifugal compressor with double-sided impeller so that air enters from both sides.

double-row radial engine A **radial engine** where the cylinders are arranged in two planes, and operate on two crank pins, 180° apart.

double stars A pair of stars appearing close together as seen by telescope. They may be at different distances (*optical double*) or physically connected as in a **binary star**.

double-wedge aerofoils See **wedge aerofoil**.

downdraught The downward draught of air occurring with the approach of a thunderstorm and due to evaporative cooling of descending air by heavy rain.

down locks See **up, down, locks**.

downwash The angle through which the airflow is deflected by the passage of an aerofoil measured parallel with the plane of symmetry.

drag Resistance to motion through a fluid. As applied to an aircraft in flight it is the component of the resultant force due to relative airflow measured along the drag axis, i.e. parallel to the direction of motion.

drag axis A line through the centre of mass of an aircraft parallel with the relative air flow, the positive direction being downwind.

drag hinge The pivot of a rotorcraft's blade which allows limited angular displacement in azimuth.

drag struts Structural members designed to brace an aerofoil against air loads in its own plane. Also, landing gear struts resisting the rearward component of impact loads.

drag wires Streamlined wires or cables bracing an aerofoil against drag (rearward) loads. Applicable to biplanes and some early monoplanes.

drift (1) The motion of an aircraft in a horizontal plane due to crosswind. (2) Slow unidirectional error of instrument or gyroscope.

drift angle The angle between the planned course and the track. Also sometimes used for angle between heading and track made good.

drift currents Ocean currents produced by prevailing winds.

drift sight A navigational instrument for measuring **drift angle**.

drogue A sea anchor used on seaplanes and flying-boats; it is a conical canvas sleeve, open at both ends, like a bottomless bucket. Used to check the way of the aircraft.

drogue parachute A small parachute used to (1) slow down a descending aircraft or spacecraft, (2) extract a larger parachute or (3) extract cargo from a hold or wing mounting.

drone Pilotless guided aircraft used as a target or for reconnaissance.

droop snoot Cockpit section hinged on to main fuselage to provide downward visibility at low speeds. Colloq. for *droop nose*.

drop tank A fuel tank designed to be jettisoned in flight. Also *slipper tank*.

drosometer An instrument for measuring the amount of dew deposited.

drought A marked deficiency of rain compared to that usually occurring at the place and season under consideration.

dry adiabatic A curve on an aerological diagram representing the temperature changes of a parcel of dry air subjected to an adiabatic process.

dry adiabatic lapse rate The *temperature lapse rate* of dry air which is subjected to adiabatic ascent or descent. This lapse rate also applies to moist air which remains unsaturated. Its magnitude is 9.76°C per km. Abbrev. *DALR*.

dry mass, weight See panel on **weight and mass**.

Duchemin's formula An expression giving the normal wind pressure on an inclined area in terms of that on a vertical area. It states that:

$$N = F\frac{2\sin\alpha}{1+\sin^2\alpha}$$

where F = pressure of wind in N/m^2 of vertical surface; a= angle of the inclined surface with the horizontal; N = normal pressure in N/m^2 of inclined surface.

ducted cooling A system in which air is constrained in ducts that convert its kinetic energy into pressure for more efficient cooling of an aero-engine or of its radiator.

ducted fan A gas turbine aero-engine in which part of the power developed is harnessed to a fan mounted inside a duct. Also *turbofan*.

duct height Height above the Earth's surface of the lower effective boundary of a tropospheric radio duct.

duct width, thickness Difference in height between the upper and lower boundaries of a tropospheric radio duct.

dump valve (1) An automatic safety valve which drains the fuel manifold of a gas turbine when it stops, or when the fuel pressure fails. (2) A large capacity valve to release residual pressure in any fluid system for emergency or operational reasons after landing, or to release all cabin pressure in an in-flight emergency.

Duperry's lines Lines on a magnetic map showing the direction in which a compass needle points, i.e. the direction of the magnetic meridian.

duplex burner A gas turbine fuel injector with alternative fuel inlets, but a single outlet nozzle.

duplexer (1) In radar, a system which takes advantage of the time delay between transmission of a pulse and reception of its echo to allow the use of the same aerial for transmission and reception. *Transmit-Receive* (TR), or sometimes TR and *Anti-Transmit-Receive* switches are used to isolate the delicate receiver during the high-power pulse transmission. (2) More generally, in radio, any system or network allowing simultaneous transmission and reception on a single aerial, although separation is normally achieved by using different frequencies for transmission and reception.

dust counter An instrument for counting the dust particles in a known volume of air.

Dutch roll Lateral oscillation of an aircraft, in particular an oscillation having a high ratio of rolling to yawing motion. Dutch roll can be countered by yaw dampers.

dwarf star The name given to a low-luminosity star of the **main sequence**. See **Hertzspring-Russell diagram, white dwarf**.

dynamic balance The condition wherein centrifugal forces due to any rotating mass, e.g. a propeller, produce neither couple nor resultant force in the shaft and hence a reduction of vibration and noise. *Dynamic balancing* is the method by which such couples and resultant forces are removed.

dynamic heating Heat generated at the surface of a fast-moving body by the bringing to rest of the air molecules in the boundary layer either by direct impact or by viscosity.

dynamic model A free-flight aircraft model in which the dimensions, inertia and masses are such as to duplicate full-scale behaviour.

dynamic noise suppressor One which automatically reduces the effective audio band-width, depending on the level of the required signal to that of the noise.

dynamic pressure The pressure resulting from the instantaneous arresting of a fluid stream, the difference between total and static pressure.

E

e Symbol for the elementary charge; 1.6022 ×10^{-19}coulomb.

η Symbol for co-efficient of viscosity.

[E] One of the Fraunhofer lines in the green of the solar spectrum. Its wavelength is 526.9723 nm, and it is due to iron.

early-warning radar A system for the detection of approaching aircraft or missiles at greatest possible distances. See **BMEWS**.

Earth See panel on p. 57.

earthing tyres Tyres for aircraft having an electrically conductive surface in order to discharge static electricity upon landing.

Earth observation See panel on p. 58.

earthshine Close to new moon, the entire disk of the Moon is often bathed in a faint light, which is sunlight reflected from the Earth. Also *ashen light* or, picturesquely, *the old moon in the new moon's arms*.

earth thermometer A thermometer used for measuring the temperature of the earth at depths up to a few metres. *Symon's earth thermometer* (the most commonly used) consists of a mercury thermometer, with its bulb embedded in paraffin wax, suspended in a steel tube.

EAS Abbrev. for **Equivalent AirSpeed**.

EBM Abbrev. for *Electron Beam Machining*.

EBW Abbrev. for *Electron Beam Welding*.

ECAC Abbrev. for *European Civil Aviation Conference*.

echelon grating A form of interferometer resembling a flight of glass steps, light travelling through the instrument in a direction parallel to the treads of the steps. The number of interfering beams is therefore equal to the number of steps. Owing to the large path difference, $t(\mu-1)$, where t is the thickness of a step and μ is the **refractive index**, the order of interference and therefore the resolving power are high, making the instrument suitable for studying the fine structure of spectral lines.

echo (1) Return signal in radar, whether from wanted object, or from side or back lobe radiation. Similarly, (2) reflected acoustic wave, which is distinct from a directly received wave because it has travelled a greater distance due to the reflection.

echo chamber Same as **reverberation chamber.**

echo flutter A rapid sequence of reflected radar (or sound) pulses arising from one initial pulse.

echo ranging sonar Determination of distance and direction of objects, such as submarines, by the reception of the reflection of a sound pulse under water. See **asdic.**

eclipse See panel on p. 59.

eclipse year The interval of time between two successive passages of the Sun through the same node of the Moon's orbit; it amounts to 346.620 03 days.

eclipsing binary A binary whose orbital plane lies so nearly in the line of sight that the components pass in front of each other in the course of their mutual revolution.

ecliptic The great circle in which the plane containing the centres of the Earth and Sun cuts the celestial sphere; hence the apparent path of the Sun's annual motion through the fixed stars. See **obliquity of the ecliptic.**

ECM Abbrev. for **Electronic Counter Measures**.

eddy diffusion The transport of quantities such as heat and momentum by eddies in regions of the atmosphere which are in a state of turbulent motion. Eddy diffusion is roughly 10^5 times as effective as molecular diffusion which for meteorological purposes can normally be ignored.

edge See **leading edge, trailing edge.**

edge effect In acoustic absorption measurements, the variations which arise from the size, shape or division of the areas of material being tested in a reverberation room.

E-display Display in which target range and elevation are plotted as horizontal and vertical co-ordinates of the blip.

effective temperature The temperature which a given star would have if it were a perfect radiator, or a **black body**, with the same distribution of energy among the different wavelengths as the star itself.

efflux The mixture of combustion products and cooling air which forms the propulsive medium of any jet or rocket engine.

EGT Abbrev. for *Exhaust Gas Temperature*.

ehp Abbrev. for *total Equivalent brake HorsePower*. Also *tehp*.

Eiffel wind tunnel An open jet, non-return flow wind tunnel.

eigenfrequency Frequency of vibration of a system which vibrates freely. See **acoustic resonance**.

Einstein shift The **redshift** of spectral lines caused when electromagnetic radiation is emitted from an object with a significant gravitational field. Also *gravitational redshift*.

ejection capsule (1) A cockpit, cabin or portion of either, in a high-altitude and/or high-speed military aircraft which can be fired clear in emergency and which, after

Earth

The third planet in order from the Sun with a mean equatorial diameter of 6378.17 km, mass 5.977×10^{24} kg and mean density 5.517. From the astronomical perspective, Earth belongs to the group of terrestrial planets, which also includes **Mercury, Venus** and **Mars**. It is with this group, and also the **Moon**, that its origin, structure and evolution are often compared. Earth has an atmosphere intermediate in thickness between those of Venus and Mars. It is unique in possessing vast oceans of liquid water. The complex interaction between the oceans, the atmosphere and the continental surfaces determines the energy balance, the temperature regime and hence the climate. Cloud cover is typically 50% and heat trapped within the atmosphere by the **greenhouse effect** raises the average temperature by more than 30° C above that expected for the Earth's distance from the Sun.

The present composition of the atmosphere is 77% molecular nitrogen, 21% molecular oxygen, 1% water vapour and 0.9% argon, with the balance in trace components. The high concentration of oxygen, which dates from 2000 million years ago, is a direct result of the presence of plants. The presence of oxygen allowed the formation of the high-level ozone layer, which shields the surface from solar ultraviolet radiation damaging to higher forms of life.

Earth is the most geologically active of the major planets. Its large-scale features have all been determined by the creation, destruction, relative movement and interaction of a dozen or so crustal plates – the *lithosphere* – which slide over the less rigid *asthenosphere* below. Collisions between the plates produce folded mountains, and zones of seismic activity are concentrated along the plate boundaries.

Seismic waves, such as are generated during earthquakes, reveal the internal structure of the Earth. At the centre, there is a molten metallic core of iron and nickel, possibly with a solid core at the very centre at a temperature around 4000°C. The silicate mantle overlies the core. The outermost crust is about 10 km thick under the oceans and 30 km thick where there are continents.

In planetary terms, the surface rocks of the Earth are very young. The basaltic rocks forming the ocean floors are among the youngest. The Precambrian shields, which occupy about 10% of the surface are the oldest and the nearest approximation to the cratered terrain that forms a large part of other planetary surfaces. Weathering and erosion processes mean that few traces of whatever impact craters there were now remain.

The molten metallic core gives rise to the Earth's magnetic field and **magnetosphere**. A layer of electrically charged particles (from the Sun) at a height of 200–300 kilometres forms the *ionosphere*. The funnelling of charged particles by the magnetic field to regions between latitudes of 60° and 75° create the phenomenon of the **aurora**. Satellite measurements have shown that the Earth is also an intense source of radio waves at kilometric wavelengths.

being slowed down, descends by parachute. (2) Container of recording instruments ejected and parachute recovered.

ejector seat A crew seat for high-speed aircraft which can be fired, usually by slow-burning cartridge, clear of the structure in emergency. Automatic releases for the occupant's safety harness and for parachute opening are usually incorporated. Also *ejection seat*.

Ekman spiral The theoretical path traced by the end of the horizontal wind velocity vector, plotted with a fixed origin on a wind **hodograph,** as the wind varies with height up through the **atmospheric boundary layer** on the assumption that density, pressure gradient and eddy viscosity are constant. It shows the approach of the wind velocity from zero near the ground to the *free atmosphere* value at a height of about 1 km and often gives a good approxiamtion to reality. Wind-driven currents on the surface layers

Earth observation

The observation of the Earth's atmosphere and surface from space by means of remote sensing techniques, thereby providing information on factors influencing terrestrial problems. Earth observation (or remote sensing) satellites can provide detailed geological and meteorological data which particularly benefit the petroleum and mining industries, fishing, weather forecasting and climatology, forestry and agriculture. Such diverse factors as the movement of ice floes and locust swarms, the location of mineral deposits, urban development, the detection of forest fires, pollution distribution and sea temperatures can be monitored, leading to a better understanding and management of the Earth's resources.

Remote sensing techniques rely on sensors that detect and measure one or several regions of the electromagnetic spectrum. The radiation received can be direct, reflected or scattered by an intervening medium and can thus provide detailed knowledge both of the source observed and the intervening medium. The sensors may be passive, e.g. a **radiometer**, for general surveillance, or active, e.g. **synthetic aperture radar**, where reflections of a signal generated on-board are received and analysed. Visual photographs of the Earth's surface and of clouds play an important role in Earth observation, but the use of microwaves provides an all-weather and 24-hour capability. In general, low-Earth orbit satellites are employed but meteorological satellites provide continuous observation of the same part of the surface from a **geosynchronous** orbit. Polar orbiters in **Sun-synchronous orbits** are particularly useful because all of the Earth's surface can be viewed eventually, and the satellite can pass over the same place at the same local time and with the same lighting conditions.

It is likely that the need for Earth observations is the greatest single motivation for a country's space ativities. This is particularly true of developing countries and of those with widely dispersed and difficult-to-access resources. Examples of remote sensing satellites are Europe's ERS, the US Landsat, France's SPOT and India's IRS.

of the ocean have a similar variation with depth.

Elara The seventh natural satellite of **Jupiter**.

elasticity of gases If the volume V of a gas is changed by $\delta\varsigma$ when the pressure is changed by $\delta\pi$, the modulus of elasticity is given by

$$-V\frac{\delta p}{\delta V}.$$

This may be shown to be numerically equal to the pressure p for isothermal changes, and equal to $\gamma\pi$ for adiabatic changes, γ being the ratio of the specific heats of the gas.

E-layer Most regular of the ionized regions in the ionosphere, which reflects waves from a transmitter back to Earth. Its effective maximum density increases from zero before dawn to its greatest at noon, and decreases to zero after sunset, at heights varying between 110 and 120 km. There are at least two such layers. Also *Heaviside layer* or *Kennelly-Heaviside layer*.

electrical analogy The correspondence between electrical and acoustical systems, which assists in applying to the latter procedures familiar in the former.

electrical dischargers Devices for discharging static electricity, e.g. earthing tyres, static wick dischargers.

electrical power distribution The provision, conditioning and supply of electrical power to satisfy the needs of a spacecraft and its payload. Continuous sources of power may be the Sun (e.g. **solar cells**, thermal devices) or carried on board (e.g. **fuel cells**, radio-isotopes). During certain mission phases of space-flight, such as launch and re-entry, an auxiliary power source (APU) must be used.

electric propulsion The use of electrostatic or electromagnetic fields to accelerate ions or plasma thereby producing propulsive thrust. See **ion propulsion**.

electric storm A meteorological disturbance in which the air becomes highly charged with static electricity. In the

eclipse

Astronomically an eclipse occurs when a body moves across the line of sight from the observer to a more distant body, thus cutting off some or all of the light from the more distant object.

Solar eclipses occur through a remarkable coincidence: the Sun is 400 times further from the Earth than the Moon and 400 times larger; both subtend an angle of about $\frac{1}{2}°$ in the sky. About once each month, at new Moon, the Moon is close to the apparent position of the Sun. Because the lunar orbit is inclined with respect to the Earth's orbit, the Moon usually swings above or below the Sun. About every 18 months on average, alignment does occur, leading to a solar eclipse. In a *total eclipse* the shadow of the Moon barely reaches the Earth. The eclipse track is no more than 300 km wide but may be many thousands of km long. Each side of totality observers see a *partial eclipse*. At totality, which lasts a maximum of 7 minutes, the solar **chromosphere**, **corona**, **Baily's beads** and **prominences** may be seen. If the Moon is at its **apogee**, the angle is too small to cover the Sun and an *annular eclipse* is seen instead.

Lunar eclipses occur when the Moon enters the Earth's shadow. These are more common and can be seen from anywhere on Earth where the Moon has risen. They are of little scientific value. At totality the Moon is still visible, shining a dull red colour as it reflects sunlight that has been refracted through the Earth's atmosphere.

The pattern of eclipses repeats over the **saros cycle**, an interval of 6585.32 days (18 years), after which the Sun and Moon return to very nearly the same positions in the sky as seen from Earth. The predictive power of the saros has been known since ancient times.

Binary star systems may be eclipsing, if the orbit is parallel to the line of sight. In those cases the total amount of light received from the binary star system will show regular variations. These eclipsing binaries are our main source of data on stellar masses and radii.

---

presence of clouds this leads to thunderstorms.

electro-acoustics The branch of technology dealing with the interchange of electric and acoustic energy, e.g. as in a transducer.

electromagnetic units Any system of units based on assigning an arbitrary value to μ_0, the permeability of free space. $\mu_0 = 4\pi\lambda$ 10^{-7} Hm^{-1} in the SI system; μ_0 is unity in the CGS electromagnetic system.

electromagnetic wave A wave comprising two interdependent mutually perpendicular transverse waves of electric and magnetic fields. The velocity of propagation in free space for all such waves is that of the velocity of light, 2.997 924 58×10^8 ms^{-1}. The electromagnetic spectrum ranges from wavelengths of 10^{-15} m to 10^3 m, i.e. from γ-rays through X-rays, ultraviolet, visible light, infrared, microwave, short-, medium- and long-wave radio waves. Electromagnetic waves undergo reflection and refraction, and exhibit interference and diffraction effects, and can be polarized. The waves can

be channelled by e.g. waveguides for microwaves or fibre optics for light.

electron A fundamental particle with negative electric charge of 1.602×10^{-19} coulombs and mass 9.109×10^{-31} kg. Electrons are a basic constituent of the atom; they are distributed around the nucleus in *shells* and the electronic structure is responsible for the chemical properties of the atom. Electrons also exist independently and are responsible for many electric effects in materials. Due to their small mass, the wave properties and relativistic effects of electrons are marked. The *positron*, the antiparticle of the electron, is an equivalent particle but with a positive charge. Either electrons or positrons may be emitted in β-decay. Electrons, muons and neutrinos form a group of fundamental particles called *leptons*.

electron charge/mass ratio The ratio of the electric charge to mass of an elementary particle. For slow moving electrons $e/m = 1.759×10^{11}$ C kg^{-1}. This value decreases with increasing velocity because of the relativistic increase in mass. Also *specific charge*.

electron density The number of electrons per gram of a material. Approx. 3×10^{23} for most light elements. In an ionized gas the equivalent electron density is the product of the ionic density and the ratio of the mass of an electron to that of a gas ion.

electronic charge The unit in which all nuclear charges are expressed. It is equal to 1.602×10^{-19} coulombs.

electronic countermeasures An offensive or defensive tactic using electronic systems and reflectors to impair the effectiveness of enemy guidance, surveillance or navigational equipment, which depend on electromagnetic signals. Also *electronic warfare, EW*. Abbrev. *ECM*.

electron mass A result of relativity theory, that mass can be ascribed to kinetic energy, is that the effective mass (m) of the electron should vary with its velocity according to the experimentally confirmed expression:

$$m = \frac{m_0}{\sqrt{1-\left(\dfrac{v}{c}\right)^2}},$$

where m_0 is the mass for small velocities, c is the velocity of light, and v that of the electron.

elements of an orbit The 6 data mathematically necessary to determine completely a planet's orbit and its position in it: (1) longitude of the ascending node, (2) inclination of the orbit, (3) longitude of perihelion, (4) semi-axis major, (5) eccentricity, (6) epoch, or date of planet's passing perihelion. Analogous elements are used in satellite and double star orbits.

elevated duct Tropospheric radio duct which has both upper and lower effective boundaries elevated.

elevator An aerodynamic surface, operated by fore-and-aft movement of the pilot's control column, governing motion in pitch.

elevons Hinged control surfaces on the wing trailing edge of tailless or delta aircraft which are moved in unison to act as elevators and differentially as ailerons.

elliptical galaxy Common type of galaxy of symmetrical form but having no spiral arms; the nearer elliptical galaxies have been resolved into stars, but contain no dust or gas. Cf. **spiral galaxy**. See **galaxy**.

elliptical orbit The orbit of a space vehicle about a primary body in the shape of an ellipse. The primary centre of mass is one of the foci and the nearest and farthest points from it are the *pericentre* and *apocentre* respectively.

El Niño, El Niño southern oscillation El Niño – The Child – is the name originally given locally to a weak warm ocean current flowing south along the coast of Ecuador at Christmas time. The El Niño southern oscillation is the term now applied to a more intense, extensive and prolonged warming of the eastern tropical Pacific occurring every few years which is associated with major meteorological anomalies. Extreme cases have serious effects on fisheries, bird life and mainland weather. See **southern oscillation**.

elongation The angular distance between the Moon or planets and the Sun. The planets Mercury and Venus have maximum elongations of about 28° and 48° respectively.

ELV See expendable launch vehicle.

e/m See **electron charge/mass ratio**.

emagram An aerological diagram, the name being derived from *energy per unit mass diagram*. The axes, rectangular or oblique, are temperature and log(pressure).

emersion The exit of the Moon, or other body, from the shadow which causes its eclipse.

emission spectrum Wavelength distribution of electromagnetic radiation emitted by self-luminous source.

emissivity The ratio of emissive power of a surface at a given temperature to that of a black body at the same temperature and with the same surroundings. Values range from 1.0 for lampblack down to 0.02 for polished silver. See **Stefan-Boltzmann law**.

empennage See **tail unit**.

Enceladus The second natural satellite of **Saturn**, 500 km in diameter.

encoding altimeter An **altimeter** designed for automatic reporting of altitude to **air-traffic control**. A special encoding disk within the instrument rotates in response to movement of the aneroid capsules, and transmits a signal which is amplified, fed to the aircraft's transponder and thence automatically to ATC.

end-fire array *End-fire aerial array*. A linear array of radiators in which the maximum radiation is along the axis of the array; the antenna may be uni- or bi-directional. A Yagi array is an example, though in most end-fire arrays, each radiator is fed from a transmission line, the relative phase of each element being determined by its position along the line.

end-plate fins Fins mounted at or near the tips of the tailplane or wing to increase its efficiency.

end speed Naval term for the speed of an aircraft relative to its aircraft carrier at the moment of release from catapult or accelerator.

endurance The maximum time that an aircraft can continue to fly without refuelling, under any agreed conditions.

energy The capacity of a body for doing work. Mechanical energy may be of two kinds: *potential energy*, by virtue of the position of the body, and *kinetic energy*, by virtue of its motion. Energy can take a wide variety of forms. Both *mechanical* and *electrical* energy can be converted into *heat* which is itself a form of energy. Electrical energy can be stored in a capacitor to be recovered on the discharge of the capacitor. *Elastic potential energy* is stored when a body is deformed or changes its configuration, e.g. in a compressed spring. All forms of wave motion have energy; in electromagnetic waves it is stored in the electric and magnetic fields. In any closed system, the total energy is constant – the *conservation of energy*. *Units of energy*: SI unit is the **joule** (symbol J) and is the work done by a force of 1 newton moving through a distance of 1 metre in the direction of the force. The CGS unit, the *erg* is equal to 10^{-7} joules and is the work done by a force of 1 dyne moving through 1 cm in the direction of the force. The foot-pound force (ft-lb f) of the British system equals 1.356 J. See **kinetic energy**, **potential energy**, **mechanical equivalent of heat**.

energy density of sound Sum of potential and kinetic sound energy per unit volume.

energy equivalent sound pressure level Sound pressure level, where the squared sound pressure is averaged over a long time, typically more than 15 minutes. Used to characterize strongly fluctuating sound levels such as those of traffic noise.

energy management Operational technique of minimizing energy loss by advanced automatic flight and engine monitoring and control systems. The actual method takes several forms but includes e.g. the measurement of an individual aircraft and engine performance in flight, and adjustment of height and Mach number to suit the monitored conditions.

engaging speed The relative speed of a carrier-borne aircraft to its ship at the moment when the **arrester gear** is engaged.

engine pod A complete turbojet power unit, including cowlings, supported on a pylon, usually under the wings of an aircraft, an installation method commonly adopted on most types of multi-engined high subsonic speed aircraft.

enstrophy Half the square of the vorticity. It is conserved in two-dimensional, adiabatic, non-dissipative flow.

enthalpy Thermodynamic property of a working substance defined as $H = U + PV$ where U is the internal energy, P the pressure and V the volume of a system. Associated with the study of heat of reaction, heat capacity and flow processes. SI unit is the joule.

entropy In thermal processes, a quantity which measures the extent to which the energy of a system is available for conversion to work. If a system undergoing an infinitesimal reversible change takes in a quantity of heat dQ at absolute temperature T, its entropy is increased by $dS = dQ/T$. The area under the absolute temperature-entropy graph for a reversible process represents the heat transferred in the process. For an adiabatic process, there is no heat transfer and the temperature-entropy graph is a straight line, the entropy remaining constant during the process. When a thermodynamic system is considered on the microscopic scale, equilibrium is associated with the distribution of molecules that has the greatest probability of occurring, i.e. the state with the greatest degree of disorder. *Statistical mechanics* interprets the increase in entropy in a closed system to a maximum at equilibrium as the consequence of the trend from a less probable to a more probable state. Any process in which no change in entropy occurs is said to be *isentropic*.

envelope The gas-bag of a balloon, or of a non-rigid or semi-rigid airship.

environmental control The provision and control of the environment of a space vehicle, or part of it, so that its payload (including man) can operate efficiently. It can involve control of temperature, humidity, atmosphere and contamination.

ephemeris A compilation, published at regular intervals, in which are tabulated the daily positions of the Sun, Moon, planets and certain stars, with other data necessary for the navigator and observational astronomer. See **Astronomical Ephemeris, Nautical Almanac**.

ephemeris time Uniform or Newtonian time, as used in the calculation of future positions of Sun and planets. The normal measurement of time by observations of stars includes the irregularities due to the changes in the Earth's rate of rotation. The difference between ephemeris time and universal time is adjusted to zero at an epoch in 1900; it amounted to about 40 sec in 1970. The *ephemeris second* is the fundamental unit of time adopted by the International Committee of Weights and Measures, its defined value being 1/31 556 925.974 7 of the tropical year for 1900 January, 0 at 12^h ET. Ephemeris time is now replaced by *dynamical time*. See **time**. Abbrev. *ET*.

epicycle The term applied in Ptolemaic or geocentric astronomy to a small circle, described uniformly by the Sun, Moon or planet, the centre of that circle itself describing uniformly a larger circle (the *deferent*), concentric with the Earth.

Epimetheus The eleventh natural satellite of **Saturn**, discovered in 1980.

epoch The precise instant to which the data of an astronomical problem are referred; thus the elements of an orbit when referred to a specific epoch also implicitly define the obliquity of the ecliptic, the rate of precession and other conditions obtaining only at that instant.

equal-signal system One in which two signals are emitted for radio-range, an aircraft receiving equal signals only when on the indicated course.

equation of time The difference between the right ascensions of the true and mean Sun, and hence the difference between apparent and mean time. In the sense mean time minus apparent time, it has a maximum positive value of nearly 14 ½ min in February, and a negative maximum of nearly 16 ½ min in November, and vanishes 4 times a year.

equator See **celestial-**, **terrestrial-**.

equatorial The name given to an astronomical telescope which is so mounted that it revolves about a polar axis parallel to the Earth's axis; when set on a star it will keep that star in the field of view continuously, without adjustment. It has two graduated circles reading *Right Ascension* and *Declination* respectively.

equatorial horizontal parallax See **horizontal parallax**.

equilibrium The state of a body at rest or moving with constant velocity. A body on which forces are acting can be in equilibrium only if the resultant force is zero and the resultant torque is zero.

equinoctial points The two points, diametrically opposite each other, in which the celestial equator is cut by the ecliptic; called respectively the *First Point of Aries* and *First Point of Libra*, from the signs of the Zodiac of which they are the beginning.

equinox (1) Either of the two points on the **celestial sphere** where the **ecliptic** intersects the **celestial equator**. Physically they are the points at which the Sun, in its annual motion,

crosses the celestial equator: the *vernal equinox* as the Sun crosses the south to north, and the *autumnal equinox* as it crosses from north to south. The vernal equinox is the zero point in celestial co-ordinate systems. (2) Either of the two instants of time at which the Sun crosses the celestial equator, being about 21 March and 23 September.

equivalent airspeed Indicated airspeed corrected for position error (angle of incidence) and air compressibility. Abbrev. *EAS*.

equivalent potential temperature The equivalent potential temperature of an air sample is the **equivalent temperature** of the sample when brought adiabatically to a pressure of 1000 mb. It is a conservative property for both dry and saturated adiabatic processes.

equivalent temperature The equivalent temperature of a sample of moist air is the temperature which would be attained by condensing all the water vapour in the sample and using the latent heat thus released to raise the temperature of the sample.

Ertel potential vorticity A rigorous formulation by Ertel of the idea of **potential vorticity** for any compressible, thermodynamically active, inviscid fluid in adiabatic flow. If S is some conservative thermodynamic property of the fluid (e.g. the potential temperature), Ω is the angular velocity of the co-ordinate system, ρ is the density, and V is the velocity of the fluid relative to the co-ordinate system, then the Ertel potential vorticity Π is defined by

$$\Pi = \nabla S \cdot \left(\frac{2\Omega + \text{curl } V}{\rho} \right)$$

Π is a conservative property for all individual fluid particles. Approximations to the Ertel potential vorticity are useful in dynamical studies of the general circulation of the atmosphere and in **numerical forecasting**.

ESA Abbrev. for *European Space Agency*, formed in 1975 combining the activities of the European Space Research Organization (ESRO) and the European Launcher Development Organization (ELDO). ESA, an intergovernmental agency, co-ordinates European space activities and related technologies; in particular, it instigates and manages international space programs on behalf of its 13 member states. These are Austria, Belgium, Denmark, France, West Germany, Ireland, Italy, The Netherlands, Norway, Spain, Sweden, Switzerland and the United Kingdom. Finland is an associate member and Canada has a special relationship with the Agency.

escape velocity V_e. The minimum velocity necessary for an object to travel in a parabolic orbit about a massive primary body, and thus to escape its gravitational attraction. An object which attains this or any greater velocity will coast away from the primary. For the surface of the Earth this velocity is 11.2 km s^{-1}, for the Moon it is 2.4 km s^{-1}, and for the Sun 617.7 km s^{-1}. The formal relation is

$$V_e = \sqrt{2GM/r}$$

where G is the Newtonian constant of gravitation, M is the mass of the object and r its radius. The escape velocity of a **black hole** exceeds the speed of light.

ETA Abbrev. for *Estimated Time of Arrival*, as forecast on a **flight plan**, for a civil aircraft or the time of arrival over a target for a military aircraft.

ETD Abbrev. for *Estimated Time of Departure*. See **departure time**.

ether A hypothetical, non-material entity supposed to fill all space whether 'empty' or occupied by matter. The theory that electromagnetic waves need such a medium for propagation is no longer tenable.

EUMETSAT Abbrev. for *European Meteorological Satellite Organisation*, an intergovernmental agency which provides operational meteorological data for its member states.

EURECA An acronym for *European Retrievable Carrier*, a free-flying platform with a mission duration of six months, launched and recovered by the *Space Shuttle Orbiter*.

eureka See rebecca-eureka.

Europa The second natural satellite of **Jupiter**, discovered by Galileo, and encased in a mantle of ice.

EUTELSAT Abbrev. for *European Telecommunications Satellite Organisation*, a European intergovernmental agency which provides satellite communications for its participating countries.

EVA Abbrev. for *Extra-Vehicular Activity*, i.e. operations performed outside the 'living environment' of a space vehicle. To accomplish EVA, it is necessary to wear a space (or pressure) suit provided with pressure control and life support systems.

evanescent waves Non-decaying surface waves.

evaporative cooling The process of evaporating part of a liquid by supplying the necessary latent heat from the main bulk of liquid which is thus cooled. Used for some piston aero-engines in the 1930s, some current types of turbine and rocket components, and also for cooling purposes in cabin air conditioning systems. Cf. **sweat cooling**.

evaporimeter An instrument used for measuring the rate of natural evaporation.

evection The largest of the four principal periodic inequalities in the mathematical expression of the Moon's orbital motion; due to the variable eccentricity of the Moon's orbit, with a maximum value of 1° 16′ and a period of 31.81 days.

evening star The name given in popular language to a planet, generally Venus or Mercury, seen in the western sky at or just after sunset. Also used loosely to describe any planet which transits before midnight.

event horizon The boundary of a **black hole**; inside this boundary no light can escape.

EW Abbrev. for *Electronic Warfare*. See **electronic countermeasures**.

exhaust cone In a turbojet or turboprop, the duct immediately behind the turbine and leading to the *jet pipe*, consisting of an inner conical unit behind the *turbine disk* and an outer unit of frustum form connecting the *turbine shroud* to the jet pipe.

exhaust-driven supercharger A piston-engine supercharger driven by a turbine motivated by the exhaust gases. Also *turbo-supercharger*.

exhaust stator blades An assembly of stator blades, usually in sections to allow for thermal expansion, mounted behind the turbine to remove residual swirl from the gases.

exhaust velocity The velocity at which a propellant gas leaves a rocket motor. It is related to the *specific impulse*, I_{sp} by the expression: $v_e = I_{sp}g$ where v_e is the exhaust velocity and g the acceleration due to gravity at the Earth's surface.

Exner function If p is the atmospheric pressure, p_0 a reference pressure, and γ is the ratio of the specific heats of a perfect gas, then the Exner function P of p is given by

$$P = (p/p_0)^{(\gamma - 1)/\gamma}.$$

It is useful in studies of compressible adiabatic flow.

exosphere Region of the Earth's atmosphere beyond about 500 km.

expanding universe The view, based on the evidence of the redshift, that the whole universe is expanding; supported by relativity theory, in which a static universe would be unstable.

expansion See **adiabatic change**, **coefficient of expansion**.

expansion of gases All gases have very nearly the same co-efficient of expansion, namely 0.003 66 per kelvin when kept at constant pressure. See **absolute temperature, gas laws**.

expansion ratio The ratio between the gas pressure in a rocket combustion chamber, or a jet pipe, and that at the outlet of the propelling nozzle.

expendable launch vehicle A launch system which is made up of throw-away stages and has no recoverable parts. Abbrev. *ELV*.

experimental mean pitch The distance of travel of a propeller along its own axis, while making one complete revolution, assuming the condition of its giving no thrust.

exploding star See **nova**.

Explorer A series of American artificial satellites used to study the physics of space cosmic rays, temperatures, meteorites etc; responsible for the discovery of the **Van Allen radiation belts**. Explorer I was the first Earth satellite launched by the US (January 1958).

exposure The method by which an instrument is exposed to the elements. The exposure in meteorological stations is standardized so that records from different stations are comparable.

extension flap A landing flap which moves rearward as it is lowered so as to increase the wing area. See **Fowler flap**.

extragalactic nebula A **nebula** external to the Galaxy.

extraterrestrial Used to describe (hypothetical) intelligent life anywhere in the universe.

eye The central calm area of a cyclone or hurricane, which advances as an integral part of the disturbed system.

eyelids Jet engine thrust-reverser nozzle deflectors, shaped and closed like eyelids.

eyepiece In an optical instrument, the lens or lens system to which the observer applies his eye in using the instrument.

F

F Symbol used, following a temperature (e.g. 41°F) to indicate the **Fahrenheit scale**.

F Symbol for faraday.

[F] A Fraunhofer line in the blue of the solar spectrum of wavelength 486.1527 nm. It is the second line in the Balmer hydrogen series, known also as Hβ.

FAA Abbrev. for *Federal Aviation Administration*, a US Government agency responsible for all aspects of US civil aviation. Cf. *CAA*.

Fabry-Pérot inteferometer An instrument in which multiple-beam circular **Haidinger fringes** are produced by the passage of monochromatic light through a pair of plane parallel half-silvered glass plates, one of which is fixed while the other can be moved by an accurately calibrated screw. In transmission, the fringes appear as sharp bright fringes on a dark background. By observing the fringes as the separation of the plates is changed, the wavelength of the light can be determined.

faculae The name given to large bright areas of the photosphere of the Sun. They can be seen most easily near sunspots and at the edge of the Sun's disk; they are at a higher temperature than the average for the Sun's surface.

Fahrenheit scale The method of graduating a thermometer in which freezing point of water is marked 32° and boiling point 212°, the fundamental interval being therefore 180°. Fahrenheit has been largely replaced by the Celsius (Centigrade) and Kelvin scales. To convert °F to °C subtract 32 and multiply by 5/9. For the *Rankine* equivalent add 459.67 to °F; this total multiplied by 5/9 gives the *Kelvin* equivalent.

fail-operational System designed so that it can continue to function after a single failure, warning being indicated.

failure modes and effects analysis The method used in the design of aircraft systems to assess the reliability of all components in all stages of flight, identify inadequate parts or system design, and then rectify by replacement or redesign. Abbrev. *FMEA*.

fairing A secondary structure added to any part of an aircraft to reduce drag by improving the streamlining.

fan Rotating bladed device for moving air in ducts or in e.g. wind tunnels. See **aeroengine**. Cf. **propeller**.

fan marker beacon A form of marker beacon radiating a vertical fan-shaped pattern.

faraday Quantity of electric charge carried by one mole of singly charged ions, i.e. 9.6487×10⁴ coulombs. Symbol *F*.

far field Sound field a long distance from the source. Every sound source has a *far field* and a *near field*, e.g. a monopole source has a far field which decays as $1/r$ (r is the distance from the source) and a near field which decays as $1/r^2$ so that the far field dominates at a large distance.

fast In acoustics, the measuring mode of a **sound-level meter** with a time constant of 0.125 s.

fast Fourier transform See **real time analyser**.

fata morgana A complicated mirage caused by the existence of several layers of varying refractive index, resulting in multiple images, possibly elongated. Especially characteristic of the Strait of Messina and Arctic regions.

F-display Type of radar display, used with directional antenna, in which the target appears as a bright spot which is off-centre when the aim is incorrect.

feathering hinge A pivot for a rotorcraft blade which allows the angle of incidence to change during rotation.

feathering pitch The blade angle of a propeller giving minimum drag when the engine is stopped.

feathering propeller See **propeller**.

feathering pump A pump for supplying the necessary hydraulic pressure to turn the blades of a *feathering propeller* to and from the feather position.

feedback Phenomenon in which part of an output signal is fed back into the input of the system. If the feedback signal is in phase with the primary input signal, the system can become unstable (positive feedback). This often occurs in electro-acoustic systems in which microphone and loudspeaker are in the same room. Negative feedback occurs if the feedback signal decreases the input signal.

Ferrel cell A mid-latitude mean atmospheric circulation cell for weather, proposed by Ferrel in the 19th century, in which air flows poleward and eastward near the surface and equatorward and westward at higher levels. This disagrees with reality. The term is now sometimes used to describe a mid-latitude circulation identifiable in mean meridional wind patterns.

fetch The length of the traverse of an airstream of fairly uniform direction across a sea or ocean area.

fibre-optics gyro Instrument for measuring angular rotation by passing two beams of coherent light in both directions round a closed loop of optical fibre. Rotation affects the phase shift at the output of the two beams. The loops are often of triangular shape, each side being between 20 and 200 mm long. The 'gyro' is fixed to the aircraft, has no rotating parts and is strictly not a gyro, but is so called because it provides equivalent information.

fiducial temperature The temperature at which a sensitive barometer reads correctly, the maker's calibration holding for latitude 45° at the temperature 285 K (12°C) only.

field (1) The interaction between bodies or particles is explained in terms of fields. For example, the *potential energy* of a body may depend on its position and then is represented by a *scalar field* with magnitude only. Other physical quantities carry direction as well as magnitude and they are represented by *vector fields*, e.g. electric, magnetic, gravitational fields. (2) Space in which there are electromagnetic oscillations associated with a radiator; the *induction field* which represents the interchange of energy between the radiator and space is within a few wavelengths of the radiator; *radiation field* represents the energy lost from the radiator to space. Where components radiated by antenna elements are parallel is called the *Fraunhofer region* and, where not, the *Fresnel region*. The latter will exist between the antenna and the Fraunhofer region and is usually taken to extend a distance $2D^2/\lambda$, where λ is the wavelength of the radiation and D is the aerial aperture in a given aspect.

field of force Principle of *action at a distance*, i.e. mechanical forces experienced by an electric charge, a magnet or a mass, at a distance from an independent electric charge, magnet or mass, because of fields established by these and described by uniform laws.

field star An individual star which is not a member of any *cluster* or *association*. Field stars are numerous on all astronomical photographs.

fifth freedom traffic Passengers or freight carried between two countries by an airline of a third country.

fillet A fairing at the intersection of 2 surfaces intended to improve the airflow by reducing breakaway and turbulence.

filtered equations Modified forms of the equations of motion, esp. of derived forms such as the **vorticity equation**, which exclude certain solutions such as fast moving sound and gravity waves that are irrelevant to the types of atmospheric motion producing phenomena of meteorological interest. Such filtering is effected by use of the **hydrostatic approximation**, and the judicious use of the **geostrophic approximation** and the **balance equation**.

filtered model A numerical weather-forecasting model which makes use of the **filtered equations**. Such models use much less computer time than those based on the **primitive equations**.

fin A fixed vertical surface, usually at the tail, which gives directional stability to a fixed wing aircraft in motion and to which a rudder is usually attached. In an airship, any fixed stabilizing surface. See **stabilizer**.

final approach The part of the landing procedure from the time when the aircraft turns into line with the runway until the flare-out is started. Colloq. *finals*.

finder A small auxiliary telescope of low power, fixed parallel to the optical axis of a large telescope for the purpose of finding the required object and setting it in the centre of the field; also used in stellar photography for guiding during an exposure.

fineness ratio The ratio of the length to the maximum diameter of a streamlined body, or flying boat planing bottom.

FIR Abbrev. for **Flight Information Region**.

fireball See **bolide**.

firewall See **bulkhead**.

firing time (1) The interval between applying a d.c. voltage to the trigger electrode of a thyristor or switching tube and the beginning of conduction. (2) In radar, the time required to establish an RF discharge in a switching tube (*transmit-receive* or *anti-transmit-receive*) after the application of RF power.

First Point of Aries The point in which the ecliptic intersects the celestial equator, crossing it from south to north; the origin from which both right ascension and celestial longitude are measured. See **equinoctial points**, **equinox**.

First Point of Libra See **equinoctial points**.

'fir tree' roots A certain type of fixing adopted for turbine blades, the outline form of the root resembling that of a fir tree.

FIS Abbrev. for **Flight Information Service**.

fissile Capable of nuclear fission, i.e. breakdown into lighter elements of certain heavy isotopes (^{232}U, ^{235}U, ^{239}Pu) when these capture neutrons of suitable energy. Also *fissionable*.

FitzGerald-Lorentz contraction The contraction in dimensions (or time scale) of a body moving through the ether with a velocity approaching that of light, relative to the frame of reference (Lorentz frame) from which measurements are made. Also *Lorentz contraction.*

fix The exact geographical position of an aircraft, as determined by terrestrial or celestial observation, or by radio cross-bearing. Cf. **pinpoint.**

fixed-loop aerial A loop aerial, used with a homing receiver, which is fixed in relation to the aircraft's centreline.

fixed-pitch propeller See **propeller.**

fixed points Temperature which can be accurately reproduced and used to define a temperature scale and for the calibration of thermometers. The temperature of pure melting ice and that of steam from pure boiling water at one atmosphere pressure define the Celsius and Fahrenheit scales. The *International Practical Temperature Scale* defined ten fixed points ranging from the triple point of hydrogen (13.81 K) to the freezing point of gold (1337.58 K). See **triple point, Kelvin thermodynamic scale of temperature.**

flame damper See **damper.**

flame trap A gauze or grid of wire, or coiled corrugated sheet, placed in the air intake to a carburettor to prevent the emission of flame from a 'pop-back'.

flame tube The perforated inner tubular 'can' of a gas turbine combustion chamber in which the actual burning occurs. Cf. **combustion chamber.**

flap Any surface attached to the wing, usually to the trailing edge, which can be adjusted in flight, either automatically or through controls, to alter the lift as a whole; primarily on fixed-wing aircraft, but occasionally on rotor systems.

flap angle The angle between the chord of the flap, when lowered or extended, and the wing chord.

flapping angle The angle between the tip-path plane of a helicopter rotor and the plane normal to the hub axis.

flapping hinge The pivot which permits the blade of a helicopter to rise and fall within limits, i.e. variation of zenithal angle in relation to the rotor head.

flap setting The flap angle for a particular condition of flight, e.g. take-off, approach, landing.

flare An energetic outburst in the lower atmosphere of the **Sun.**

flare-out Controlled approach path of aircraft immediately prior to landing.

flaring A term applied to the end of a pipe etc., when it is shaped out so as to be of increasing diameter towards the end.

flash spectrum A phenomenon seen at the first instant of totality in a solar eclipse; the dark lines of the Fraunhofer spectrum formed in the chromosphere flash out into bright emission lines as soon as the central light of the Sun is cut off.

flat four Four cylinder, horizontally opposed piston engine.

flat random noise See **white noise.**

F-layer Upper ionized layer in the ionosphere resulting from ultraviolet radiation from the Sun and capable of reflecting radio waves back to Earth at frequencies up to 50 MHz. At a regular height of 300 km during the night, it falls to about 200 km during the day. During some seasons, this remains as the F_1 layer while an extra F_2 layer rises to a maximum of 400 km at noon. Considerable variations are possible during particle bombardment from the Sun, the layer rising to great heights or vanishing. Also *Appleton layer.*

Fletcher-Munson curves Equal-loudness curves for aural perception, measured just outside the ear, extending from 20 to 20 000 Hz, and from the threshold of hearing to the threshold of pain. They are the basis of the **phon** scale.

flex-wing A collapsible single surface fabric wing of delta planform investigated first for the return of space vehicles as gliders; later applied to army low-performance tactical aircraft of collapsible type; now for **microlights** and hang gliders.

flick roll See **roll.**

flight control Control of vehicle, e.g. spacecraft, missile or module, so that it attains its target, taking all conditions and corrections into account. Generally done by computer, controlled by signals representing actual and intended path.

flight deck (1) Upper part of an aircraft carrier. (2) Crew compartment of a large aircraft.

flight director An aircraft instrument (i.e. blind) flying system in which the dials indicate what the pilot must do to achieve the correct flight path as well as the actual attitude of the aircraft. The dial display is usually a **gyro horizon** with a spot or pointer which must be centred. The equipment can be coupled to a radio- or **gyro-compass** to bring the aircraft on to a desired heading, preset on the compass, or it can be coupled to the **instrument landing system** to receive these signals. In all cases, the flight director can be made automatic by switching its signals into the autopilot.

flight engineer A member of the flying crew of an aircraft responsible for engineering duties i.e. management of the engines, fuel consumption, power systems etc.

flight envelope Plot of altitude versus speed defining performance limits within which an aircraft and/or its equipment can operate.

flight fine pitch The minimum blade angle, held by a removable stop, which the propeller of a turboprop engine can reach while in the air, and which provides braking drag for the landing approach.

flight information A *flight-information centre* provides a *flight-information service* of weather and navigational information within a specified **flight-information region**.

flight-information region An airspace of defined dimensions within which information on air-traffic flow is provided according to the types of airspace therein. Abbrev. *FIR*.

flight-information service One giving advice and information to assist in the safe, efficient conduct of flights. Abbrev. *FIS*.

flight level **Air-traffic control** instructions specify heights at which controlled aircraft must fly and these are given in units of 100 ft (30 m) for altitudes of 3000 or 5000 ft (900 or 1500 m) and above.

flight Mach number Ratio of true air speed of an aircraft to speed of sound under identical atmospheric conditions.

flight management system Computer controlled automatic flight control system allowing the pilot to select specific modes of operation. These could include: standard instrument departure; autothrottle; standard terminal arrival system; Mach hold.

flight path The path in space of the centre of mass of an aircraft or projectile. (Its *track* is the horizontal projection of this path).

flight plan A legal document filed with *air-traffic control* by a pilot before or during a flight (by radio), which states his destination, proposed course, altitude, speed, ETA and alternative airfield(s) in the event of bad weather, fuel shortage etc.

flight recorder A device which records data on the functioning of an aircraft and its systems on tape or wire. (1) The *Flight Data Recorder* (*FDR*) should be in a crashproof, floatable box which may be ejected in case of an accident, and is usually fitted with a homing radio beacon and flashlight. (2) The *Cockpit Voice Recorder* (*CVR*) stores all speech between crew on the flight deck, and between crew and ground ATC. (3) The *Maintenance Data Recorder* receives data from hundreds of inputs from engineering systems. Popular name, *black box* (frequently a misnomer).

flight time See **block time**.

FLIR See **forward-looking infrared**.

float (1) The distance travelled by an aircraft between flattening-out and landing. (2) A watertight buoyancy unit which is of combined streamline and hydrodynamic form to reduce air and water resistance; *main floats* are the principal hydrodynamic support of floatplanes, while *wing-tip floats*, often retractable, give lateral stability to flying-boats.

floated rate-integrating gyro An electrically-driven single degree-of-freedom gyro whose cylindrical or spherical case floats with neutral buoyancy in a fluid within an outer casing. Precession of the gyro is detected electrically and these signals combined with the viscous torque set up by relative motion between rotor and case are used to measure the integral of the angular motion. Cf. **tuned rotor gyro**. Abbrev. *RIG*, *MIG* if miniaturized.

float seaplane An aircraft of the sea-plane type, in which the water support consists of floats in place of the main undercarriage, and sometimes at the tail and wing tips. It may be of the *single-* or *twin-float type*.

flocculi See **plage**.

flow noise Acoustic signal caused by a flow process, e.g. siren, ventilator noise, jet noise.

fluctuation noise Noise produced in the output circuit of an amplifier by shot and flicker effects.

flutter Rapid fluctuation of frequency or amplitude.

flutter Sustained oscillation, usually on wing, fin or tail, caused by interaction of aerodynamic forces, elastic reactions and inertia, which rapidly break the structure. *Asymmetrical flutter* occurs where the port and starboard sides of the aircraft simultaneously undergo unequal displacements in opposite directions, as opposed to *symmetrical flutter*, where the displacements and their direction are the same; *classical* or *coupled flutter* is due solely to the inertial, aerodynamic or elastic coupling of two or more degrees of freedom.

flutter echo See **multiple echo**.

flutter speed The lowest **equivalent airspeed** at which flutter can occur.

flyback The return of the scanning beam to its starting point at the completion of a radar trace or a line of a TV picture, the line being blanked out during the process. Also *retrace*.

fly before buy Process of procuring new military aircraft by flying a prototype prior to giving the production order. An alternative is to order 'off the drawing board', which shortens delivery time at the risk of inadequate performance or delays in fixing inadequacies found during flight test.

fly-by A type of space mission where the spacecraft passes close to the target but does not rendezvous with it, orbit around it or land on it.

fly-by-light Flight control system in which the signalling is performed by coherent light beams travelling in optic fibres.

fly-by-wire Flight control system using electric/electronic signalling.

flying-boat A seaplane wherein the main body or hull provides water support.

flying speed The *maximum flying speed* is the highest attainable speed in level flight, under specified conditions and corrected to standard atmosphere. The *minimum flying speed* is the lowest speed at which level flight can be maintained.

flying tail See **all-moving tail**.

FMEA See **failure modes and effects analysis**.

focus A point to which rays converge after having passed through an optical system, or a point from which such rays appear to diverge. In the first case the focus is said to be *real*; in the second case, *virtual*. The *principal focus* is the focus for a beam of light rays parallel to the principal axis of a lens or spherical mirror.

fog Minute water droplets with radii in the range 1 to 10 µm suspended in the atmosphere and reducing visibility to below 1 km (1100 yd in UK).

fogbow A bow seen opposite the Sun in fog. The bow is similar to the rainbow, but the colours are faint, or even absent, owing to the smallness of the drops, which causes diffraction scattering of the light.

föhn wind A warm dry wind which blows to the lee of a mountain range. It is prevalent on the northern slopes of the Alps.

foot thumper Stall warning device that vibrates the rudder pedals when the detector senses that a stall is imminent.

forbidden lines Spectral lines which cannot be reproduced under laboratory conditions. Such lines correspond to transitions from a metastable state, and occur in extremely rarefied gases, e.g. in the solar corona and in gaseous nebulae.

force That which, when acting on a body which is free to move, produces an acceleration in the motion of the body, measured by rate of change of momentum of body. The unit of force is that which produces unit acceleration in unit mass. See **newton**. Extended to denote loosely any operating agency. Electromotive force, magnetomotive force, magnetizing force etc. are

strictly misnomers.

forebody strake Low aspect-ratio extension of the wing at the root along sides of the forebody. These create powerful vortices during high incidence flight, thereby improving handling and increasing lift.

forecast A statement of the anticipated weather conditions in a given region, for periods of from 1 hour to 30 days in advance, the longer term being less reliable; made from a study of current synoptic charts, or by carrying out a **numerical forecast**.

foreplane Horizontal aerofoil mounted on front fuselage for pitch control. In a tail-first or *canard* configuration it replaces the function of the tailplane. In a delta wing design it may assist slow speed behaviour. Never used for roll control, a foreplane may be fixed or retractable and have **slats**, **flaps** or **elevators**.

forked lightning A popular name given to a lightning stroke; the name derives from the branching of the stroke channel which is commonly observed.

form drag The difference obtained when the *induced drag*, i.e. the fraction of the total drag induced by lift, is subtracted from the **pressure drag**. See **drag**.

former A structural member of a fuselage, nacelle, hull or float to which the skin is attached, and having the primary purpose of preserving form or shape. It generally carries structural loads. Cf. **frame**.

Fortin's barometer A type of mercury barometer suited for accurate readings of the pressure of the atmosphere. The zero of the scale is indicated by a pointer inside the mercury cistern, the bottom of which is flexible and may be moved by an adjusting screw until the mercury surface just touches the pointer.

forward-looking infrared Sensor systems in the nose of aircraft or guided weapon for target detection and vehicle guidance.

forward sweep See **sweep**.

Foucault's measurement of the velocity of light One of the first successful attempts to obtain an accurate result for this important constant. Foucault, in 1862, made use of a rapidly rotating mirror sending light to a distant fixed concave mirror which reflected it back. Measurement of the displacement of the reflected image gave a value of 2.986×10^8 m/s for the velocity of light *in vacuo*.

Foucault's pendulum An instrument devised by Foucault, in 1851, to demonstrate the rotation of the Earth; it consists of a heavy metal ball suspended by a very long fine wire. The plane of oscillation slowly changes through $15°$ sin (latitude) per sidereal hour.

Fourier analysis The determination of the harmonic components of a complex waveform (i.e. the terms of a Fourier series that represents the waveform) either mathematically or by a wave-analyser device.

Fowler flap A high-lift trailing-edge flap that slides backward as it moves downward, thereby increasing the wing area, also leaving a slot between its leading edge and the wing when fully extended.

frame A transverse structural member of a fuselage, hull, nacelle or float, which follows the periphery and supports the skin, or the skin-stiffening structure. See **spar**.

Fraunhofer lines Sharp narrow absorption lines in the spectrum of the Sun, 25 000 of which are now identified. The most prominent lines are due to the presence of calcium, hydrogen, sodium and magnesium. Most of the absorption occurs in cool layers of the atmosphere, immediately above the incandescent **photosphere**. See [A], [B], [C] etc.

frazil ice Ice, in the form of small spikes and plates, formed in rapidly flowing streams, where the formation of large slabs is inhibited.

free atmosphere The atmosphere above the friction layer where motion is determined primarily by the large-scale pressure field.

free balloon Any balloon floating freely in the air, not propelled or guided by any power or mechanism, either within itself or from the ground.

Freedom The name of the international space station being developed jointly by NASA, ESA, Japan and Canada. See **space station**.

free fall The motion of an unpropelled body in a gravitational field. In orbit beyond the Earth's atmosphere, free fall produces near-weightlessness. See **microgravity**.

free field Sound field which is radiated directly from a source, without being reflected elsewhere.

free-flight wind tunnel (1) A wind tunnel, usually of up-draught type, wherein the model is not mounted on a support, but flies freely. (2) One in which three pilots each control one axis of a free-flying model, as used by NASA in the large Ames 80 ft tunnel. (3) Ballistic shape fired into airflow of a wind tunnel for shock wave or re-entry experiments.

free turbine A power take-off turbine mounted behind the main turbine/compressor rotor assembly and either driving a long shaft inside the main rotor to a gearbox at the front of the engine, or a short shaft to a gearbox at the rear of the engine. It can also be a separate unit fed by a remotely pro-

duced gas supply.

free vortex flow A vortex persisting away from a solid surface as in a natural tornado. A bound vortex is one attached to the body creating e.g. a wing tip vortex.

freezing point The temperature at which a liquid solidifies, which is the same as that at which the solid melts (the *melting point*). The freezing point of water is used as the lower fixed point in graduating a thermometer. Its temperature is defined as 0°C (273.15 K). Abbrev. *fp*. See **triple point**, **water**, **depression of freezing-point**.

Freons TN for compounds consisting of ethane or methane with some or all of the hydrogen substituted by fluorine, or by fluorine and chlorine. More commonly *chlorofluorocarbons* (*CFCs*), *fluorocarbons*.

frequency distortion In sound reproduction, variation in the response to different notes solely because of frequency discrimination in the circuit or channel. Generally plotted as a decibel response on a logarithmic frequency base.

frequency-division multiplex A method of *multiplex* transmission in which individual speech or information channels are modulated to separate channels and then transmitted simultaneously over a cable or microwave link. Single-sideband suppressed carrier methods are normally employed.

Fresnel's biprism An isosceles prism having an angle of nearly 180°, used for producing interference fringes from the two refracted images of an illuminated slit.

friction The resistance to motion which is called into play when it is attempted to slide one surface over another with which it is in contact. The frictional force opposing the motion is equal to the moving force up to a value known as the *limiting friction*. Any increase in the moving force will then cause slipping. *Static friction* is the value of the limiting friction just before slipping occurs. *Kinetic friction* is the value of the limiting friction after slipping has occurred. This is slightly less than the static friction. The *coefficient of friction* is the ratio of the limiting friction to the normal reaction between the sliding surfaces. It is constant for a given pair of surfaces.

friction layer The atmospheric layer, extending to a height of about 600 m, in which the influence of surface friction is appreciable. Sometimes referred to as the '*planetary boundary layer*'.

front (1) Surface of separation of two air masses. (2) Line of intersection of the surface of separation of two air masses with another surface or with the ground. If warm air replaces cold, it is a warm front; if cold

---

fundamental dynamical units

The basic equations of dynamics are such as to be the same for any system of fundamental units. Unit force acting on unit mass produces unit acceleration; unit force moved through unit distance does unit work; unit work done in unit time is unit power. Four systems are, or have been, in general use, the SI system now being the only one employed in scientific work.

	System				Dimensions
	ft lb sec	gravitational	CGS	SI (MKSA)	
length	foot (ft)	foot	centimeter (cm)	metre (m)	L
mass	pound (lb)	slug	gram (g)	kilogram (kg)	M
time	second (s)	second	second	second	T
velocity	ft/s	ft/s	cm/s	m/s	LT^{-1}
acceleration	ft/s^2	ft/s^2	cm/s^2	m/s^2	LT^{-2}
force	poundal (pd)	pound force (lbf)	dyne	newton (N)	MLT^{-2}
work	ft pdl	ft lbf	erg	joule (J)	ML^2T^{-2}
power	ft pdl/s	ft lbf/s	erg/s	watt (W)	ML^2T^{-3}

Notes: (1) There is no name for the unit of power except in the SI system. It is possible to express power in the ft lb sec system and in the gravitational system by the horse-power (550 ft lbf s^{-1}) and in the CGS system by the watt (10^7 erg s^{-1}). (2) The unit of force (lbf) in the gravitational system is also known as the pound weight (lb wt).

---

replaces warm, it is a cold front.

frontogenesis, frontolysis Respectively, the intensification or realization of a front, and its weakening or disappearance.

frost A *frost* is said to occur when the air temperature falls below the freezing point of water (0°C or 32°F). See **hoar frost**.

frost hollow A hollow in hilly ground into which cold air drains, stagnates and becomes exceptionally cold owing to radiation at night.

frostpoint The temperature at which air becomes saturated with respect to ice if cooled at constant pressure.

fuel accumulator A reservoir which augments the fuel supply when the critical fuel pressure is reached during the starting cycle of a gas turbine.

fuel cell A galvanic cell in which the oxidation of a fuel (e.g. methanol) is utilized to produce electricity.

fuel-cooled oil cooler A compact oil cooler for high-performance gas turbines in which heat is transferred to fuel passing in the counter bores of the device, instead of to air.

fuel cut-off A device which shuts off the fuel supply of an aero-engine. Also *slow-running cut-out*.

fuel grade The quality of piston aero-engine fuel as expressed by its **knock rating**.

fuel jettison Apparatus for the rapid emergency discharge of fuel in flight.

fuel manifold The main pipe, or gallery, with a series of branch pipes, which distributes fuel to the burners of a gas turbine.

fuel tanks These may be of many forms, for which the names vary. The *main tanks* are normally all those carried permanently and are usually composed of either flexible, self-sealing bags, or cells in wing or fuselage, or are integral with the wing structure; *auxiliary tanks* can be mounted additionally to increase range. See **drop tank**.

fuel trimmer A variable-datum device for resetting in flight the automatic fuel regulation, by *barostat*, of a gas turbine to meet changes in ambient temperature.

full mass See **weight and mass**.

full Moon The instant when the geocentric longitudes of the Sun and Moon differ by 180°; the Moon is then opposite the Sun, and therefore fully illuminated.

fundamental dynamical units See panel on this page.

funnelling The strengthening of a wind blowing along a valley, esp. when the valley narrows.

fuselage The name generally applied to the main structural body of a heavier-than-air craft, other than the hull of a flying-boat or amphibian.

G

g See **load factor**.

g Symbol for acceleration due to gravity.

[G] A pair of Fraunhofer lines in the deep blue of the solar spectrum. One, of wavelength 430.8081 nm is due to iron; the other, of wavelength 430.7907 nm, is due to calcium.

galactic circle The great circle of the celestial sphere in which the latter is cut by the galactic plane: hence the primary circle to which the galactic co-ordinates are referred.

galactic clusters Same as **open clusters**.

galactic co-ordinate Two spherical co-ordinates referred to the galactic plane; the origin of galactic longitude is at RA $17^h42.4^m$, dec. $-28°55'$ (1950); galactic latitude is measured positively from the galactic plane towards the north galactic pole.

galactic halo An almost spherical aggregation of stars, gas and dust, which is concentric with the Galaxy. It contains stars of Population II, and is responsible for much of the background of radio emission from the sky. Similar halos surround other galaxies.

galactic plane The plane passing as nearly as possible through the centre of the belt known as the Milky Way or Galaxy.

galactic rotation The rotation of the Galaxy about its centre. All stars, gas and dust within the Galaxy share this rotation, which is fastest towards the centre. In the Sun's vicinity the velocity due to galactic rotation is about 250 km/sec. The Sun takes about 250 million years to complete one orbit round the Galaxy.

galaxy (1) See panel on p. 73. (2) Automatic star-plate analyser for measuring brightness and position of high density photographic images of portions of the Galaxy (*General Automatic Luminosity And XY measuring engine*).

gale A wind having a speed of 34 knots (63 km/h) or more, at a height of 33 ft (10 m) above the ground. On the Beaufort scale, a gale is a wind of force 8.

gamma-ray astronomy The study of radiation from celestial sources at wavelengths shorter than 0.01 nm. Gamma-rays have been detected from the *gamma-ray background*, a few energetic galaxies and **quasars**, and from certain highly-evolved stars.

gantry The servicing tower beside a rocket on its launching pad.

Ganymede The third natural satellite of **Jupiter**, discovered by Galileo. It is 5200 km in diameter, making it the largest moon in the Solar System, and larger than Mercury. It is the brightest of the Galilean satellites.

gap The distance from the leading edge of a biplane's upper wing to the point of its projection on to the chord line of a lower wing.

GAPAN Abbrev. for *Guild of Air Pilots and Air Navigators*.

gap filler Radar to supplement long range surveillance radar.

gas-bag Any separate gas-containing unit of a rigid airship.

gas generator The high-pressure compressor/combustion/turbine section of a gas turbine which supplies a high-energy gas flow for turbines which drive propellers, fans or compressors.

gas laws Boyle's law, Charles's law and the pressure law which are combined in the equation $pV = RT$, where p is the pressure, V the volume, T the absolute temperature and R the gas constant for 1 mole. A gas which obeys the gas laws perfectly is known as an *ideal* or *perfect* gas. Cf. **van der Waals' equation**.

gas producer A turbo-compressor of which the power output is in the form of the gas energy in the efflux, sometimes mixed with air from an auxiliary compressor. Essentially, the gas producer is mounted remotely from the point of utilization of its energy, e.g. helicopter **rotor-tip jets** or a **free turbine**.

gassing See **inflation**.

gas temperature The temperature of the gas stream resulting from the combustion of fuel and air within a turbine engine. For engine performance monitoring, the temperature may be measured at either of two points signified by the abbrevs. *JPT* and *TGT*. *JPT* (jet pipe temperature) is the measured temperature of the gas stream in the exhaust system, usually at a point behind the turbine. *TGT* (turbine gas temperature) is the measured temperature of the gas stream between turbine stages. *EGT* (exhaust gas temperature), frequently used, is synonymous with *JPT*.

gated throttle A supercharged aero-engine throttle quadrant with restricting stop(s) to prevent the throttle being wrongly used. See **boost control**.

Gaussian noise See **white noise**.

GCA Abbrev. for **Ground-Controlled Approach**.

GCI Abbrev. for **Ground-Controlled Interception**.

galaxy

A galaxy is a large family of stars, interstellar gas and dust, held together by its mutual gravitational force, and generally isolated by almost empty space from its neighbour galaxies. Galaxies are the basic large scale components from which our universe is constructed. Their masses range from $10^7 - 10^{12}$ solar masses. Our own *Milky Way* Galaxy has about 10^{11} solar masses altogether. Since galaxies come in a bewildering variety of forms, it is useful to distinguish three major classes: spiral, elliptical and irregular.

The *spiral* galaxies are flat disks of stars with two spiral arms emerging from the nuclear region. In 30% of spirals a central bar of stars links the arms. They account for 25% of all galaxies. Diameters are 25 – 800 kiloparsec (kpc). The arms are rich in interstellar matter, and they play a key role in star formation. The spiral pattern does not rotate rigidly, like a wheel. Theorists believe a compression wave propagates through the galaxy, and the spiral pattern delineates its location. At the leading edge of the arm, compression of the interstellar gas triggers vigorous star formation.

Most galaxies are *elliptical*, with diameters up to 100 kpc. They range from almost spherical up to ratio of about 3 for length relative to diameter. They contain hardly any dust or gas, and so there is very little star formation. Elliptical galaxies are thus characterized by an ageing population of stars. Irregular galaxies have no clear morphology and they embrace many of the more active subtypes.

There are no clear explanations as to why galaxies come in spiral and elliptical shapes. It appears rather unlikely that spirals change into ellipticals, but beyond that not much is known. Dust may hold the key: in a collapsing protogalaxy with dust, the formation of a flat disk is inevitable. Most galaxies are extremely old: the Milky Way is at least 12 billion years old. The epoch of galaxy formation was perhaps one billion years after the onset of the **Big Bang**.

Distances to the galaxies are mainly derived from the **Hubble law** and they are very large. The *Magellanic Clouds*, satellites of the Milky Way, are 55 kpc away. Our own Local Group, which includes the Andromeda galaxy, will fit in a sphere a few Mpc across. The next significant group of galaxies is tens of Mpc away. The Virgo cluster, for example, is 2 Mpc across, 20 Mpc distant, and it has hundreds of members. Clusters of galaxies sit at the apex of the hierarchical structure of the universe: as the universe expands, clusters and the galaxies within them are unchanged, but the spacing between clusters increases monotonically with time. Very remote galaxies are of considerable importance in cosmology. At distances of 5000 Mpc, for example, they bring information on the universe as it was some 15 billion years ago. The **Hubble space telescope** has the investigation of very distant galaxies as one of its major tasks.

Active galaxies are numerically quite rare, but have been much observed on account of the exotic phenomena within them. About one galaxy per million is a giant **radio galaxy**. These have two clouds of radio emission extending to 1 Mpc from the nucleus, and their total luminosity (10^{38} W) is around a million times brighter than a normal galaxy. This suggests that stars cannot be the source of their energy. **Quasars** (see separate panel) are similar to radio galaxies as regards their radio emissions, but they have higher **redshifts** (up to $z = 5$) and much smaller angular sizes than galaxies. Possibly they are extremely active galaxies that are so far away that we can only see their brilliant nuclei. If the Hubble law holds out to large distances, these are the remotest objects known to us.

continued on next page

galaxy (contd)
Individual astronomers have studied and classified particular sorts of galaxies that have taken the name of their investigator. For example, *Seyfert* galaxies are spirals with brilliant nuclei and faint arms; *Markarian* galaxies have unusual blue spectra; the *Maffei* galaxies are two under-luminous members of our Local Group; and *Zwicky* galaxies are very compact, only just distinguishable from stars on photographs.

An important goal of theoretical work is to provide a unified picture of galactic activity on all scales. The commonest model invokes a central **black hole** of up to a billion solar masses. Matter falling into this black hole behaves in bizarre ways: a gigantic *accretion disk* is formed and relativistic jets can emerge along its rotation axis. In a qualitative way many of the observed features of galactic nuclei can be explained. Alternatives to such models include the *starburst galaxy* in which it is envisaged that a temporarily enhanced rate of star formation causes much higher luminosity for a while. A few astronomers have chosen to challenge the Hubble law, and argued that the active galaxies are much closer than supposed, in which cases their energy requirements would not be as extreme.

G-display Similar to **F-display** but indicating increasing or diminishing range of target by increasing or diminishing lateral extension of the spot.

gegenschein Ger. *counter-glow*. A term applied to a faint illumination of the sky sometimes seen in the ecliptic, diametrically opposite the Sun, and connected with the zodiacal light.

general aviation Private, agricultural and survey aviation.

general circulation of the atmosphere See separate panels on the troposphere (p. 75) and the middle atmosphere (p. 77).

general theory of relativity Einstein's extension of the special theory of relativity to deal with accelerating frames of reference. Based on the *principle of equivalence*, it shows that the laws of physics must be in such a form that it is impossible to distinguish between a uniform gravitational field and an accelerated frame of reference.

GEO Abbrev. for *GEosynchronous Orbit*. See **orbit**.

geocentric The term applied to any system or mathematical construction which has as its point of reference the centre of the Earth.

geocentric parallax The apparent change of position of a heavenly body due to a shift of the observer by the rotation of the Earth; hence only observed in bodies (e.g. the Moon and Sun) sufficiently close for the Earth's radius to subtend a measurable angle when seen from the body. Also *diurnal parallax*.

geodetic construction A redundant space frame whose members follow diagonal geodesic curves to form a lattice structure, such that compression loads induced in any member are braced by tension loads in crossing members. Does not need stress-carrying covering.

geopotential height The height of a point in the atmosphere expressed in units (geopotential metres) proportional to the geopotential at that height. One geopotential metre is numerically equal to $(9.8/g)$ of a geometric metre, where g is the local acceleration of gravity. All reported measured heights of pressure levels are given in geopotential metres thus obviating the necessity for considering variations of gravity when making dynamical calculations.

George Colloq. for *automatic pilot*.

geostationary Of an orbit lying above the equator, in which an artificial satellite moves at the same speed as the Earth rotates, thus maintaining position above the Earth's surface. Such a satellite would have an altitude of 35 800 km above the Earth's surface.

geostrophic approximation Use of the *geostrophic wind* as an approximation to the actual wind, either in operational forecasting or as a replacement for certain terms in the equations of motions.

geostrophic force A virtual force used to account for the change in direction of the wind relative to the Earth's surface, arising from the Earth's rotation and the **Coriolis force**.

geostrophic wind The theoretical wind arising from the **pressure gradient** force and the **geostrophic force**.

geosynchronous Of an orbit where the period of the satellite is equal to the period of rotation of the Earth; usually used synonymously with **geostationary** but the latter is always circular and in the equatorial plane.

general circulation of the troposphere

The way in which the excess of solar heat energy received at the surface of the Earth at low latitudes is distributed upwards and polewards by the wind until it is lost by radiation out to space. The Sun's radiative energy is contained mainly in the visible region of the spectrum to which the atmosphere, apart from clouds, is largely transparent. This energy is absorbed by soil and vegetation, and the upper layers of the seas and oceans, and it then either heats the lowest part of the atmosphere or evaporates large quantities of water. Outward radiation by the atmosphere takes place at infrared wavelengths and is inadequate to compensate for received solar radiation roughly between latitudes 40°N. and 40°S. The excess of latent and sensible heat in the lowest atmospheric layer between these latitudes induces systematic windfields, vertical as well as horizontal, which dominate global weather patterns.

Because surface winds are subject to friction and loss of kinetic energy at the Earth's surface, they exchange angular momentum with the solid rotating Earth. There are also a number of constraints on their general circulation: (1) the need to conserve the angular momentum of the global wind system; (2) all winds are subject to the **Coriolis force** because the Earth rotates; (3) total energy must be conserved, however it is transformed between kinetic and potential energy, the latent heat of water etc. The net effect of all these processes is roughly as follows.

Over the tropics, air rises by vertical convection causing heavy showers and thunderstorms with much release of latent heat up to heights of 12 km and then it spreads north and south. In the subtropics, air subsides gently to levels near the surface where some returns equatorwards to form the trade winds. Polewards of this subtropical high-pressure belt, the potential energy of thermal contrast between warm tropical air and cold polar air causes the formation of typical travelling mid-latitude depressions and anticyclones in which vast quantities of warm moist air are carried upwards, polewards and eastwards, with cold air drawn down equatorwards as a replacement. These mid-latitude weather systems are not minor eddies superimposed on a much greater general circulation pattern. Where they occur they *are* the general circulation, and charts of mean values can be very misleading. In polar regions tropospheric winds are generally much lighter again and air subsides with the strong radiational cooling which dominates the energy balance.

Schematic representation of sea-level winds and pressure systems on an idealized Earth. The mean upper-level flow in a meridional vertical cross-section is also indicated.

On the real Earth, continental land masses and major mountain ranges cause much divergence from this simple picture, esp. in the northern hemisphere; there are also seasonal variations. NP and SP indicate the poles and H and L, high and low pressure systems.

continued on next page

general circulation of the troposphere (contd)
The various constraints and physical laws already mentioned give rise to two important tropospheric **jet streams** in each hemisphere. One in the subtropics at about 12 km altitude and fairly constant latitude of 30°, and one associated with the polar front and the formation of mid-latitude depressions. The polar frontal jet is much stronger in winter than in summer and varies considerably in position and orientation, often forming a wave-like pattern around the hemisphere. See diagram.

Systematic complications to all these patterns are caused by the geographical features of mountain masses and chains, and the relative distribution of land and sea, esp. in the northern hemisphere; the monsoons are typical examples.

The oceans, which cover 71% of the Earth's surface, absorb twice as much solar energy as the land. This energy is eventually given up to the atmosphere as sensible and latent heat, much of it in higher latitudes. Probably about 40% of the total heat which is transported out of the tropics and subtropics is provided directly by the oceans and there is good evidence that anomalies of sea-surface temperature have far-reaching effects on weather patterns.

See next page for discussion of the middle atmosphere

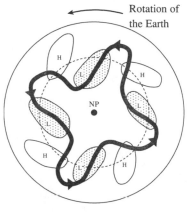

Rotation of the Earth

Mean upper tropospheric flow

Schematic diagram of planetary waves (indicated by the upper tropospheric flow) and alternating low (L) and high (H) pressure systems. The SW–NE orientation of the axes of the pressure systems gives rise to poleward transport of heat and momentum. NP, North Pole.

giant star A star which is more luminous than the main sequence stars of the same spectral class. Smaller groups of subgiants and supergiants are recognized.

gibbous The word applied to the phase of the Moon, or of a planet, when it is between either quadrature and opposition, and appears less than a circular disk but greater than a half disk.

gills Controllable flaps which vary the outlet area of an air-cooled engine cowling or of a radiator. Also *cooling gills, cowl flaps, radiator flaps.*

GIOTTO Name of the **ESA** spacecraft which flew through the coma of Halley's comet and approached within 600 km of the nucleus. The **fly-by** took place at a relative velocity of 68 km/sec in March 1986. The encounter was supported by data from two Japanese probes (*Suisei* and *Sakigake*), two USSR

spacecraft (*Vega 1* and *2*) and the US International Cometary Explorer (*ICE*). Co-ordination was performed by the truly international *Inter-Agency Consultative Group* (IACG).

GLAVCOSMOS The Soviet civilian agency for marketing space expertise and facilities.

glazed frost A smooth layer of ice which is occasionally formed when rain falls and the temperature of the air and the ground is below freezing point.

glide path The approach slope (usually 3.5° or 5°) along which large aircraft are assumed to come in for a landing. The term is imprecise because such aircraft do not glide, but are brought in with a considerable amount of power.

glide-path beacon A directional radio beacon, associated with an ILS, which provides an aircraft, during approach and landing,

general circulation of the middle atmosphere

The motions described thus far are largely confined to the troposphere which accounts for about 80% of the total atmospheric mass. Above the troposphere is the stratosphere, a region where temperature increases with height and hence, unlike the tropsphere, a region of great static stability. Deviations from simple axially-symmetric zonal atmospheric flow may be described in terms of linear wave dynamics to a much greater extent than in the troposphere which is dominated by the highly non-linear flows due to polar-front depressions, tropical storms, topographic forcing and so on. The stratosphere accounts for most of the remaining atmospheric mass and is bounded at about 50 km by the *stratopause*. Above this lies the *mesosphere* which together with the stratosphere is known as the *middle atmosphere*. The middle atmosphere is a region where complex photochemical reactions, including the formation and destruction of ozone, take place under the influence of solar radiation and where man-made pollutants, including chlorine and chlorofluorocarbons (*CFCs*), are able to exert a widespread and deleterious influence by the effect they have on these reactions. In consequence, there has been a greatly increased interest in the middle atmosphere in recent years.

Most of the middle atmosphere lies well above the range of conventional observational techniques, and measurements of wind and temperature involve the use of devices dropped from rockets (*rocket sondes*), radiometers carried on satellites and specially adapted high-flying aircraft.

The chief energy inputs to the middle atmosphere are (1) the difference between the absorption of short-wave radiation (particularly ozone) and the emission of long-wave radiation (chiefly by carbon dioxide and ozone); (2) tropospheric motions that propagate energy upwards through the tropopause, with the tropospheric motions which have significant effects being the very long, quasi-stationary planetary waves occurring in the winter hemisphere and gravity waves including atmospheric tides and mountain lee-waves.

The main characteristics (see figure) of the general stratospheric circulation are:

1. A strong cyclonic vortex surrounding the winter pole with a maximum of windspeed in middle latitudes (the *polar night jet*).

2. A moderate anticyclonic vortex centered on the summer pole, covering much of the summer hemisphere.

3. Relatively weak circulations at the equinoxes.

Schematic diagram of meridional mean circulation in the middle atmosphere at the solstices.

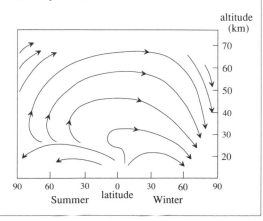

altitude (km)

70
60
50
40
30
20

90   60   30   0   30   60   90
Summer   latitude   Winter

continued on next page

---

general circulation of the middle atmosphere (contd)

These characteristics are complemented by additional features:

4. A semi-annual oscillation of the zonal wind field which is especially marked at the equator.

5. A strong oscillation of the zonal wind in the tropics with a somewhat irregular period of about 26 months, the Quasi Biennial Oscillation (*QBO*). It propagates downward through the stratosphere, taking about 2 years to do so, but decays rapidly below about 22 km.

6. In winter in the northern hemisphere, the polar vortex occasionally, but not every year, breaks down completely with mean zonal winds becoming easterly and temperatures rising by up to 50 K in a few days. This phenomenon is called *sudden warming*.

In mid-winter, sudden warmings are usually followed by a re-establishment of the normal cyclonic vortex, but in the late winter they mark the beginning of the summer circulation regime. In the southern hemisphere, the polar winter vortex is more nearly circular than in the northern, has stronger winds, is colder, is not subject in the same way to violent distortions, such as reversible sudden warmings, and lasts until well after the equinox. These differences are due to the widely different distributions of land, sea and mountains in the two hemispheres.

Explanations for these and related phenomena are being sought by theoreticians. It has become clear that the temperature structure of the middle atmosphere differs widely from that appropriate to radiative equilibrium, with the winter stratosphere and mesosphere being very much warmer, and the summer mesopause much colder, than the calculated equilibrium state. Indeed the tropopause and mesopause are substantially colder over the equator than over the poles which is especially surprising for the winter pole where there is no radiative warming at all. It seems probable that dynamically induced motions yield a 'refrigerator' effect whereby the vertical flux of energy from tropospheric waves, made available through a 'breaking' phenomenon, produces a poleward flux of heat against the temperature gradient. Average motions in the vertical meridional plane, at the solstice, are probably as follows: there is rising motion in the summer troposphere and mesosphere, cross-equatorial flow in the mesosphere and descent in the winter middle atmosphere. In the lower stratosphere this merges with a type of **Hadley cell** circulation with equatorial ascent and extratropical descent in both hemispheres.

---

with indications of its vertical position relative to the desired approach path.

glide-path landing beam Radio signal pattern from a radio beacon, which aids the landing of an aircraft during bad visibility.

glider A heavier-than-air craft not power driven within itself, although it may be towed by a power driven aircraft. Cf. **sailplane**.

gliding (1) Flying a heavier-than-air craft without assistance from its engine, either in a spiral or as an approach glide before flattening-out antecedent to landing. (2) The sport of flying *gliders* which are catapulted into the air, or launched by accelerating with a winch, or towed by car, or towed to height by an aircraft.

gliding angle The angle between the flight path of an aircraft in a glide and the horizontal.

glint Pulse-to-pulse variation in the amplitude of a reflected signal; may be due to reflection from different surfaces of a rapidly moving target, or from propellors or rotor blades.

Global Positioning System *Navstar*. System for providing an extremely accurate three-dimensional position (to 10 m) and velocity information to users anywhere in the world. Data is derived from the transit time of radio-frequency signals from four staellites out of a total of 24.

globular cluster Densely packed family of stars which is arranged characteristically as a compact sphere of stars. They contain tens of thousands to millions of stars, all thought to have been formed at the same time. Over 100 are known as members of our Galaxy, distributed through the **galactic halo**.

globule Small dark nebula composed of opaque gas and dust, representing an early stage of star formation.

glory A small system of coloured rings surrounding the shadow of the observer's head, cast by the Sun on a bank of mist, as in the *spectre of the Brocken*. The *glory* is produced by diffraction caused by the water droplets in the mist.

glow plug In a gas turbine, an electrical igniting plug which can be switched on to ensure automatic relighting when the flame is unstable, e.g. under icing conditions. Also used in diesel engines.

golden number A term derived originally from medieval church calendars, and still used to signify the place of a given year in the Metonic cycle of 19 years.

Gold slide A slide, named after its designer, which is attached to a marine mercury barometer to make the corrections for index error, latitude, height and temperature mechanically.

gore One of the sector-like sections of the canopy of a parachute.

gothic wing A very low **aspect ratio** wing with the double curvature plan form, similar to a gothic arch of the perpendicular period, used for supersonic aircraft to combine low **wave drag** with **separated lift**. See **ogee wing**.

Göttingen wind tunnel A return flow wind tunnel in which the working section is open. See **open-jet wind tunnel**.

GPS Abbrev. for *Global Positioning System*.

gradient wind A theoretical wind which, when blowing along curved isobars with no tangential acceleration, represents a balance between the pressure gradient, the Coriolis force and the centrifugal force. It is less than the **geostrophic wind** round a **depression**, and greater round an **anticyclone**.

Grand Unified Theory See panel on p. 80.

granulation The mottled appearance of the Sun's **photosphere** when viewed at very high resolution.

graphecon A double-ended storage tube used for the integration and storage of radar information, and as a translating medium.

grass Irregular deflection from the timebase of a radar display, arising from electrical interference or noise.

grating An arrangement of alternate reflecting and non-reflecting elements, e.g. wire screens or closely spaced lines ruled on a flat (or concave) reflecting surface, which, through diffraction of the incident radiation analyses this into its frequency spectrum. An *optical grating* can contain a thousand lines or more per cm. A *standing-wave* system of high-frequency sound waves with their alternate compressive and rarefied regions can give rise to a diffraction grating in liquids and solids. With a *criss-cross* system of waves, a 3-dimensional grating is obtainable.

grating spectrum An optical spectrum produced by a **diffraction grating** instead of by *refraction* in a prism.

gravipause That point or border in space in which the gravitational force of one body is matched by the counter-gravity of another. Also *neutral point*.

gravitation The name given to that force of nature which manifests itself as a mutual attraction between masses, and whose mathematical expression was first given by Newton, in the law which states: 'Any two particles of matter attract one another with a force directly proportional to the product of their masses and inversely proportional to the square of the distance between them.' This may be expressed by the equation:

$$F = G\frac{m_1 m_2}{d^2},$$

where F is the force of gravitational attraction between bodies of mass m_1 and m_2, separated by a distance D. G is the constant of gravitation and has the value $6.670{\times}10^{-11}$ N m^2 kg^{-2}.

gravitational astronomy See **celestial mechanics**.

gravitational field That region of space in which at all points a gravitational force would be exerted on a test particle.

graviton A *gauge boson* that is the agent for gravitational interactions between particles. The graviton has not been detected but is believed to have zero mass and zero charge. It bears the same relationship to the gravitational field as the photon does to the electromagnetic field. Gravitational interactions between particles are so weak that no quantified effects have been observed.

gravity assist Use of a body to provide an energy boost to a spacecraft on a trajectory planned to pass close to the body for this purpose.

gravity feed tanks Fuel tanks, situated above the point of delivery to the engine, which feed the engine solely by gravity.

gravity stabilization A way of stabilizing the orientation of a spacecraft with reference to a primary body (such as Earth) by using

Grand Unified Theories

Modern theoretical physics aims to describe the nature of the physical universe with as few assumptions and laws as possible. In a complete framework of physical theory parameters, which we currently have to measure experimentally, such as the speed of light or the fine structure constant, would emerge naturally from the equations. Just as Newton was able to explain **Kepler's laws of planetary motion** through his more powerful theory of gravitation, so in modern physics there is a desire to find all-embracing physical laws.

Physics recognizes four distinct interactions that affect matter. These are the *electromagnetic* interaction, the *strong* and *weak* interactions affecting particles and nuclei, and the gravitational interaction. Although gravitation is by far the weakest of these, it is always an attractive force and it acts over immense distances; that is why it is so important in **cosmology**. Each of these interactions has its own theoretical formalism. *Gravitation* is now described through the **general theory of relativity**, for example. Electromagnetic, weak and strong interactions are also described through different *gauge theories*. (Gauge group theory is a powerful branch of mathematics.)

In the 1960s and 1970s, physicists such as S.Weinberg, A.Salam and S.Glashow found ways of achieving partial unification of these different classes of interaction. *Electroweak theory* could explain most features of electromagnetism and the weak interaction. The more elaborate quantum chromodynamics described the interactions between quarks and heavy particles such as neutrons. The unified theories achieved well-publicized successes at the time by predicting new elementary particles that were eventually found by the CERN accelerator.

Grand unified theories (or GUTs) are particularly associated with H.M.Georgi and they aim to fold the electromagnetic, weak and strong interactions into a single gauge group. This theory predicts that the *proton* is not stable, although the half-life is extremely long: 10^{29} years. Experiments to measure the proton decay are being attempted as an important test of the theory. The theory also attempts to describe how elementary particles with energies of 10^{21} electron volts will behave. This energy is so colossal, with enough to run a light bulb for a minute residing on a single particle, that they will never be made in accelerators. They were however abundant in the first 10^{-32} seconds of the Big Bang and it is possible that they could be found in cosmic rays. It is because of these astrophysical connections that astronomers, as well as physicists, are interested in GUTs. A more distant goal is the unification of gravitation also, to produce a viable theory of **quantum gravity**.

---

the gravity gradient, in which the long axis of the spacecraft is directed toward the centre of the body.

gravity waves Liquid surface waves controlled by gravity and not surface tension. In meteorology, they are atmospheric transverse waves in which the restoring force is due to the effect of gravity on pressure and density fluctuations. Examples are **lee waves**.

green flash A phenomenon sometimes seen in clear atmospheres at the instant when the upper rim of the Sun finally disappears below the horizon as a bright blob of light; green is the last apparent colour from the Sun, because the more greatly refracted blue is dispersed.

greenhouse effect Phenomenon by which

thermal radiation from the Sun is trapped by water vapour and carbon dioxide on a planet's surface, thus preventing its re-emission as long-wave radiation. This leads to the temperature at the planet's surface being considerably higher than would otherwise be the case. The effect is most pronounced for **Venus** and the **Earth**.

Greenwich Mean Time   *GMT. Mean solar time* referred to the zero meridian of longitude, i.e. that through Greenwich. GMT is the basis for scientific and navigational purposes. See **universal time**.

Gregorian calendar The name commonly given to the civil calendar now used in most countries of the world. It is the Julian calendar as reformed by decree of Pope Gregory XIII in 1582. The Gregorian reform omitted certain leap years, and brought the length of the year on which the calendar is based nearer to the true astronomical value.

Gregorian telescope A form of reflecting telescope, very similar in principal to the Cassegrainian, in which the large mirror is pierced at the centre and the light is reflected back into an eyepiece in the centre by a small concave mirror on the principal axis and a little outside the focus. See **astronomical telescope**.

grid navigation A navigational system in which a grid instead of true North is used for the measurement of angles and for heading reference. The grid is of parallel lines referenced to the 180° meridian which is taken as Grid North. Used principally in polar route flying and, as the system is independent of convergence of meridians, headings remain the same from departure to destination.

gross mass, weight See **weight and mass**.

grosswetterlage The sea-level pressure distribution averaged over a period during which the essential characteristics of the atmospheric circulation over a large region remain nearly unchanged.

gross wing area The full area of the wing, including that covered by the fuselage and any nacelles.

ground clutter The effect of unwanted ground-return signals on the screen pattern of a radar indicator.

ground control Control of an aircraft or guided missile in flight from the ground.

ground-controlled approach Aircraft landing system in which information is transmitted by a *ground controller* from a ground radar installation at end of runway to a pilot intending to land. Also *talk-down*. Abbrev. *GCA*.

ground-controlled interception Radar system whereby aircraft are directed on to an interception course by a station on the ground. Abbrev. *GCI*.

ground engineer An individual, selected by the licensing authorities, who has power to certify the safety for flight of an aircraft, or certain specified parts of it. Term now superseded by *licensed aircraft engineer*.

ground fine pitch A very flat blade angle on a turboprop propeller, which gives extra braking drag and low propeller resistance when starting the engine. Colloq. *disking*.

ground frost A temperature of 0°C (32°F) or below, on a horizontal thermometer in contact with the shorn grass tips on a turf surface.

ground loop An uncontrollabe and violent

swerve or turn by an aircraft while taxiing, landing or taking off.

ground noise See **background noise**.

groundplot Method of calculating the position of an aircraft by relating groundspeed and time on course to starting position.

ground-position indicator An instrument which continuously displays the dead-reckoning position of an aircraft.

ground reflection The wave in radar transmission which strikes the target after reflection from the Earth.

ground resonance A sympathetic response between the dynamic frequency of a rotorcraft's rotor and the natural frequency of the alighting gear which causes rapidly increasing oscillations.

ground return The aggregate sum of the radar echoes received after reflection from the Earth's surface. May include **clutter**.

ground safety lock See **retraction lock**.

groundspeed Speed of aircraft or missile relative to the ground and not to the surrounding medium. Cf. **airspeed**.

ground support equipment All the handling facilities employed to service aircraft on the airport, e.g. tractors, steps, fuelling tankers, food and cleaning supplies. Also weapon trolleys and installation check-out gear for military aircraft.

GSE See **ground support equipment**.

g-suit See **anti-g-suit**.

GTO Abbrev. for *Geosynchronous Transfer Orbit*. See **orbit**.

g-tolerance The tolerance of an object or person to a given value of g-force.

guided atmospheric flight See panel on p. 82.

guided missile See **guided atmospheric flight** panel.

guided wave Electromagnetic or acoustic wave which is constrained within certain boundaries as in a wave guide.

guide-vanes A general term for aerofoils which guide the airflow in a duct. Also *cascades*. See **impeller-intake-**, **nozzle-**, **toroidal-intake-**.

guide wavelength Wavelength in a guide, operated above the cut-off frequency. $1/\lambda_g^2 = 1/\lambda_0^2 - 1/\lambda_{0c}^2$, where λ_g is the guide wavelength, λ_0 the wavelength in the unbounded medium at the same frequency, and λ_{0c} the wavelength in the unbounded medium at the cut-off frequency for the mode in question.

Guppy Aircraft modified by substituting a very large diameter section for a major part of the fuselage. Nose and tail hinge sideways for carrying large indivisible loads as in Boeing 377. *Airbus* major assemblies are delivered this way. Also *Super Guppy*,

guided atmospheric flight

This term refers to any automatic (unpiloted) and *aeronautical* flight of which the major subdivisions are: guided missiles, recoverable pilotless aircraft, drones and radio-controlled model aircraft.

There are various categories of guided missiles: a *guided bomb* is an unpowered bomb with aerodynamic stearing; a *smart bomb* is unpowered but has terminal guidance; a *stand-off-bomb* is an air-launched long-range cruise missile, e.g. Blue Steel; *ground to air* is an anti-aircraft missile; *ground to ground* is ground launched to attack ground targets, e.g. the Atlas Intercontinental Ballistic Missile (ICBM) or the Thor Intermediate Range Ballistic Missile; *air to sea* is e.g. an aerial torpedo for attacking ships; *air to ground* is for anti-tank or anti-radar use for destroying hard targets like submarine pens; *underwater to ground* is a submarine launched nuclear ballistic missile like Trident or Polaris.

Guided missiles can be propelled by solid or liquid rocket, airbreathing turbojet, ramjet or combinations of these. Their body contains propulsion fuel, warhead, control and guidance systems and they have to be launched from a ground system or carried and released from an aircraft or naval vessel. They also contain safety systems to prevent premature operation and to allow external destruction by radio command.

Guidance systems vary greatly: *direct command guidance* is control entirely from the launcher by radio or wire signals; *radar command guidance* is radio-signal guidance from a lock-on radar/computer system on the launcher; *beam rider* is a missile which follows a radar beam directly from launcher to target; *semi-active homing* is the radar 'illumination' of the target, on the reflection from which the missile 'homes'; *collision course homing* is similar, but employs an offset missile aerial to bring it onto a converging course with the target; *fully-active homing* is self-contained, the missile generating its own radar signals and carrying a lock-on or collision-course computer device; *passive homing* is by a sensitive detector on to infrared, heat, sound, static electricity, magnetic or other wave emissions from the target.

Long-range guidance devices include the *celestial*, wherein the missile's automatic astronavigation equipment is given its starting and target co-ordinates, and its present position is determined from stellar angular measurements so permitting its on-board trajectory program to set speed, altitude and direction to reach the target in an optimum manner. See **inertial navigation system, Strategic Defence Initiative**.

Remotely piloted aircraft (RPV) are small military aircraft, propeller-driven, which usually carry out photographic reconnaissance flights over enemy territory, guided by radio and internal programs and returning to be collected in catch-nets for re-use. UMA are *unmanned aircraft*, while ROV are *remotely operated vehicles* and UMV *unmanned vehicles*.

Drones are expendable radio-controlled unmanned aircraft or guided missiles for development of experimental aircraft or missile systems or training of operational crews.

Model aircraft are radio-controlled to fly for leisure or aerial photography.

Pregnant Guppy.

GUT Abbrev. for **Grand Unified Theory**.

gyro Commonly used abbrev. for **gyroscope**.

gyrocompass A gyroscope, electrically rotated, controlled and damped either by gravity or electrically so that the spin axis settles in the meridian.

gyrodyne A form of rotorcraft in which the rotor is power-driven for take-off, climb, hovering and landing, but is in **autorotation** for cruising flight, there usually being small wings further to unload the rotor.

gyro horizon An instrument which employs a gyro with a vertical spin axis, so arranged that it displays the attitude of the aircraft about its pitch and bank axes, referenced against an artificial horizon. It is normally electrically, but sometimes pneumatically, operated.

gyromagnetic compass A magnetic compass in which direction is measured by gyroscopic stabilization.

gyroplane Rotorcraft with unpowered rotor(s) on a vertical axis. Also *autogyro*.

gyroscope *Gyro*. Spinning body in a **gimbal mount** or similar, which resists torques altering the alignment of the spin axis, and in which **precession** or **nutation** replace the direct response of static bodies to such applied torques. Used as a compass, or as a controlling device for servos which reduce the misalignment and thus correct the course or e.g. stabilize a ship. Also *gyro, gyrostat*. See **fibre-optic gyro, floated rate-integrating gyro, laser gyro, rate gyro, tuned-rotor gyro.**

Gyrosyn TN for a remote-indicating compass system employing a directional *gyroscope* which is monitored by and *synchronized* with signals from an element fixed in azimuth and designed to sense its angular displacement from the Earth's magnetic meridian by **fluxgate**. The element is located at some remote point, e.g. wing tips, away from extraneous magnetic influences.

H

h Symbol for (1) Planck's constant. See **Planck's law**. (2) Specific enthalpy.
H Symbol for enthalpy.
[H] One member of the strongest pair (H and K) of Fraunhofer lines in the solar spectrum, almost at the limit of visibility in the extreme violet. Their wavelengths are [H], 396.8625 nm; [K], 393.3825 nm; and the lines are due to ionized calcium.
Ha, Hb, Hc etc. The lines of the Balmer series in the hydrogen spectrum. Their wavelengths are: Ha, 656.299; Hb, 486.152; Hc, 434.067; Hd 410.194 nm. The series continues into the ultraviolet, where about 20 more lines are observable.
haar Local term for a wet sea-fog advancing in summer from the North Sea upon the shores of England and Scotland.
Hadley cell A meridional circulation of the atmosphere consisting of low-level equatorward movement of air from about 30° to the equator, rising air near the equator, poleward flow aloft, and descending motion near 30°. It was suggested by Hadley in the 18th century, and is at least partially confirmed by observation.
hadron An elementary particle that interacts *strongly* with other particles. Hadrons include *baryons* and *mesons*.
Haidinger fringes Optical interference fringes produced by transmission and reflection from two parallel, partly reflecting surfaces, e.g. a plate of optical glass. The fringes are produced by division of amplitude of the wave front and are circular fringes formed at infinity (cf. **contour fringes**). Used extensively in interferometry, e.g. Fabry and Pérot, and Michelson interferometers.
hail, hailstones Precipitation in the form of hard pellets of ice, which often fall from cumulonimbus clouds and accompany thunderstorms. They are formed when raindrops are swept up by strong air-currents into regions where the temperature is below freezing point. In falling, the hailstone grows by condensation from the warm moist air which it encounters.
hail stage That part of the condensation process taking place at a temperature of 0°C so that water vapour condenses to water liquid which then freezes.
hair hygrometer A form of *hygrometer* which is controlled by the varying length of a human hair with humidity. It is not an absolute instrument, but it can be used at temperatures below freezing point, and it can be made self-recording.
half-power Condition of a resonant system, electrical, mechanical, acoustical etc. when amplitude response is reduced to $1/\sqrt{2}$ of maximum, i.e. by 3 dB.
half-roll See **roll**.
half-width A measure of sharpness on any function $y = f(x)$ which has a maximum value y_m at x_0 and also falls off steeply on either side of the maximum. The half-width is the difference between x_0 and the value of x for which $y = y_m/2$. Used particularly to measure the width of spectral lines or of a response curve.
halo (1) A spherical region of space surrounding our Galaxy (and others), and permeated by very hot transparent gas. (2) A bright ring or system of rings often seen surrounding the Sun or Moon. The *large halo*, of radius 22°, is due to light refracted at minimum deviation by ice crystals in high cirrostratus clouds. See **corona**.
hangar A special construction for the accommodation of aircraft.
hang glider Original manned glider as used by Lilienthal in Germany in 1890s. Revived in the 1960s as the **Rogallo wing**, employing flexible wing surfaces and now a major class of ultra-light aircraft. Both flexible and fixed hard wings are now used.
hardware The physical material (e.g. structure, electrical harness, computers) produced for aerospace systems as opposed to nontangible aspects such as computer software and operating procedures.
harness (1) The entire system of engine ignition leads, particularly those which are screened to prevent electromagnetic interference with radio equipment. (2) The parallel combination of leads interconnecting the thermocouple probes of a turbine engine exhaust gas temperature indicating system. See **gas temperature**. (3) Prefabricated electrical connections for any electrical or electronic system. (4) Straps by which aircrew are held in their seats.
Hartman oscillator A device, consisting basically of a conical nozzle, supersonic gas jet and cylindrical cavity (resonator), used for generating high-intensity ultrasound in fluids or gases.
Harvard classification A method of classifying stellar spectra, employed by the compilers of the Draper Catalogue of the Harvard Observatory and now in universal use. See **spectral types**.
harvest moon The name given in popular

language to the full Moon occurring nearest to the autumnal equinox, at which time the Moon rises on several successive nights at almost the same hour. This retarded rising, due to the small inclination of the Moon's path to the horizon, is most noticeable at the time of the full Moon, although it occurs for some phase of the Moon each month.

haze A suspension of solid particles of dust and smoke etc. reducing visibility above 1 km.

HDD See **head-down display**.

H-display Modified **B-display** to include angle of elevation. The target appears as two adjacent bright spots and the slope of the line joining these is proportional to the sine of the angle of elevation.

head-down display Usually a visual display mounted inside the cockpit to supplement the **head-up display**.

heading indicator The development of supersonic fighters, which climb at extremely steep angles, made it essential to have an even more comprehensive instrument than the *attitude indicator*. A sphere enables compass heading to be included, thereby giving complete 360° presentation about all three axes.

head-up display The projection of instrument information on to the windscreen or a sloping glass screen in the manner of the **reflector sight**, so that the pilot can keep a continual lookout and receive flight data at the same time. Used originally with fighter interception and attack radar, later adapted for blind approach, landing and general flight data. Abbrev. *HUD*.

hearing The subjective appreciation of externally applied sounds.

heat Heat is energy in the process of transfer between a system and its surroundings as a result of temperature differences. However, the term is still used also to refer to the energy contained in a sample of matter. Also colloq. for **temperature**, e.g. forging or welding *heat*. For some of the chief branches in the study of heat, see **calorimetry, internal energy, latent heat, mechanical equivalent of heat, radiant heat, radiation, specific heat capacity, temperature, thermal conductivity**.

heating muff A device for providing hot air, consisting of a chamber surrounding an exhaust pipe or jet pipe.

heat pipe A means of cooling where heat is transferred along a tube from a heat source to a heat sink of small temperature difference. Heat transfer is effected by a liquid which vaporizes at a desired temperature.

heavier-than-air aircraft See **aerodyne**.

Heaviside layer See **E-layer**.

helical rising, helical setting The rising or setting of a star or planet, simultaneously with the rising or setting of the Sun. It was much observed in ancient times as a basis for a solar calendar for agricultural purposes.

helical tip speed The resultant velocity of the tip of a propeller blade, which is a combination of the linear speed of rotation and the flight speed.

helicopter A **rotorcraft** whose main rotor(s) are power driven and rotate about a vertical axis, and which is thus capable of vertical take-off and landing.

heliocentric parallax See **annual parallax**.

heliometer An instrument for determining the Sun's diameter and for measuring the angular distance between two celestial objects in close proximity. It consists of a telescope with its object glass divided along a diameter, the two halves being movable, so that a superposition of the images enables a value of the angular separation to be deduced from a reading of the micrometer.

heliostat An instrument designed on the same principle as the coelostat, but with certain modifications that make it more suitable for reflecting the image of the Sun than for use on a larger region of the sky; hence used, in conjunction with a fixed instrument, esp. for photographic and spectroscopic study of the Sun.

helium stars Those stars, of spectral type B in the Harvard classification, whose spectrum shows only dark lines, in which those due to the element helium predominate.

Helmert's formula An empirical formula giving the value of g, the acceleration due to terrestrial gravity, for a given latitude and altitude: $g = 9.806\ 16 - 0.025\ 928\ \cos 2\lambda + 0.000\ 069\ \cos^2 2\lambda - 0.000\ 003\ 086H$, where λ is the latitude and H is the height in metres above sea level, g in ms^{-2}.

Helmholtz resonance The type of acoustic resonance arising in a **Helmholtz resonator**.

Helmholtz resonator An air-filled cavity with an opening. The resonance frequency depends on the stiffness of the cavity and the mass of air which oscillates in the opening.

henry SI unit of self and mutual inductance. (1) A circuit has an inductance of 1 henry if an e.m.f. of 1 volt is induced in the circuit by a current variation of 1 ampere per second. (2) A coil has a self-inductance of 1 henry when the magnetic flux linked with it is 1 weber per ampere. (3) The mutual inductance of two circuits is 1 henry when the flux linked with one circuit is 1 weber per ampere of current in the other. Symbol H.

HERMES A manned spaceplane for space transportation and servicing the Columbus Free-Flying Laboratory. Visits to the space stations 'Freedom' and 'MIR' are possibilities.

hertz SI unit of frequency, indicating number of cycles per second (c/s). Symbol Hz.

Hertzian waves Electromagnetic waves from, e.g. 10^4 to 10^{10} Hz, used for communication through space, covering the range from very low to ultra-high frequencies, i.e. from audio reproduction, through radio broadcasting and television to radar.

Hertzsprung-Russell diagram See panel on p. 87.

heterogeneous radiation Radiation comprising a range of wavelengths or particle energies.

HF Abbrev. for *High Frequency*. Radio transmissions occurring between 3 000 and 30 000 kHz.

high aspect ratios See **aspect ratio**.

high by-pass ratio Applied to a *turbofan* **aero-engine** in which the **air mass flow** ejected directly as propulsive thrust by the fan is more than twice the quantity passed internally through the *gas generator* section.

high-energy ignition A gas-turbine ignition system using a very high voltage discharge.

high-level RF signal A radiofrequency signal having sufficient power to fire a switching tube.

high-pressure compressor In a gas-turbine engine with two or more compressors in series, the last is the high-pressure one. In a dual-flow turbojet this feeds the combustion chamber(s) only. Abbrev. *HP compressor*.

high-pressure turbine The first turbine after the combustion chamber in a gas-turbine engine with two or more turbines in series. Abbrev. *HP turbine*.

high-pressure turbine stage The first stage in a **multistage turbine**. Abbrev. *HP stage*.

high-speed wind tunnel A high **subsonic** wind tunnel in which compressibility effects can be studied.

high-wing monoplane An aircraft with the wing mounted on or near the top of the fuselage.

hi/lo stages Refers to the high and low compression stages of the compressor of a gas turbine. See **aero-engine**.

Himalia The sixth natural satellite of **Jupiter**. Its diameter is about 100 km.

HiMAT Abbrev. for *Highly Manoeuvrable Aircraft Technology*.

hinge moment The moment of the aerodynamic forces about the hinge axis of a control surface, which increases with speed, necessitating **aerodynamic balance**.

hoar frost A deposit of ice crystals formed on objects, esp. during cold clear nights when the dewpoint is below freezingpoint. The conditions favouring the formation of hoarfrost are similar to those which produce *dew*.

hodograph A curve used to determine the acceleration of a particle moving with known velocity along a curved path. The hodograph is drawn through the ends of vectors drawn from a point to represent the velocity of the particle at successive instants.

Hohmann orbit A space trajectory tangential to, or *osculating*, two co-planar planetary orbits at its **perihelion** and **aphelion** respectively: it is the most energetically economical transfer orbit.

holding altitude The height at which a controlled aircraft may be required to remain at a **holding point**.

holding pattern A specified flight track, e.g. *orbit* or figure-of-eight, which an aircraft may be required to maintain about a holding point.

holding point An identifiable point, such as a radio beacon, in the vicinity of which an aircraft under **air-traffic control** may be instructed to remain.

homing aid Any system designed to guide an aircraft to an airfield or aircraft carrier.

honeycomb A gridwork across the duct of a wind tunnel to straighten the airflow. Also *straighteners*.

honeycomb structure Lightweight, very rigid, material for aircraft skin or floors, usually made from thin light-alloy plates with a bonded foil interlayer of generally honeycomb-like form. For high supersonic speeds, a heat-resistant material is made by brazing together stainless steel or nickel-alloy skins and honeycomb core.

hood jettison Mechanism, often operated by explosive bolts or cartridge, for releasing the pilot's canopy in flight. Confined to military aircraft.

hop Distance along Earth's surface between successive reflections of a radio wave from an ionized region; also called a *skip*.

horizon That great circle, of which the zenith and the nadir are the poles, in which the plane tangent to the Earth's surface, considered spherical at the point where the observer stands, cuts the celestial sphere.

horizontal parallax The value of the geocentric parallax for a heavenly body in the solar system when the body is on the observer's horizon. In astronomy, the *equatorial horizontal parallax* of a planet or of

Hertzsprung-Russell diagram

Two properties that can be measured with relative ease for a great many stars are the **spectral type** and the **luminosity**. The first of these tells us what type of star we are looking at and its surface temperature. The second is a measure of the absolute quantity of energy being radiated by the star. The Hertzsprung-Russell diagram (or HR diagram) is named for E. Hertzsprung and H.N. Russell who independently (1911 and 1913) discovered that the two quantities are correlated.

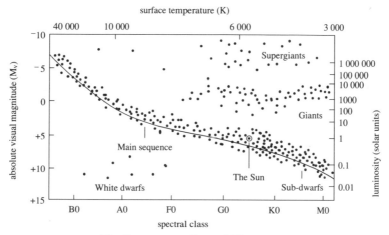

The Hertzsprung-Russell Diagram

The dots, which are schematic, represent the location of stars to show the scatter which is in practice observed.

For the stars this correlation is very striking. About 90% of all stars lie on a narrow diagonal band known as the **main sequence**. Stars on the left side of this diagram are hot, those to the right cool. There is also a line of cool luminous stars. These are giants on the giant red branch. *White dwarf* stars form a separate population at the lower left: they are hot but dim. The fact that not all combinations of temperature and luminosity are found means that the study of stars is drastically simplified compared to the study of galaxies or the interstellar medium. In particular, this branch of astrophysics is amenable to investigation via rigorous mathematical physics.

The explanation of the HR diagram has been one of the great triumphs of **stellar evolution** theory. Essentially, when new stars form, the only parameter that distinguishes one from another is mass. For most of its life a star burns hydrogen in the core to form helium. The more massive stars have higher central temperatures and burn hydrogen faster, have hotter surfaces and are more luminous. These stars burning hydrogen in the core are all found on the main sequence. The Sun is one such star. While on the main sequence the temperature and luminosity scarcely change, so a given star will be at more or less the same place on the main sequence.

Eventually hydrogen burning in the core stops. Structural changes occur and the star moves off the main sequence on to the giant branch. Although the surface temperature falls, the surface area expands greatly as the star swells. The luminosity therefore goes up temporarily. As the star nears the end of its life it contracts again, crosses the main sequence from right to left and eventually settles as a white dwarf.

continued on next page

Hertzsprung-Russell diagram (contd)

The HR diagram is a valuable tool for investigating discrete populations of stars, such as those found in **star clusters**. All stars in a cluster are at more or less the same distance. For such a population the HR diagram can be plotted in relative terms without knowing the distance exactly. The point at which the main sequence breaks over to the red giant branch is an indication of cluster age. These vary from 20 million years for a young *open cluster* such as the Pleiades, to 10,000 million years for highly evolved *globular clusters*. The HR diagram can also be used to gauge a star's distance. The spectral type is a proxy indicator of absolute **magnitude**, which can be compared to the directly measured relative magnitude to yield a distance.

Related to the HR diagram are the **colour-luminosity array** and *colour-colour* diagrams. In these plots the colour is the difference in magnitude when the star is examined through two or more colour filters. The advantage of this cruder measure of star type is that large numbers of colours can be measured from a series of survey photographs taken through filters. This technique can be automated. Traditional spectroscopy is, by contrast, very much slower.

---

the Moon is the angle subtended at the centre of that body by the equatorial radius of the Earth. See **solar parallax**.

horizontal stabilizer Tailplane. See **stabilizer**.

horn balance An **aerodynamic balance** consisting of an extension forward of the hinge line at the tip of a control surface; it may be *shielded* (i.e. screened by a surface in front) or *unshielded*.

horn feed The feed to an antenna in the form of a horn.

Horsehead Nebula A famous dark nebula in the constellation Orion; it bears a plausible resemblance to the silhouette of a horse's head.

horse latitudes See **tradewinds**.

HOTOL Abbrev. for *Horizontal Take-Off and Landing*, a proposed single-stage-to-orbit launch vehicle burning oxygen and hydrogen propellants; a revolutionary engine uses atmospheric oxygen below Mach 5.

hot-wire anemometer An instrument which measures wind velocities by using their cooling effect on a wire carrying an electric current, the resistance of the wire being used as an indication of the velocity.

hour angle The angle, generally measured in hours, minutes and seconds of time, which the hour circle of a heavenly body makes with the observer's meridian at the celestial pole; it is measured positively westwards from the meridian from 0 to 24 hr.

hour circle (1) The great circle passing through the celestial poles and a heavenly body, cutting the celestial equator at 90°. (2) The graduated circle of an equatorial telescope which reads sidereal time and right ascension.

housekeeping data A term used to denote

information on the working of a space system, subsystem or component (as opposed to scientific data *per se*).

Hovmüller diagram A diagram, with one axis representing time and the other longitude, on which are shown isopleths of an atmospheric variable such as pressure or *thickness*, usually averaged over a band of latitude. It demonstrates, very effectively, the movement over weeks or months of large-scale atmospheric features.

howl See **screeching**.

howl A high-pitched audio-tone due to unwanted acoustic (or electrical) feedback.

H-radar Navigation system in which an aircraft interrogates two ground stations for distance.

Hubble classification A scheme for the classification of galaxies according to morphology, the principle types being **elliptical** and **spiral**.

Hubble diagram A plot of redshifts of galaxies against their inferred distances. See **redshift-distance relation**.

Hubble parameter Originally known as the *Hubble constant* of proportionality. It is a measure of the rate at which the expansion of the universe varies with distance, and is expressed in the units km/sec per megaparsec, and has the value 50–100 km/sec/Mpc. It is determined from the observed recession velocities or *redshifts* of distant **galaxies**. See **redshift-distance relation**.

Hubble relation *Hubble law*. The relationship first described by E. Hubble in 1925, which states that the recession velocity (or *redshift*) of a distant galaxy is directly proportional to the distance from the observer.

Hubble space telescope An optical telescope of 2.5 m diameter carried by satellite above the Earth's atmosphere which will permit seeing up to 100 times fainter objects than is possible with Earth-bound instruments.

HUD Abbrev. for **Head-Up Display.** Cf. **head-down display.**

hull The main body (structural, flotation and cargo-carrying) of a flying-boat or boat amphibian.

hum Objectionable low-frequency components induced from power mains into sound reproduction, caused by inadequate smoothing of rectified power supplies, induction into transformers and chokes, unbalanced capacitances, or leakages from cathode heaters.

humidity See **vapour concentration, relative-, specific-humidity.**

humidity mixing ratio The ratio of the mass of water vapour in a sample of moist air to the mass of dry air with which it is associated.

hump speed The speed, on the water, at which the water resistance of the floats or boat body of a seaplane or flying-boat is a maximum. After this is past the craft begins to plane over the water.

hunting (1) An uncontrolled oscillation, of approximately constant amplitude, about the flight path of an aircraft. (2) The angular oscillation of a rotorcraft's blade about its drag hinge. (3) The oscillation of instrument needles.

hurricane (1) A wind of force 12 on the *Beaufort Scale.* A mean wind speed of 75 mph (120 km/h). (2) A *tropical revolving storm* in the North Atlantic, eastern North Pacific and the western South Pacific.

hyades The name of a star cluster, of the 'open' type, situated in the constellation Taurus; visible to the naked eye.

hydraulic accumulator A weight-loaded or pneumatic device for storing liquid at constant pressure, to steady the pump load in a system in which the demand is intermittent. In aircraft hydraulic systems an accumulator also provides fluid under pressure for operating components in an emergency, e.g. failure of an engine-driven pump.

hydraulic reservoir In an aircraft hydraulic system, the header tank which holds the fluid; not to be confused with the *accumulator*, which is a pressure vessel wherein hydraulic energy is stored.

hydrofoil An immersed aerofoil-like surface to facilitate the take-off of a seaplane by increasing the hydrodynamic lift.

hydrogen I, II The hydrogen of interstellar space, known in two clearly defined states: H I (*neutral hydrogen*) and H II (*ionized*

hydrogen). The H I regions, confined mainly to the spiral arms of the galaxy, emit no visible light, and are detected solely by their emission of the 21-cm radio line. The H II regions, found in the gaseous nebulae, emit both visible and radio radiations.

hydrogen bomb Uses the nuclear fusion process to release vast amounts of energy. Extremely high temperatures are required for the process to occur and these temperatures are obtained by an *atomic bomb* around which the fusion material is arranged. Lithium deuteride can initiate a number of fusion processes involving the hydrogen isotopes, deuterium and tritium. These reactions also produce high-energy neutrons capable of causing *fission* in a surrounding layer of the most abundant isotope of uranium, ^{238}U so that further energy is released.

hydrolapse The rate of decrease with height of atmospheric water vapour as measured by humidity mixing ratio, dew point or other suitable quantity.

hydrological cycle The evaporation and condensation of water on a world scale.

hydrometeor A generic term for all products of the condensation or sublimation of atmospheric water vapour, including: ensembles of falling particles which may either reach the Earth's surface (rain and snow) or evaporate during their fall (virga); ensembles of particles suspended in the air (cloud and fog); particles lifted from the Earth's surface (blowing or drifting snow and spray); particles deposited on the ground or on exposed objects (dew, hoar frost etc.).

hydroskis Hydrofoils, proportioned like skis and usually retractable, fitted to seaplanes without a **planing bottom** as the sole source of hydrodynamic lift. They are also fitted to aircraft landing gears to make them amphibious (*pantobase*), in which case a minimum taxiing speed is necessary to keep the aircraft above water.

hydrostatic approximation The assumption that the atmosphere is in hydrostatic equilibrium in the vertical which is equivalent to ignoring the vertical components of acceleration and **Coriolis force** in the equations of motion. The approximation is valid for atmospheric disturbances on horizontal scales of not less than about 10 km.

hydrovane See **hydrofoil, hydroskis, sponson.**

hyetograph An instrument which collects, measures and records the fall of rain (Gk. *hyetos*, rain).

hygrometry The measurement of the hygrometric state, or **relative humidity,** of the atmosphere.

hyperbarism An agitated bodily condition when the pressure within the body tissues, fluids and cavities is countered by a greater external pressure, such as may happen in a sudden fall from a high altitude.

hypergolic A rocket propellant mixture (fuel and oxidizer) which ignites spontaneously upon mixing.

Hyperion The seventh natural satellite of **Saturn**.

hypersonic Velocities of **Mach number** 5 or more. *Hypersonic flow* is the behaviour of a fluid at such speeds, e.g. in a shock tube.

hypersound Sound of a frequency over 10^9 Hz; not audible.

Hz Symbol for **hertz**.

I

Iapetus The eighth natural satellite of **Saturn**.

IAS Abbrev. for **Indicated Air Speed**.

IATA Abbrev. for **International Air Transport Association**.

ICAO Abbrev. for **International Civil Aviation Organization**.

Icarus A minor planet, discovered in 1949, with a small eccentric orbit approaching the Sun to within 30 million kilometres, inside the orbit of Mercury. In June 1968 Icarus passed within four million miles of the Earth.

ice Ice is formed when water is cooled below its freezing point. It is a transparent crystalline solid of rel. d. 0.916 and specific heat capacity 0.50. Because water attains its maximum density at 4°C, ice is formed on the surface of ponds and lakes during frosts, and thickens downwards.

iceberg A large mass of ice, floating in the sea, which has broken away from a glacier or ice barrier. Icebergs are carried by ocean currents for great distances, often reaching latitudes of 40° to 50° before having completely melted. Approximately one-tenth of an iceberg shows above the surface.

iceblink A whitish glare in the sky over ice which is too distant to be visible.

ice guard A wire-mesh screen fitted to a piston aero-engine intake so that ice will form on it and not inside the intake; a *gapped ice guard* is mounted ahead of the intake so that air can pass round it, while a *gapless ice guard* is inside the intake and an alternative air path comes into use when it ices up.

ice pellets Precipitation of transparent or translucent pellets of ice with diameters of 5 mm or less.

I-display Display representing the target as a full circle when the antenna is pointed directly at it, the radius being in proportion to the range.

IFF Abbrev. for World War II system whereby vessels and aircraft carried **transponders** capable of indicating to a 'friendly' radar system that they were not hostile. *IFF* is used now as a general description of such radar identification systems. See **ATCRBS**.

IFR Abbrev. for **Instrument Flight Rules**.

igniter plug An electrical-discharge unit for lighting up gas turbines.

IGV Abbrev. for **Inlet Guide Vanes**.

illumination, illuminance The quantity of light or luminous flux falling on unit area of a surface. Illumination is inversely proportional to the square of the distance of the surface from the source of light, and proportional to the cosine of the angle made by the normal to the surface with the direction of the light rays. The unit of illumination is the *lux*, which is an illumination of 1 lumen/m². Symbol *E*.

ILS Abbrev. for **Instrument Landing System**.

image Optical images may be of two kinds, real or virtual. A *real image* is one which is formed by the convergence of rays which have passed through the image-forming device (usually a lens) and can be thrown on to a screen, as in the camera and the optical projector. A *virtual image* is one from which rays appear to diverge. It cannot be projected on to a screen or a sensitive emulsion.

imaging system An instrument with its supporting **hardware** and **software** for obtaining remote images of the Earth and objects in space. The types of image may be optical, microwave or obtained by using some other selected part of the electromagnetic spectrum. The images may be recorded, e.g. on film or digitally and recovered later, or transmitted in real-time to an Earth receiving station.

IMC Abbrev. for *Instrument Meteorological Conditions*, wherein aircraft must conform to **instrument flight rules**.

immersed pump An electrical pump mounted inside a fuel tank.

immersion The entry of the Moon, or other body, into the shadow which causes its eclipse.

impact accelerometer An accelerometer which measures the deceleration of an aircraft while landing.

impeller The rotating member of a **centrifugal-flow compressor** or supercharger.

impeller-intake guide-vanes The curved extension of the vanes of a centrifugal impeller which extend into the intake eye or throat, and which thereby give the airflow initial rotation.

impulse starter A mechanism in a magneto which delays the rotor against a spring so that, when released, there is a strong and retarded spark to help starting.

impulsive sound Short sharp sound, the energy spectrum of which spreads over a wide frequency range.

incidence, angle of See **angle of incidence**.

indicated airspeed The reading of an airspeed indicator which when corrected for instrument errors, reads low by a factor equal to the square root of the relative air density as the latter falls with altitude. Abbrev. *IAS*.

induced drag The portion of the **drag** of an aircraft attributable to the derivation of lift.

induction flame damper See **flame trap**.

inequality The term used to signify any departure from uniformity in orbital motion; it may be (1) *periodic*, i.e. completing a full cycle within a specific time and then repeating it; or (2) *secular*, i.e. increasing steadily in magnitude with time.

inertia The property of a body, proportional to its mass, which opposes a change in the motion of the body. See **inertial force**, **inertial reference frame**.

inertial damping That which depends on the acceleration of a system and not velocity.

inertial force Newton's laws of motion do not apply in accelerating and rotating frames of reference. Newton's laws may still be used in these frames if *inertial forces* are introduced to preserve the second law of motion. In the case of rotating frames these forces are the centrifugal and Coriolis forces.

inertial navigation system An assembly of highly accurate gyros to stabilize a platform supported on gimbals on which are mounted a group of similarly accurate accelerometers (typically one for each of the three rectilinear axes) which measure all accelerations imparted. With one automatic time integration, these measurements give a continuous read-out of velocity and with another a read-out of present position related to the start. Accuracy is typically 1 in 10^9.

inertial reference frame In mechanics, a reference frame in which Newton's first law of motion is valid.

inferior conjunction See **conjunction**.

inferior planets See **planet**.

inflatable aircraft A small aircraft of low performance for military use, in which the aerofoil surfaces (and sometimes the fuselage) are inflated so that it can be compactly packed for transport.

inflation The process of filling an airship or balloon with gas. Sometimes *gassing*.

inflationary universe Model of the very early universe (10^{-44} seconds after the **Big Bang**) in which the universe expands momentarily much faster than the speed of light. This model is able to account for the flatness of spacetime and the isotropy of the observed universe.

infrared astronomy The study of radiation from celestial objects in the wavelength range, 800 nm – 1 mm. Absorption by water vapour in our atmosphere poses severe difficulties, some of which are overcome at high altitude observatories such as Mauna Kea in Hawaii at 4000 m. There are many thousands of infrared sources in our Galaxy, principally cool giant stars, nascent stars and the Galactic centre itself.

infrared countermeasures Means of deceiving missiles guided by infra-red sensors by (1) flares deployed as decoys from aircraft or (2) in design, by reducing heat output from jet exhaust or shielding hot parts.

infrared radiation Electromagnetic radiation in the wavelength range from 0.75 to 1000 µm approximately, i.e. between the visible and microwave regions of the spectrum. The *near* infrared is from 0.75 to 1.5 µm, the *intermediate* from 1.5 to 20 µm and the *far* from 20 to 1000 µm.

infrared spectrometer An instrument similar to an optical spectrometer but employing non-visual detection and designed for use with infrared radiation. The infrared spectrum of a molecule gives information as to the functional groups present, and is very useful in the identification of unknown compounds.

infrasound Sound of frequencies below the usual audible limit, viz. 20 Hz.

injection carburettor A pressure carburettor in which the fuel delivery to the jets is maintained by pressure instead of by a float chamber. It is unaffected by negative *g* in aerobatics or severe atmospheric turbulence.

inlet guide vanes Radially positioned aerofoils in the annular air intake of axial-flow compressors which direct the airflow on to the first stage at the most efficient angle. The vanes are often rotatable about their mounting axes so that different entry airspeeds can be accomodated; they are then called *variable inlet guide vanes*. Abbrev. *IGV*.

INMARSAT Abbrev. for *International Maritime Satellite Organisation*, an intergovernmental agency concerned with ship-to-ship and ship-to-shore communications via satellite. INMARSAT's activities now include mobile communications on land and in the air as well as at sea.

inner marker beacon A vertically directed radio beam which marks the airport boundary in a beam-approach landing system, such as the **instrument landing system**.

insolation The radiation received from the Sun. This depends on the position of the Earth in its orbit, the thickness and transparency of the atmosphere, the inclination of the intercepting surface to the Sun's rays, and the **solar constant**.

instability (1) An aircraft possesses *instability* when any disturbance of its steady motion tends to increase, unless it is over-come by a movement of the controls by the pilot. (2) Vibration with an exponentially growing amplitude. It occurs if there is some sort of positive **feedback**, e.g. thermo-acoustic instability, screech.

instantaneous automatic gain control A rapid action system in radar for reducing the clutter.

instrument approach Aircraft approach made using **instrument landing system**.

instrument flight rules The regulations governing flying in bad visibility under strict flight control. Abbrev. *IFR*.

instrument landing system A localizer for direction in the horizontal plane and glideslope in the vertical plane, usually in-clined at 3° to the horizontal. Two markers are used for linear guidance. Abbrev. *ILS*.

integral stiffeners The stiffening ridges left when an aircraft skin panel is machined from a solid billet.

INTELSAT The *INTernational TELecom-munications SATellite organisation*. Body responsible for the design, construction, de-velopment, operations and maintenance of the worldwide satellite communications sys-tem. Organized by government telecom-munication authorities, who have shares in the organization in proportion to their use of it.

intensity level Level of power per unit area as expressed in decibels above an arbitrary zero level, e.g. sound intensity level.

intensity of sound The magnitude of a sound wave, measured in terms of the power transmitted, in watts, through unit area nor-mal to the direction of propagation.

intercooler (1) A heat exchanger on the de-livery side of a supercharger which cools the charge heated by compression. (2) A second-ary heat exchanger used in a cabin air con-ditioning system for cooling the charge air from the compressor to the turbine of a **boot-strap cold-air unit**.

INTERCOSMOS A consortium of Soviet al-lied nations which carries out co-operative efforts in space matters.

interface Relationship between parts of a system or subsystem which ensures that their eventual meeting will be harmonious; the interface may be physical (e.g. mechanical, thermal, electrical) or non-physical (e.g. software, organizational) and all conditions of the eventual union are controlled as part of the system documentation.

interference (1) Interaction between two or more waves of the same frequency emitted from coherent sources. The wavefronts are combined, according to the *principle of*

superposition, and the resulting variation in the disturbances produced by the waves is the interference pattern. See **interference fringes**. (2) Mutual aerodynamic interactions between solid bodies in airflow, the drag of the com-bined bodies exceeding that of their separate drags by the interference drag. Thus the lift of the lower wing of a biplane is reduced by the flow under the upper wing.

interference colours The *colours of thin films*. When light is reflected from a thin film, such as a soap bubble or a layer of oil on water, coloured effects are seen which are due to optical interference between light re-flected from the upper surface and that from the lower. The colours are not bright unless the film is less than about 10^{-3} mm thick.

interference filter If a beam of light is inci-dent normally on a Fabry-Pérot etalon with a plate spacing d of between 2×10^{-7} and 6×10^{-7}, then there will be only one wavelength in the visible region for which there is an inter-ference maximum ($n\lambda=2d$, where λ is the wavelength and $n=$ 0,1,2,...). The highly re-flecting surfaces of the etalon ensure that only a narrow band of wavelengths around λ is transmitted. Interference filters can also be made by successively evaporating dielectric and silvered films of suitable thickness on a glass plate.

interference fringes Alternate light and dark bands formed when two beams of monochromatic light, having a constant phase relation, overlap and illuminate the same portion of a screen. The method of producing fringes is either by division of wavefront (see **Fresnel's biprism, Lloyd's mirror**), or by division of amplitude (see **Haidinger fringes, Fabry-Pérot interfe-rometer, Newton's rings**).

interference pattern See **interference**.

interferometer Instrument in which an acoustic, optical or microwave interference pattern of *fringes* is formed and used to make precision measurements, mainly of wave-length.

intergalactic medium General term for any material which might exist isolated in space far from any galaxy. Theoretical con-siderations suggest that an enormous quan-tity of undetected material exists, the total mass of which could far exceed the visible galaxies.

intermediate pressure compressor The section of the axial compressor of a turbofan or by-pass turbojet between the *LP* and *HP* sections. It may be on a shaft of its own with a separate turbine, or it may be mounted on either of the other shafts.

intermittent jet See pulse jet.

internal energy For a thermodynamic system, the difference between the heat absorbed by the system and the external work done by the system, is the change in its internal energy (the *first law of thermodynamics*). The internal energy takes the form of the kinetic energy of the constituent molecules and their potential energies due to the molecular interactions. The internal energy is manifest as the temperature of the system, latent heat, as shown by a change of state, or the repulsive forces between molecules, seen as expansion. Symbol U; SI unit is the joule.

International Air Transport Association Association founded in Havana (1945) for the promotion of safe, regular and economic air transport, to foster air commerce and collaboration among operators, and to co-operate with ICAO and other international organizations. Abbrev. *IATA*.

International Civil Aviation Organization An intergovernmental co-ordinating body, which has its headquarters at Montreal, Canada, for the regulation and control of civil aviation and the co-ordination of **airworthiness** requirements and other safety measures on a worldwide basis. Abbrev. *ICAO*.

International Space Station The former name for 'Freedom', the space station jointly being developed by NASA, ESA, Japan and Canada. See **space station**.

International Standard Atmosphere A standard fixed atmosphere, adopted internationally, used for comparing aircraft performance; mean sea-level temperature 15°C at 1013.2 millibars, lapse rate 6.5°C/km altitude up to 11 km (ISA tropopause), above which the temperature is assumed constant at −56.5°C in the stratosphere. Abbrev. *ISA*.

interplane struts In a multiplane structure, those struts, either vertical or inclined, connecting the spars of any pair of planes, one above the other.

interplanetary matter Material in the solar system other than the planets and their satellites. This matter includes the streams of charged particles from the **solar wind**, the resulting tenuous gas being termed the *interplanetary medium*. Interplanetary matter also includes dust, **meteorites** and **comets**.

interplanetary space See panel on **space**.

INTERSPUTNIK The Soviet space communication system.

interstellar hydrogen The **interstellar medium** is predominantly hydrogen. Cool hydrogen clouds *H I regions* are diffuse clouds of atomic hydrogen detected by the radio emission at a frequency of 1420.4 Mhz ('21 centimetre line') associated with a *forbidden*

transition between two ground state energy levels. Ionized hydrogen, or *H II regions*, exists as discrete clouds often associated with star formation or other sources of ultraviolet radiation. See **forbidden lines**.

interstellar medium The gaseous and dusty matter pervading interstellar space and amounting to one-tenth of the mass of the Galaxy. Where cold gas and dust conglomerate, regions of star formation occur. The medium comprises hydrogen, helium, **interstellar molecules** and dust. It is replenished by stellar winds and the ejecta of **nova** outbursts and **supernova** explosions.

interstellar molecule More than fifty species of molecule are found within the gas nebulae of the **interstellar medium**, particularly in cold dense clouds. The most common molecules include: CO, H_2O, NH_3, $HCHO$, CH_4 and CH_3OH.

intertropical convergence zone *ITCZ*. Air masses originating in the northern and southern hemispheres converge. Over the Atlantic and Pacific oceans it is the boundary between the northeast and southeast trade winds. The mean position is somewhat north of the equator but over the continents the range of movement is considerable. It is a zone of generally cloudy, showery weather. Sometimes the *intertropical front (ITF)*.

interval Ratio of the frequencies of two sounds; or in some cases its logarithm. See **semitone**.

invariable plane A certain plane which remains absolutely unchanged by any mutual action between the planets in the solar system; defined by the condition that the total angular momentum of the system about the normal to this plane is constant. The plane is inclined at 1°35' to the ecliptic.

inverse square law The intensity of a field of radiation is inversely proportional to the square of the distance from the source. Applies to any system with spherical wavefront and negligible energy absorption.

inversion Reversal of the usual temperature gradient in the atmosphere, the temperature increasing with height. Inversions are of frequent occurrence near the ground on clear nights and in anticyclones, often causing dense smoke fogs over cities.

inverted engine An in-line engine having its cylinders below the crankshaft. Adopted in certain types of aircraft to improve the forward view of the pilot.

inverted loop An *aerobatic* manoeuvre consisting of a complete revolution in the vertical plane with the upper surface of the aircraft outside, which is started from the inverted position. Also *outside loop*.

Io The first natural satellite of **Jupiter**, discovered by Galileo. It is characterized by very active volcanoes, and stimulates radio emission in the ionosphere of **Jupiter**.

ion Strictly, any atom or molecule which has resultant electric charge due to loss or gain of valency electrons. Free electrons are sometimes loosely classified as *negative ions*. Ionic crystals are formed of ionized atoms and in solution exhibit ionic conduction. In gases, ions are normally molecular and cases of double or treble ionization may be encountered. When almost completely ionized, gases form a fourth state of matter, known as a **plasma**. Since matter is electrically neutral, ions are normally produced in pairs.

ion engine A device for propulsion or attitude control of spacecraft or satellites, esp. in outer space. Ions and electrons are expelled at high velocity from *combustion chamber*. Outside, they recombine, thus preventing any space-charge effect which would counteract thrust.

ionization temperature A critical temperature, different for different elements, at which the constituent electrons of an atom will become dissociated from the nucleus; hence a factor in deducing stellar temperatures from observed spectral lines indicating any known stage of ionization.

ionizing radiation Any electromagnetic or particulate radiation which produces ion pairs when passing through a medium.

ionosphere That part of the Earth's atmosphere (the upper atmosphere) in which an appreciable concentration of ions and free electrons normally exist. This shows daily and seasonal variations. See **E-layer, F-layer, hop**.

ionospheric disturbance An abnormal variation of the ion density in part of the ionosphere, commonly produced by solar flares. Has a marked effect on radio communication. See **ionospheric storm**.

ionospheric prediction The forecasting of ionospheric conditions relevant to communication.

ionospheric regions These are: D region, between 90 and 150 km; F region, over 150 km; all above surface of Earth. Internal effective layers are labelled E, *sporadic E*, E_2, F, F_1, $F_{1.5}$, F_2.

ionospheric storm Turbulence in parts of the ionosphere, probably connected with sunspot activity, causing dramatic changes in its reflective properties and sometimes totally disrupting short-wave communications.

ion propulsion Method of rocket propulsion in which charged particles (e.g. lithium or caesium ions) are accelerated by an electrostatic field, giving a small thrust but a high specific impulse; the thrust-to-weight ratio is small so that the main use is for station keeping.

IR Abbrev. for *InfraRed*.

IRCM See **infrared countermeasures**.

iridescent clouds High clouds which show colours, generally delicate pink and green, in irregular patches. It is thought that the effect is caused by the diffraction of sunlight by supercooled water droplets.

irregular galaxies Small galaxies showing no symmetry; they contain little dust or gas. See **Magellanic Clouds**.

irregular variables See **variable star**.

irreversible controls A flying control system, hydraulically or electrically operated, wherein there is no feed-back of aerodynamic forces from the control surfaces.

ISA Abbrev. for **International Standard Atmosphere**.

isallobar The contour line on a weather chart, signifying the location of equal changes of pressure over a specified period.

isallobaric high and low Centres, respectively, of rising and falling **barometric tendency**.

isallobaric wind Theoretical component of the wind arising from the spatial non-uniformity of local rates of change of pressure.

ISAS Abbrev. for *Institute of Space and Aeronautical Science* of the University of Tokyo, mainly reponsible for the Japanese scientific satellites.

island universe The name applied to an extragalactic nebula.

isobar A line drawn on a map through places having the same atmospheric pressure at a given time.

isobarometric, isobaric, charts Maps on which isobars arc drawn. See **isobar**.

isocline Line on a map, joining points where the angle of dip (or inclination) of the Earth's magnetic field is the same.

isoclinic A wing designed to maintain a constant angle of incidence even when subject to dynamic loads.

isodynamic lines Lines on a magnetic map which pass through points having equal strengths of the Earth's field.

isogonic line Line on a map joining points of equal magnetic declination, i.e. corresponding variations from true north.

isogrivs Lines drawn on a **grid navigation** system chart joining points at which convergency of meridians and magnetic variation are equal.

isohel A line drawn on a map through places having equal amounts of sunshine.

isohyet A line drawn on a map through places having equal amounts of rainfall.

isolation Prevention of sound transmission by walls, mufflers, resilient mounts etc.

isolux Locus, line or surface where the light intensity is constant. Also *isophot*.

isomagnetic lines Lines connecting places at which a property of the Earth's magnetic field is a constant.

isopleth *Isogram*, esp. one on a graph showing variations of a climatic element as a function of two variables.

isopycnic A line on a chart joining points of equal atmospheric density.

isostere A line on a chart joining points of equal atmospheric *specific volume*, i.e. the volume of unit mass..

isotach A line on a chart joining points of equal windspeed.

isotherm A line drawn on a chart joining points of equal temperature.

ISRO Abbrev. for *Indian Space Research Organisation*, which oversees all Indian space activities.

ISS Abbrev. for **International Space Station.**

J K

jansky A unit in radio astronomy to measure the power received at the telescope from a cosmic radio source. 1 jansky (Jy) = 10^{-26} W m^{-2} Hz^{-1} sr^{-1}.

Janus The tenth natural satellite of **Saturn**, discovered in 1980.

JATO Abbrev. for *Jet-Assisted Take-Off*. See **take-off rocket**.

J-display A modified *A-display* with circular time base. See **R-display**.

JEM Abbrev. for *Japanese Experimental Module*, part of the international space station. See **space station**.

Jeppesen chart Airway charts, airport maps and information, named after Ebroy Jeppesen, who built up the basic format from 1926 to 1940.

jet co-efficient The basic non-dimensional thrust-lift relationship of the **jet flap**;

$$C_J = J/\tfrac{1}{2}e^{V^2 s}$$

where J = jet thrust, e = air density, V = speed and S = wing area.

jet deflection A jet-propulsion system in which the thrust can be directed downwards to assist take-off and landing.

jet flap A high-lift flight system in which (1) the whole efflux of turbojet engines is ejected downward from a spanwise slot at the wing trailing edge or (2) a large surplus efflux from turboprop engines is so ejected, with the propellers providing a relative airflow over the wing. The downward ejection of the jet forms a barrier to the passage of air under the wing and induces more air to flow across the upper surface, so giving very high lift co-efficients of the order of 10 and even higher. See **NGTE rigid rotor**.

jet lag Delayed bodily effects felt after long flight by fast jet aircraft.

jet noise The noise of jet efflux varies as the eighth power of its velocity.

jet pipe shroud A covering of heat-insulating material, usually layers of bright foil, round a jet pipe.

jet propulsion Propulsion by reaction from the expulsion of a high-velocity jet of the fluid in which the machine is moving. It has been used for the propulsion of small ships by pumping in water and ejecting it at increased velocity, but the principal application is to aircraft. See **aero-engines**.

jet stream A fairly well-defined core of strong wind, perhaps 200–300 miles (320–480 km) wide with wind speeds up to perhaps 200 mph (320 km/h) occurring in the vicinity of the **tropopause**.

joint Permanent or semi-permanent connection between two lengths of waveguide. It may consist of plain flanges with direct metallic contact and no discontinuity in the waveguide walls, or it may be a **choke flange**.

Joint Tactical Information Distribution System *JTIDS*. A full-scale tactical, jam-resistant, command and control system developed under the leadership of the US Air Force. It integrates the hitherto separate functions of communication, navigation and identification.

joule The SI unit of work, energy and heat. 1 joule is the work done when a force of 1 newton moves its point of application 1 metre in the direction of the forces. Abbrev. *J*. 1 erg = 10^{-7}J, 1 k Wh = $3.6{\times}10^6$J, 1 eV = $1.602{\times}10^{-19}$J, 1 calorie = 4.18 J, 1 Btu = 1055 J.

joystick Colloq. for *control column*.

JP- Nomenclature for jet fuels. See **aviation fuels**.

JPT Abbrev. for *Jet Pipe Temperature*. See **gas temperature**.

JTIDS Abbrev. for **Joint Tactical Information Distribution System.**

Julian calendar The system of reckoning years and months for civil purposes, based on a tropical year of 365.25 days; instituted by Julius Caesar in 45 BC and still the basis of our calendar, although modified and improved by the Gregorian reform.

Julian date The number of days which have elapsed since 12.00 GMT on 1 January 4713 BC. This consecutive numbering of days gives a calendar independent of month and year which is used for analysing periodic phenomena, esp. in astronomy. This system devised in 1582 by J. Julius Scaliger has no connection whatsoever with the Julian calendar.

Jupiter See panel on p. 98.

jury strut A strut giving temporary support to a structure. Usually required for folding-wing biplanes, sometimes for naval monoplanes with folding parts.

K The symbol for kelvin.

[K] A very strong Fraunhofer line in the extreme violet of the solar spectrum. See **[H]**.

Karman vortex street Regular vortex pattern behind an obstacle in a flow where vortices are generated and travel away from the object. The frequency of vortex generation is determined by the **Strouhal number**.

katabatic wind A local wind flowing down

Jupiter

The largest planet in the solar system and the fifth in order out from the Sun. After **Venus**, it is the second brightest planet as seen from Earth.

Jupiter is ten times the size of the Earth and one tenth of the Sun's diameter. Its composition (by number of molecules) is very similar to the Sun: 90% hydrogen and 10% helium. The most significant trace gases are water vapour, methane and ammonia. Jupiter has no solid surface: there is a gradual transition from gas to liquid that takes place as the pressure increases with depth below the outermost layers, followed by an abrupt change to metallic liquid hydrogen, in which the atoms are stripped of their electrons. At the very centre there may be a small core of rock and perhaps ice. Jupiter radiates 1.5–2 times as much heat as it absorbs from the Sun; the source of this internal energy is heat generated as Jupiter continues to contract gravitationally.

Observed visually, the disk of Jupiter is crossed by alternating light zones and dark belts. These are easily visible in the smallest telescope. Results from four space probes that have passed by Jupiter between 1973 and 1981 (Pioneers 10 and 11, Voyagers 1 and 2) have revealed the full complexity of the flow pattern within these bands. There are five or six in each hemisphere, correlating with wind currents. White or coloured ovals appear as relatively long-lived features. The most well-known and conspicuous is the *Great Red Spot*, which has been observed for around 300 years. The origin of this feature, which is as wide as the Earth, is uncertain. One popular theory is that it is a huge hurricane. A similar dark spot occurs in Neptune.

The coloured clouds in the highest layers of Jupiter occupy a zone only 0.1–0.3% of the total radius. The origin of their colour is uncertain though it must involve trace constituents of the atmosphere and is evidence of complex chemistry. Cloud colour correlates with altitude: blue features are the deepest, followed by brown, then white, with red being the highest.

The existence of a faint ring around Jupiter was first suggested by results from Pioneer 11 in 1974 and confirmed by direct Voyager images. It lies between 1.81 and 1.72 Jupiter radii from the centre of the planet. The ring must have many particles with dimensions measured in micrometres. A constant source of replenishment is also required, which may be a population of boulder-sized objects, constantly bombarded by high-velocity particles.

There are sixteen known natural satellites orbiting Jupiter. They fall into four distinct groups. The four small inner satellites (*Adrastea, Metis, Amalthea* and *Thebe*) and the four large Galilean satellites (*Io, Europa, Ganymede* and *Callisto*) are in circular orbits in the equatorial plane. The third group (*Leda, Himalia, Lysithea* and *Elara*) are small satellites in circular orbits, inclined at angles between 25° and 29° to the equatorial plane and at distances between 11 and 12 million kilometres from Jupiter. The outermost group (*Ananke, Carme, Pasiphae* and *Sinope*) are small satellites in retrograde orbits that are relatively eccentric ellipses, inclined substantially to the equatorial plane. These orbits all lie between 21 and 24 million kilometres from Jupiter. The four Galilean satellites and their movements in orbit are easily visible with a small telescope or binoculars.

Radio emission from Jupiter, discovered in 1955, indicates the presence of a strong magnetic field which is 4000 times greater than the Earth's. The **magnetosphere** is consequently 100 times larger. The radio emission is caused by the spiralling of electrons around the field lines. Trapped electrons near the planet give rise to **synchroton** radiation at decimetre wavelengths. Decametric radiation, observed only from certain regions of the planet, is associated with the interaction between Jupiter's ionosphere and Io, whose orbit lies within a huge *plasma torus*. This interaction also creates *aurorae*. Radiation at kilometre wavelengths was discovered by the Voyager probes, originating in high latitude regions near the planet and in the plasma torus.

a slope cooled by loss of heat through radiation at night. It is caused by the difference in density between cold air in contact with the ground and the warmer air at corresponding levels in the *free atmosphere*. It is often quite strong, esp. when channelled down a narrow valley, and can reach gale-force at the edges of the Antarctic and Greenland ice-caps.

kata-front A situation at a front, warm or cold, where the warm air is sinking relative to the front.

K-band US designation of microwave band between 12 and 40 GHz, widely adopted in the absence of internationally agreed standards. Subdivided into Ku-band (*K-under*), 12–19 GHz, K-band, 18–26 GHz and Ka (*K-above*), 26–40 GHz; replacing approximately the UK designations of *J-, K-* and *Q-bands*.

K-display A form of A-display produced with a lobe-switching antenna. Each lobe produces its own peak and the antenna is directly on target when both parts of the resulting double peak have the same height.

keelson A longitudinal structural member inside the bottom of the hull of a flying-boat. It forms part of the main framework, connecting the transverse members and bulkheads.

kelvin The SI unit of thermodynamic temperature. It is 1/273.16 of the temperature of the **triple point** of water above **absolute zero**. The temperature interval of 1 kelvin (K) equals that of 1°C (degree Celsius). See **Kelvin thermodynamic scale of temperature**.

Kelvin thermodynamic scale of temperature A scale of temperature based on the thermodynamic principle of the performance of a reversible heat engine. The scale cannot have negative values so *absolute zero* is a well defined thermodynamic temperature. The temperature of the **triple point** of water is assigned the value 273.16 K. The temperature interval corresponds to that of the Celsius scale so that the freezing point of water (0°C) is 273.15 K. Unit is the **kelvin**. Symbol K.

Kennelly-Heaviside layer See **E-layer**.

Kepler's laws of planetary motion (1) The planets describe ellipses with the Sun at a focus. (2) The line from the Sun to any planet describes equal areas in equal times. (3) The squares of the periodic times of the planets are proportional to the cubes of their mean distances from the Sun.

Kew-pattern barometer The *Adie barometer*. Specially graduated so that error arising from changes in the free level in the cistern is obviated. Also *compensated scale barometer*.

kick stage A propulsive stage used to provide an additional velocity increment required to put a spacecraft on a given trajectory.

kiloparsec See **parsec**.

kilowatt-hour The commonly used unit of electrical energy, equal to 1000 watt-hours or 3.6 MJ. Often called simply a *unit*; abbrev. *kWh*. See **Board of Trade Unit**.

kinematic viscosity The co-efficient of viscosity of a fluid divided by its density. Symbol ν. Thus $\nu = \eta/\rho$. Unit in the CGS system is the stokes (cm^2 s^{-1}); in SI m^2 s^{-1}.

kinetic energy Energy arising from motion. For a particle of mass m moving with a velocity v it is $\tfrac{1}{2}mv^2$, and for a body of mass M, moment of inertia I_g, velocity of centre of gravity v_γ and angular velocity ω, it is

$$\tfrac{1}{2}Mv_g^2 + \tfrac{1}{2}I_g\omega^2.$$

kinetic heating See **dynamic heating**.

kinetic pressure See **dynamic pressure**.

Kirchhoff's law The ratio of the coefficient of absorption to the coefficient of emission is the same for all substances and depends only on the temperature. The law holds for the total emission and also for the emission of any particular frequency.

kite Any aerodyne anchored or towed by a line, not mechanically or power-driven. Derives its lift from the aerodynamic forces of the relative wind.

knot Speed of 1 nautical mph (1.15 mph or 1.85 km/h) used in navigation and meteorology.

Köppen classification A system of classifying climate by groups of letters. The general climatic type is shown by a letter or letter-pair in capitals, and further detail is given by additional lower-case letters, e.g. the climate of the British Isles is coded as Cf where C indicates 'warm or temperate rainy' and f is 'moist, no marked dry season'.

Kruger flap Leading edge flap which is hinged at its top edge so it normally lies flush with the underwing surface, and swings down and forward to increase circulation and lift for low speed flight.

Kuchemann tip Low drag wingtip shape developed by Kuchemann at **RAE** for transonic aircraft. It has a large radius curve in plan finishing with a corner at the trailing edge.

Kundt's tube Transparent tube in which standing sound waves are established, indicated by lycopodium powder, which accumulates at the nodes. Used for measuring sound velocities. See **acoustic streaming**.

L

λ Symbol for (1) wavelength, (2) linear co-efficient of thermal expansion, (3) mean free path, (4) radioactive decay constant, (5) thermal conductivity.

L Symbol for (1) angular momentum, (2) luminance, (3) self inductance.

Lagrangian points The five points associated with a binary system (particularly Earth-Moon), where the combined gravitational forces are zero (also *libration points*). A Lagrangian *point* is thus that point in the orbital plane of two objects in orbit around their common centre of gravity at which a third particle of negligible mass can remain in equilibrium.

laminar flow A type of fluid flow in which adjacent layers do not mix except on the molecular scale. See **streamline flow, viscous flow, boundary layer**.

land and sea breezes Winds occurring at the coast during fine summer weather, esp. when the general pressure gradient is small. During the day, when the land is warmer than the sea, the air over the land becomes warmer and less dense than that over the sea and a local circulation is created with air flowing from land to sea at high levels and from sea to land near the surface. At night, conditions are reversed. The sea breeze can penetrate many miles inland and can become strong at low latitudes.

landing area That part of the movement area of an unpaved airfield intended primarily for take-off and landing.

landing beacon A transmitter used to produce a landing beam.

landing beam Field from a transmitter along which an aircraft approaches a landing field during blind landing. See **instrument landing system**.

landing direction indicator A device indicating the direction in which landings and take-offs are required to be made, usually a T, toward the cross bar of which the aircraft is headed.

landing gear That part of an aircraft which provides for its support and movement on the ground, and also for absorbing the shock on landing. It comprises main support assemblies incorporating single or multi-wheel arrangements, and also auxiliary supporting assemblies such as nose-wheels, tail-wheels or skids. See **bogie-, drag struts, oleo**.

landing ground In air transport, any piece of ground that has been prepared for landing of aircraft as required; not necessarily a fully equipped airfield. An *emergency landing* *ground* is any area of land that has been surveyed and indicated to pilots as being suitable for forced or emergency landings. See **airstrip**.

landing parachute See **brake parachute**.

landing procedure The final approach manoeuvres, beginning when the aircraft is in line with the axis of the runway, either for the landing, or upon the reciprocal for the purpose of a procedural turn, until the actual landing is made, or *overshoot* action has to be taken.

landing speed The minimum airspeed, with or without engine power, at which an aircraft normally alights.

landing wires Wires or cables which support the wing structure of a biplane when on the ground and also the negative loads in flight. Also *anti-lift wires*.

lapse rate The rate of fall of a quantity with increase in height.

laser *Light Amplification by Stimulated Emission of Radiation*. A source of intense monochromatic light in the ultraviolet, visible or infrared region of the spectrum. It operates by producing a large population of atoms with their electrons in a certain high energy level. By *stimulated emission*, transitions to a lower level are induced, the emitted photons travelling in the same direction as the stimulating photons. If the beam of inducing light is produced by reflection from mirrors or *Brewster's windows* at the ends of a resonant cavity, the emitted radiation from all stimulated atoms is in phase, and the output is a very narrow beam of coherent monochromatic light. Solids, liquids and gases have been used as lasing materials. See **maser**.

laser gyro An instrument, usually triangular with internally reflecting prisms at the corners, which senses angular rotation in its plane by measuring the frequency shift of laser energy passing around the circuit.

latent heat More correctly, **specific latent heat**. The heat which is required to change the state of unit mass of a substance from solid to liquid, or from liquid to gas, without change of temperature. Most substances have a latent heat of fusion and a latent heat of vaporization. The specific latent heat is the difference in *enthalpies* of the substance in its two states.

latent instability A type of **conditional instability** of the atmosphere which exists only if a rising parcel of air reaches a critical level.

lateral axis The crosswise axis of an aircraft, particularly that passing through its centre of mass, parallel to the line joining the wing tips.

lateral instability A condition wherein an aircraft suffers increasing oscillation after a rolling disturbance. Also *rolling instability*.

lateral stability Stability of an aircraft's motions out of the plane of symmetry, i.e. sideslipping, rolling and yawing.

lateral velocity The rate of **sideslip** of an aircraft, i.e. the component of velocity which is resolved in the direction of the lateral axis.

latitude *Northing*. Distance of a point north (if positive) or south (negative) from point of origin of survey. This, together with departure (easting), locates the point in a rectangular grid orientated on true meridian.

latitude and longitude, celestial Those spherical co-ordinates which are referred to the ecliptic and its poles. *Celestial latitude* is the angular distance of a body north or south of the ecliptic. *Celestial longitude* is the arc of the ecliptic intercepted between the latitude circle and the First Point of Aries, and is measured positively eastwards from 0° to 360°.

launch pad Special area from which a launch system is fired; it contains all the necessary support facilities such as a servicing tower, cooling water, safety equipment and flame deflectors.

launch system See panel on p. 102.

launch window The time slot within which a spacecraft must be launched to best achieve its given mission trajectory; a launch window may become available a number of times during a particular *launch opportunity*.

layers Ionized regions in space which vary vertically and affect radio propagation. See **E-layer**, **F-layer**.

L-band A radar and microwave frequency band, generally accepted as being 1 to 2 GHz; not frequently referred to because of lack of international conformity in its use.

lb.s.t. Abbrev. for *static thrust* in pounds.

LCN Abbrev. for **Load Classsification Number**.

L-display A radar display in which the target appears as two horizontal pulses, left and right from a central vertical time base, varying in amplitude according to accuracy of aim.

leading edge The edge of a streamline body or aerofoil which is forward in normal motion; structurally, the member that constitutes that part of the body or aerofoil.

leading-edge flap A hinged portion of the wing leading edge, usually on fast aircraft, which can be lowered to increase the camber

and so reduce the stalling speed. Colloq. *droops*.

leap second A periodic adjustment of time signal emissions to maintain synchronism with co-ordinated universal time (UTC). A positive leap second may be inserted, or a negative one omitted, at the end of December or June.

leap years Those years in which an extra day, 29 February, is added to the civil calendar to allow for the fractional part of a tropical year of 365.2422 days. Since the Gregorian reform of the Julian calendar, the leap years are those whose number is divisible by 4, except centennial years unless these are divisible by 400.

Leda The thirteenth natural satellite of **Jupiter**, perhaps only 10 km in diameter, discovered in 1979.

lee wave Stationary wave set up in an air stream to the lee of a hill or range over which air is flowing. Special conditions of atmospheric stability and vertical wind-structure are required. Lee waves are sometimes of large amplitude and can be dangerous to aircraft.

left-handed engine An aero-engine in which the propeller shaft rotates counter-clockwise with the engine between the observer and the propeller.

left-hand propeller See **propeller**.

LEO Abbrev. for *Low-Earth Orbit*. See **orbit**.

Leonids A swarm of meteors whose orbit round the Sun is crossed by the Earth's orbit at a point corresponding to about 17 November , when a display of more than average numbers is to be expected; the radiant point is in the constellation Leo.

lepton A fundamental particle that does not interact strongly with other particles. There are several different types of lepton; the electron, the negative muon, the tau-minus particle and their associated neutrinos. See **antilepton**.

letting down The reduction of altitude from cruising height to that required for the approach to landing.

level Logarithmic ratio of two energies or two field quantities, where the nominator is the measured quantity and the denominator a reference quantity.

libration in latitude That in which, owing to the Moon's axis of rotation not being perpendicular to its orbital plane, an observer on the Earth sees alternately more of the north and south regions of the lunar surface, and so, in a complete period, more than a hemisphere.

libration in longitude That due to the uni-

launch system

An assemblage of propulsive stages capable of accelerating a space vehicle to the velocity needed to achieve a particular trajectory. The principles involved are illustrated by the following considerations.

For an engine using rocket propulsion, one can deduce from Newton's second law of motion that the relationship between burn-out velocity (V_b) and the exhaust velocity (V_e) is:

$$V_b = V_e \, \log_e \frac{M}{M_b}$$

where M and M_b are the initial and burn-out masses respectively. This relationship is known as the *rocket equation*. M/M_b is referred to as the mass ratio of the rocket. For a single stage the mass change is due to propellant consumption. If the principle of staging is used, the unwanted mass of the rocket itself can be jettisoned before ignition of the next stage. The rocket equation then becomes:

$$V_b = \left(V_e\right)_1 \log_e \left(\frac{M}{M_b}\right)_1 + \left(V_e\right)_2 \log_e \left(\frac{M}{M_b}\right)_2 + \cdots$$

where the subscripts refer to successive stages. The final velocity is now greater than that achieved by using one stage and is the sum of the velocity increases contributed by each stage. In this way, the launch velocity requirements can be met, the desired azimuth being achieved by adjusting the direction of the ascent trajectory. The thrust developed by the first stage of the launch vehicle must exceed the total weight of the launch sytem plus payload to ensure that the ensemble leaves the ground. In other words the thrust-to-weight ratio must exceed 1. This is sometimes achieved with booster rockets, which usually burn simultaneously with the first stage.

A launch system may be reusable or expendable. The former implies recovery and refurbishment but the latter has no recoverable parts. Examples of launch systems from around the world, together with their payload capabilities in kg, are given in the following table. GTO is Geosynchronous Transfer Orbit and LEO is Low-Earth Orbit. See **orbit**.

Launch system	Origin	Capability to GTO
Proton	USSR	3800
Long March CZ-3A	China	2500
Ariane 44L	Europe	4200
Atlas 2A (Atlas-Centaur)	US	2800
Titan 3 (with Transtage)	US	ca 5000
H-II (operational 1992)	Japan	3800

Launch system	Origin	Capability to LEO
Energia (four boosters)	USSR	ca 100 000
Titan 4 (Titan-Centaur)	US	18 000
Space Shuttle	US	30 000

Both the US and Soviet shuttle orbiters are recoverable and may be manned whereas the others mentioned above are expendable, although Soviet sources foresee a reusable Energia in the future. The US shuttle solid boosters are recovered from the sea and are refuelled before reuse; this is a good example of a partially reusable system. Also *launch vehicle, launcher*.

form rotation of the Moon on its axis combining with its non-uniform orbital motion to cause an observer on the Earth to see more, either on the east and or on the west, of the lunar surface than an exact hemisphere.

librations Apparent oscillations of the Moon (or other body). The actual physical librations (due to changes in the Moon's rate of rotation) are very small; the other librations are librations in latitude or in longitude and diurnal libration.

licensed aircraft engineer An engineer licensed by the airworthiness authority (in the UK the *Civil Aviation Authority*) to certify that an aircraft and/or component complies with current regulations.

lid Temperature inversion in the atmosphere which prevents the mixing of the air above and below the inversion region.

lidar Abbrev. for *LIght Detection And Ranging*. Device for detection and observation of distant cloud patterns by measuring the degree of back scatter in a pulsed laser beam.

life support The provision of the necessary conditions of health and comfort to support a human in space, either during the occupation of a space vehicle or during *extra-vehicular activity* (EVA).

lift *Aerodynamic lift* is the component of the aerodynamic forces supporting an aircraft in flight, along the lift axis, due solely to relative airflow. Lift force acts at right angles to the direction of the undisturbed airflow relative to the aircraft.

lift axis See **axis**.

lift co-efficient A non-dimensional number representing the aerodynamic lift on a body.

$$C_L = \frac{L}{1/2\ \rho V^2 S},$$

where L = lift, ρ = air density, V = airspeed, S = wing area.

lift engine An engine used on **VTOL** or **STOL** aircraft, having the primary purpose of providing lifting force. It is shut down in normal flight and the intake usually closed by a sealing flap.

lift-off (1) The speed at which a pilot pulls back on the control column to make an aircraft leave the ground. It is a carefully defined value for large aircraft depending upon the weight, runway surface, gradient, altitude and ambient temperature. It is one of the functions established from **WAT curves**. Colloq. *unstick*. (2) The point at which the vertical thrust of a spacecraft exceeds the force due to local gravity and it begins to rise.

light *Electromagnetic radiation* capable of

inducing visual sensation, with wavelengths between about 400 and 800 nm. See **illumination, speed of light**.

light aircraft One having a maximum take-off weight less than 12 500 lb (5670 kg).

light-curve The line obtained by plotting, on a graph, the apparent change of brightness of a variable star, against the observed times.

lighter-than-air craft See **aerostat**.

lightning Luminous discharge of electric charges between clouds, and between cloud and earth (or sea). A path is found by the *leader stroke*, the main discharge following along this ionized path, with possible repetition. See **thunderstorm**.

light quanta When light interacts with matter, the energy appears to be concentrated in discrete packets called *photons*. The energy of each photon $E = h\nu$ where ν is the frequency and h is Planck's constant.

light-up The period during the starting of a *turbojet* or *turboprop* engine when the fuel/air mixture has been ignited.

light-year An astronomical measure of distance, being the distance travelled by light in space during a year, which is approximately 9.46×10^{12} km (5.88×10^{12} miles).

limb Term applied to the edge or rim of a heavenly body having a visible disk; used specially of the Sun and Moon.

limb darkening The apparent darkening of the limb of the Sun due to the absorption of light in the deeper layers of the solar atmosphere near the edge of the disk.

limiting frequency (1) Frequency at which the wavelength of an air-borne sound wave coincides with the wavelength of a bending wave of a wall or plate. Important in sound transmission. (2) Name for the highest or lowest frequency transmitted by an electro-acoustic system or by a waveguide.

limiting Mach number The maximum permissible *flight Mach number* at which any particular aircraft may be flown, either because of the **buffet boundary** or for structural strength limitations.

limiting velocity The steady speed reached by an aircraft when flown straight, the angle to the horizontal, power output, altitude and atmospheric conditions all specified. See **max level speed, terminal velocity**.

limit load The maximum load anticipated from a particular condition of flight and used as a basis when designing an aircraft structure.

line broadening A term used for the increase in width of the lines of a stellar spectrum due to rotation of the star, turbulence in the stellar atmosphere, or the

Stark or *Zeeman effects.*

line of apsides See **apse line**.

line-of-sight velocity The velocity at which a celestial body approaches, or recedes from, the Earth. It is measured by Doppler shift of the spectral lines emitted by the body as observed on the Earth. Also *radial velocity.*

line spectrum A spectrum consisting of relatively sharp lines, as distinct from a **band spectrum** or a **continuous spectrum**. Line spectra originate in the atoms of incandescent gases or vapours.

line squall A system of squalls occurring simultaneously along a line, sometimes hundreds of miles long, which advances across the country. It is characterized by an arch or line of low dark cloud, a sudden drop in temperature and rise in pressure. Thunderstorms and heavy rain or hail often accompany these phenomena.

live room One which has a longer period of reverberation than the optimum for the conditions of performance and listening.

Lloyd's mirror A device for producing interference fringes. A slit, illuminated by monochromatic light, is placed parallel to and just in front of the plane of a plane mirror or piece of unsilvered glass. Interference occurs between direct light from the slit and that reflected from the mirror. See **interference fringes**.

LLTV Abbrev. for *Low Light TeleVision.*

lo Low level military flight, usually at less than 200 ft.

load classification number A number defining the load-carrying capacity of the paved areas of an airport without cracking or permanent deflection. Abbrev. *LCN.*

load factor (1) In relation to the structure, the ratio of an external load to the weight of an aircraft. Loads may be centrifugal and aerodynamic due to manoeuvring, to gravity, to ground or water reaction; usually expressed as g, e.g. 7 g is a load seven times the weight of the aircraft. (2) In aircraft operations, the actual *payload* on a particular flight as a percentage of the maximum permissible payload.

loading and cg diagram A diagram, usually comprising a side elevation of the aircraft concerned, with a scale and the location of all items of removable equipment, payload and fuel, which is used to adjust the weight of the aircraft so that its resultant lies within the forward and aft **cg limits**.

lobe switching Method of determining the precise direction of a target without resorting to impossibly narrow antenna beams. While the antenna is turning, its beam is switched

periodically to the left and right of the dead-ahead position; when equal signals are received in both positions, the antenna is accurately aimed.

local group The name of the family (or cluster) of **galaxies** to which our **Milky Way** belongs. There are two dozen members within 5 million light-years or so. Another prominent member of the group is the **Andromeda Nebula**.

localizer beacon A directional radio beacon associated with the **ILS**, which provides an aircraft during approach and landing with an indication of its lateral position relative to the runway in use.

local Mach number The ratio of the velocity of the airflow over a part of a body in flight to the local speed of sound. Usually it is concerned with a part of greater curvature, where the airflow accelerates momentarily, thereby increasing the Mach number above that of the body as a whole, e.g. over the wing or canopy.

local time Applied to any of the 3 systems of time reckoning, sidereal, mean solar or apparent solar time, it signifies the hour angle of the point of reference in question measured from the local meridian of the observer. The local times of a given instant at 2 places differ by the amount of their difference in longitude expressed in time, the local time at a place east of another being greater.

locator beacon The 'homing' beacon on an airfield used by the pilot until he picks up the localizer signals of the **ILS**.

locks, up and/or down See **up and/or down locks**.

logarithmic amplifier One with an output which is related logarithmically to the applied signal amplitude, as in decibel meters or recorders.

logarithmic array Tapered radar **end-fire array** designed to operate over a wide range of frequency.

longeron Main longitudinal member of a fuselage or nacelle.

longitude See **latitude and longitude, celestial**.

longitudinal A girder that runs fore and aft on the outside of a rigid airship frame. Longitudinals connect the outer rings of the transverse frames.

longitudinal axis See **axis**.

longitudinal wave Propagating sound wave in which the motions of the relevant particles are in line with the direction of translation of energy.

longitudinal instability The tendency of an aircraft's motion in the plane of symmetry to depart from a steady state, i.e. to pitch up or down, to rise or fall, or to vary in horizontal speed. See **stability and control**.

longitudinal oscillation A periodic variation of speed, height and angle of pitch. See **phugoid oscillation**

longitudinal stability See **stability and control**.

long-period variable See **variable star**.

long range Aircraft, ship or missile capable of covering great distances without refuelling.

looming The vague enlarged appearance of objects seen through a mist or fog, particularly at sea.

loop An aircraft manoeuvre consisting of a complete revolution about a lateral axis, with the normally upper surface of the machine on the inside of the path of the loop. See **inverted loop**.

LORAN Abbrev. for *LOng RAnge Navigation*. An early but much developed hyperbolic navigation aid using on-board aircraft systems to translate the time difference between pulsed transmissions received from two or more ground stations to provide positional information.

Lorentz transformations The relations between the co-ordinates of space and time of the same event as measured in two inertial frames of reference moving with a uniform velocity relative to one another. They are derived from the postulates of the special theory of **relativity**.

loss factor Quantity to describe the damping of structure-borne sound in materials or structures.

loudness The intensity or volume of a sound as perceived subjectively by a human ear.

loudness contour Line drawn on the audition diagram of the average ear which indicates the intensities of sounds that appear to the ear to be equally loud.

loudness level That of a specified sound is the intensity of the *reftone*, 1000 Hz, on the *phon* scale, which is adjusted to equal, in apparent loudness, the specified sound. The adjustment of equality is made either subjectively or objectively as in special sound-level meters.

low A region of low pressure, or a *depression*.

low aspect ratios See **aspect ratio**.

lower culmination See **culmination**.

lower-pitch limit The minimum frequency for a sinusoidal sound wave which produces a sensation of pitch.

lower transit Same as *lower culmination*. See **culmination**.

low light television Sensor system, capable of operating at dawn and dusk, smaller and cheaper than radar.

low-pressure compressor, stage, turbine Stages of an **axial-flow turbine**. Abbrev. *LP compressor* etc.

low-wing monoplane A monoplane in which the main planes are mounted at or near the bottom of the fuselage.

luminance Measure of brightness of a surface, e.g., candela per square metre of the surface radiating normally. Symbol *L*.

luminosity The intrinsic or absolute amount of energy radiated per second from a celestial object. Its units in astronomy are usually absolute **magnitude**, rather than watts, largely for historical reasons.

lunar bows Bows of a similar nature to *rainbows* but produced by moonlight.

lunar month See **synodic month**.

lunation See **synodic month**.

lux Unit of illuminance or illumination in SI system, 1 lm/m^2. Abbrev. *lx*.

Lyman series One of the hydrogen series occurring in the extreme ultraviolet region of the spectrum. The series may be represented by the formula

$$v = N\left(\frac{1}{1^2} - \frac{1}{n^2}\right),$$

$n = 2, 3, 4...$(see **Balmer series**), the series limit being at wave number $N = 109\,678$, which corresponds to wavelength, 91.26 nm. The leading line, called *Lyman alpha*, has a wavelength of 121.57 nm, and is important in upper atmosphere research, as it is emitted strongly by the Sun.

Lyot filter See **polarizing monochromator**.

Lysithea The tenth natural satellite of **Jupiter**.

M

Mach angle In supersonic flow, the angle between the **shock wave** and the airflow, or line of flight, of the body. The cosecant of this angle is equal to the Mach number.

Mach cone In supersonic flow, the conical **shock wave** formed by the nose of a body, whether stationary in a **wind tunnel** or in free flight through the air.

machmeter A pilot's instrument for measuring **flight Mach number**.

Mach number The ratio of the speed of a body, or of the flow of a fluid, to the speed of sound in the same medium. At *Mach 1*, speed is *sonic*; below *Mach 1*, it is *subsonic*; above *Mach 1*, it is *supersonic*, creating a *Mach* (or *shock*) *wave*. Hypersonic conditions in air are reached at Mach numbers exceeding 5. See **velocity of sound**.

mackerel sky Cirrocumulus or altocumulus cloud arranged in regular patterns suggesting the markings on mackerel.

MAD Abbrev. for (1) *Mutual Assured Destruction*, (2) *Magnetic Anomaly Detection*.

MADGE Abbrev. for *Microwave Aircraft Digital Guidance Equipment*.

Mae West Personal lifejacket designed for airmen, inflated by releasing compressed cabon dioxide.

Magellanic Clouds Two dwarf galaxies, satellites of the **Milky Way**, visible in the night sky of the southern hemisphere as cloudy patches. They were first recorded by F. Magellan in 1519 and are about 180 000 light-years away, containing a few thousand million stars. They are of immense astrophysical importance because individual stars within them can be studied and yet they are essentially all at the same distance from us. This removes a great source of uncertainty compared with the situation within our own Galaxy, where actual distances to individual stars are hard to determine.

magnetic monopole Some **grand unified theories** of matter predict the existence of point-like defects in the structure of spacetime which would behave like the isolated pole of a bar magnet. These are a major obstacle in certain theories of the expanding universe but the difficulties may be circumvented in the theory of the **inflationary universe**. See **Big Bang**.

Magnetic North The direction in which the North pole of a pivoted magnet will point. It differs from the Geographical North by an angle called the *magnetic declination*.

Magnetic South The dircction in which the South pole of a pivoted magnet will point. It

differs from the Geographical South by an angle called the *magnetic declination*.

magnetic storm Magnetic disturbance in the Earth, causing spurious currents in submarine cables; probably arises from variation in particle emission from the Sun, which affects the ionosphere.

magnetic surface wave Magnetic wave propagated along the surface of a ferromagnetic garnet substrate. Used in microwave delay lines, filters etc. Abbrev. *MSW*.

magnetic variables Stars in which strong variable magnetic fields have been detected by the Zeeman effect.

magnetic variations Both diurnal and annual variations of the magnetic elements (dip, declination etc.) occur, the former having by far the greater range. In the northern hemisphere, the declination moves to the west during the morning and then gradually back, the extreme range being nowhere more than 1°. The dip varies by a few minutes during the day. It is thought that these effects are caused by varying electric currents in the ionized upper atmosphere.

magnetohydrodynamics The study of the motions of an electrically conducting fluid in the presence of a magnetic field. The motion of the fluid gives rise to induced electric currents which interact with the magnetic field which in turn modifies the motion. The phenomenon has applications both to magnetic fields in space and to the possibility of generating electricity. If the free electrons in a plasma or high velocity flame are subjected to a strong magnetic field, then the electrons will constitute a current flowing between two electrodes in a flame. Abbrev. *MHD*.

magnetosphere The asymmetrically shaped volume round the Earth and other magnetic planets in which charged particles are subject to the planet's magnetic field rather than the Sun's. Its radius is least towards the Sun and greatest away from it.

magnetron A two-electrode valve in which the flow of electrons from a large central cathode to a cylindrical anode is controlled by crossed electric and magnetic fields; the electrons gyrate in the axial magnetic field, their energy being collected in a series of slot resonators in the face of the anode. Magnetrons, used mainly as oscillators, can produce pulsed output power at microwave frequencies, with high peak power ratings. Used in microwave and radar transmitters, and microwave cookers.

magnitude

The brightness of any celestial object when measured from Earth is the *apparent magnitude* (m). Ideally this should be expressed in units of energy flux, as is done in radio astronomy, where the **jansky** is used. The system in optical astronomy dates from 120 BC when Hipparchos divided the visible stars into six classes, from first magnitude for the brightest down to sixth magnitude for those barely visible. In fact his natural system is logarithmic and in 1854 it was shown that first magnitude represents 100 times as much energy as sixth. A difference of one magnitude corresponds to a brightness ratio of 2.512, this being the fifth root of 100.

The system is fixed by reference to a few standard stars which have been very accurately determined by photometry. Confusingly, very bright objects have negative magnitudes and the faintest large positive magnitudes.

Apparent magnitudes are estimated by eye (amateur astronomy of variable stars), photographically with a large number of stars on a single exposure, photoelectrically or with **CCD** detectors. This last is now preferred because computer reduction of the data gives exact apparent magnitudes. Magnitude measurements depend on the wavelength studied.

The conversion from measured apparent magnitude to an absolute measure of intrinsic luminosity requires knowledge of the distance of the object. By definition, the *absolute magnitude* (M) is the apparent magnitude an object would have if placed 10 parsec from Earth. The conversion between the two is: $M = m + 5 - \log D$, where D is the distance in parsec. Much of the emphasis in practical astronomy is on measuring m, then inferring M from other evidence, such as variability in **Cepheid** stars, and thus calculating D, which is the most elusive quantity in this relation. Some examples of magnitude are given in the table.

Object	Apparent magnitude	Absolute magnitude
Sun	−26.74	+4.83
Moon, full	−12.6	
Venus at maximum	−4.0	
Sirius	−1.5	+1.41
Polaris	+2.0	−4.5
faintest detectable galaxies	+29	−19
typical quasar	+19	−25/−30
bright meteor	−10	
visible in binoculars	+8	
in small telescope	+11	
in Hubble Space telescope	+30	

magnitude See panel on this page.

main float The two single, or one central, float(s) which give buoyancy to a seaplane or amphibian.

main plane The principal supporting surface, or wing, of an aircraft or glider, which can be divided into centre, inner, outer, and/or wing-tip sections.

main rotor (1) The principal assembly or assemblies of rotating blades which provide lift to a rotorcraft. (2) The assembly of compressor(s) and turbine(s) forming the rotating parts of a gas-turbine engine.

main sequence In the **Hertzsprung-Russell diagram**, the vast majority of stars lie in a broad band known as the main sequence, running diagonally from high temperature, high luminosity stars to low temperature, low luminosity stars, in a smooth progression. A star spends most of its life on this main sequence, and throughout that time it converts hydrogen to helium. The position on main sequence depends mostly on mass, the more massive and luminous stars being located higher on the sequence. Once the hydrogen in the core is consumed, the star evolves away from the main sequence, becoming first a **red giant**. The Sun is a main sequence star.

main tanks See **fuel tanks**.

majority voting system In a redundant electrical or computer system a means whereby signals from all channels (3) are continually monitored; any discrepancy in a single channel is recognized and 'voted out' of circuit so that the system can continue to function.

Maksutov telescope An optical telescope in which the image-forming surfaces are spherical, and therefore easy to make. A deeply curved meniscus lens corrects for aberrations in the primary mirror. Design published by Maksutov in 1944. Cf. *Schmidt* telescope in **astronomical telescope**.

mamma Clouds with rounded protuberances on their lower surfaces, like udders. They often occur below thunder clouds.

manifold pressure The absolute pressure in the induction manifold of a reciprocating aero-engine which, indicated by a cockpit gauge, is used together with r.p.m. settings to control engine power output and fuel consumption. See **supercharging**.

manifold pressure gauge See **boost gauge**.

manned space flight Refers to a manned presence on-board a spacecraft; the additional flexibility provided must be reconciled with the design impacts to safeguard the well-being of the individual. Pioneering efforts in manned space flight involved the Russian *Vostok* and American *Mercury* projects. More sophisticated systems are associated with Earth-orbiting space stations.

manoeuvre demand system A **fly-by-wire** auto-control system in which the pilot's action determines a required manoeuvre, e.g. a pitch up of a certain value of *g*, the automatic control system then setting the control deflections to achieve the desired result while maintaining overall stability despite changes of speed.

map comparison unit See **chart comparison unit**.

map matching Navigation by auto-correlation of terrain with data stored in aircraft, missile or RPV, often in the form of film. See **tercom**.

marangoni convection The flow resulting from gradients in surface tension giving rise to the transfer of heat and mass; it is particularly relevant to **microgravity** conditions when gravity-induced convection is absent.

maria The Latin designation of the so-called 'seas' on the lunar surface, named before the modern telescope showed their dark areas to be dry planes. Since 1959 spacecraft probes and landings (manned and unmanned) have provided much detailed information but their origin is still problematical. The sing. is *Mare* (e.g. *Mare Imbrium*), but there is a marked tendency to delatinize this picturesque terminology (e.g. Sea of Tranquility, Sea of Fertility, Ocean of Storms, Sea of Moscow etc.). See **mascon, Moon**.

Markarian galaxy Galaxy with excessive ultraviolet emission. See **galaxy**.

marker Pip on a radar display for calibration of range and direction.

marker antenna One giving a beam of radiation for marking air routes, often vertically for blind or instrument landing.

marker beacon A radio beacon in aviation which radiates a signal to define an area above the beacon. See **fan, inner-, middle-, outer-, Z-, glide-path beacon, localizer beacon, locator beacon, radar beacon, track guide**.

markers See **airport markers**.

Mars See panel on p. 109.

mascon *Mass concentrations*. Regions of high gravity occurring within certain **maria** of the Moon. Their origin is still conjectural.

maser *Microwave Amplification by Stimulated Emission of Radiation*. A *microwave* oscillator that operates on the same principle as the **laser**. Maser oscillations produce coherent monochromatic radiation in a very narrow beam. Less noise is generated than in other kinds of microwave oscillators. Materials used are generally solid-state, but masers have been made using gases and liquids.

mass balance A weight or mass attached ahead of the hinge line of an aircraft control surface to give static balance with no moment about the hinge, and to reduce to zero inertial coupling due to displacement of the control surface. Mass balancing is a precaution against control surface **flutter**.

mass control Said of mechanical systems, particularly those generating sound waves, when the mass of the system is so large that the compliance and resistance of the system are ineffective in controlling motion.

mass-energy equation $E = mc^2$. Confirmed deduction from Einstein's special theory of relativity that all energy has mass. If a body gains energy E, its inertia is increased by the amount of mass $m = E/c^2$, where c is the speed of light. Derived from the assumption that all conservation laws must hold equally in all frames of reference and using the principle of conservation of *momentum*, of *energy* and of *mass*. See

Mars

Mars is the fourth major planet from the Sun, often known as the red planet, because of its distinctive colour, noticeable even to the naked eye.

Mars is one of the terrestrial planets and has a diameter just over half that of the Earth. It had long been regarded as the planet most likely to have life, a view encouraged by the presence of polar ice-caps and observations of seasonal changes. Nineteenth century observers, notably Percival Lowell, convinced themselves that they could make out systems of straight channels or canals that might be artificially constructed. Exploration of the planet by spacecraft has produced no evidence that life exists, or has existed, on Mars.

The relatively low density of Mars (3.3 times that of water) suggests that 25% of its mass is contained in an iron core. There is a weak magnetic field, about 2% of the strength of Earth's. The crust is rich in olivine and ferrous oxide, which gives the rust colour. The tenuous atmosphere is 95.3% carbon dioxide, 2.7% molecular nitrogen and 1.6% argon, with oxygen as the major trace constituent. The pressure at the surface is only 0.7% of that at the surface of the Earth. However, strong winds can cause extensive dust storms, which occasionally engulf the entire planet.

A variety of clouds and mists occur. Early morning fog forms in valleys, and orographic clouds over the high mountains of the *Tharsis* region. In winter, the North Polar Cap is swathed in veils of icy mist and dust, known as the polar hood. A similar phenomenon is seen to a lesser extent in the south.

The polar regions are covered with a thin layer of ice, thought to be a mixture of water, ice and solid carbon dioxide. High resolution images show a spiral formation and strata of wind-born material. The north polar region is surrounded by stretches of dunes; the polar ice-caps grow and recede with the seasons which arise – as they do in the case of Earth – because the planet's rotation axis is tilted (by 25°) to the orbital plane.

The US has successfully sent six probes to Mars: Mariner 4 in 1965, Mariners 6 and 7 in 1969, Mariner 9 in 1971 and Vikings 1 and 2 in 1976. The Vikings incorporated both orbiters and landers. Mars is considered to be a realistic target for a manned landing by about 2020. Such a mission would take around two years.

There is a marked difference in the nature of the terrain between the two halves of Mars, divided roughly by a great circle tilted at 35° to the equator. The more southerly part consists largely of ancient, largely cratered terrain. The major impact basins – the *Hellas, Argyre* and *Isidis planitiae* – are located in this hemisphere. The north is dominated by younger, more sparsely cratered terrain, lying 2–3 kilometres lower. The highest areas are the large volcanic domes of the *Tharsis* and *Elysium planitiae*. Both areas are dominated by several huge extinct volcanoes.

These volcanic areas are located at the east and west ends of an immense system of canyons, the *Valles Marineris*, which stretches for more than 5000 kilometres around the equatorial region and has an average depth of 6 kilometres. It is believed to have been caused by faulting associated with the upthrust of the Tharsis dome.

There is evidence, in the form of flow channels, that liquid water once existed on the surface of Mars. Channels from the Valles Marineris appear to have been created in some kind of sudden flood. There are also sinuous, dried-up river beds with many tributaries, found only in the heavily cratered terrain.

Mars has two small natural satellites, *Phobos* and *Deimos*, which are in circular orbits in the equatorial plane, close to the planet. They are very difficult to see from Earth. Their nature is so different from Mars that it seems likely that they are captured asteroids.

rest-mass energy.

mass law Law describing the sound transmission through walls. The transmission coefficient is approximately proportional to the inverse of the mass per unit area and to the inverse of the frequency, i.e. the *transmission loss* increases by approximately 6 dB when mass or frequency are doubled. There are many exceptions.

mass-luminosity law A relationship between the mass and absolute magnitude of stars, the most massive stars being the brightest; applicable to all stars except the white dwarfs.

mass ratio The ratio of the fully-fuelled mass of a rocket-propelled vehicle or stage to that when all fuel has been consumed.

master connecting-rod The main member of the master and articulated assembly of a radial aero-engine. It incorporates the crank-pin bearing, and the *articulated rods* of the other cylinders oscillate on it by means of wrist pins.

matching Said of the insertion of *matching sections* into radio-frequency transmission lines, with the aim of minimizing power reflections at a *mismatch*. A matching section may consist of specifically chosen lengths of waveguide, stripline or co-axial cable having a different impedance from the main system and connected in series so as to cause impedance transformation; alternatively, a *matching stub* may be connected in parallel with the transmission path.

materialization Reverse of Einstein energy released with annihilation of mass. A common example is by *pair production* (electron-positron) from gamma rays.

Maunder diagram See **butterfly diagram**.

max gross See **maximum weight**.

maximum and minimum thermometer An instrument for recording the maximum and minimum temperatures of the air between 2 inspections, usually a period of 24 hr. A type widely used is *Six's thermometer*.

maximum continuous rating See **power rating**.

maximum flying speed See **flying speed**.

maximum landing weight See **weight and mass**.

maximum-reading accelerometer See **accelerometer**.

maximum safe airspeed indicator A pilot's *airspeed indicator* with an additional pointer showing the **indicated airspeed** corresponding to the aircraft's **limiting Mach number** and also having a mark on the dial for the maximum permissible airspeed.

maximum take-off rating See **power rating**.

maximum weight See **weight and mass**. Also *max take-off weight*, colloq. *max gross*.

max level speed The maximum velocity of a power-driven aircraft at full power without assistance from gravity; the altitude should always be specified.

max take-off weight See **weight and mass**.

MCRIT See **critical Mach number**.

M display Modified form of *A display* in which range is determined by moving an adjustable pedestal signal along the baseline until it coincides with the target signal; range is read off the control which moves the pedestal.

mean chord See **standard mean chord**.

mean daily motion The angle through which a celestial body would move in the course of 1 day if its motion in the orbit were uniform. It is obtained by dividing 360° by the period of revolution.

mean free path Average distance travelled by a sound wave in an enclosure between wall reflections; required for establishing a formula for reverberation calculations.

mean noon The instant at which the mean Sun crosses the meridian at upper culmination at any place; unless otherwise specified, the meridian of Greenwich is generally meant.

mean place The position of a star freed from the effects of precession, nutation and aberration, and of parallax, proper motion and orbital motion where appreciable. These corrections can be computed for any future date, and when applied to the mean place give the apparent place.

mean solar day The interval, perfectly constant, between two successive transits of the mean Sun across the meridian.

mean solar time Time as measured by the hour angle of mean Sun. When referred to the meridian of Greenwich it is called Greenwich Mean Time. Before 1925 this began at noon but, by international agreement, is now counted from midnight; it is thus the hour angle of mean Sun plus 12 hr, and is identical with *universal time*.

mean sun A fictitious reference point which has a constant rate of motion and is used in timekeeping in preference to the non-uniform motion of the real Sun. The mean sun is imagined to follow a circular orbit along the celestial equator and is used o measure **mean solar time**.

mechanical and electrical systems of aircraft See panel on p. 111.

mechanical equivalent of heat Once conceived as a conversion co-efficient

---

mechanical and electrical systems of aircraft

Mechanical power can be created in the main propulsive engines as shaft output for hydraulic pumps etc., compressors for pneumatic systems and electrical generators with, in the latter instance, batteries for back up. In addition an *auxiliary power unit* or *APU*, which is separate from the main propulsive or lift engines, can provide power for airborne systems including electrical, hydraulic, air conditioning, avionics, cabin pressurization and main engine starting. A *ram air turbine* can be used as an emergency source of power in the event of main engine failure. In these a propeller-like turbine is rotated to face into the airflow, and by shaft rotation and hydraulic pump provides essential power for flying controls to keep the aircraft airborne.

Hydraulic systems use high-pressure oil pumped from the main engines through pipes to actuate linear jacks to actuate flying controls, flaps, retractable under carriage (landing gear), entry doors etc. *Pneumatic systems* employ compressed air to operate reaction controls on hovering **VTOL** aircraft, air motors to provide shaft work, environmental cooling systems etc. *Electric systems* operate at several voltages for **avionics**, navigation sub-systems, fuel pumps, lighting and heating etc. Instruments indicating data on flight conditions, engine management, fuel available etc., for the benefit of the flight crew, can be operated mechanically, electrically and electronically.

New methods are being continually developed including electrically *powered* flying controls and mechanical fuel pumps.

Mechanical efficiency is the work delivered by a machine as a percentage of the work put into it, the difference being largely frictional losses.

---

between mechanical work and heat (4.186 joules =1 calorie) thereby denying the identity of the concepts. Now recognized simply as the specific heat capacity of water, 4.186 kJ/kg.

mechanical equivalent of light The ratio of the radiant flux, in watts, to the luminous flux, in lumens, at the wavelength for which the **relative visibility factor** is a maximum. Its value is about 0.0015 watts per lumen.

mechanics The study of forces on bodies and of the motions they produce.

megaparsec Unit used in defining distance of extragalactic objects. 1 Mpc = 10^6 parsec, = 3.26×10^6 light-years. See **parsec**.

megaton Explosive force equivalent to 1 000 000 tons of TNT. Used as a unit for classifying nuclear weapons.

mel Unit of subjective pitch in sound, a pitch of 131 mels being associated with a simple tone of frequency 131 Hz.

melting band A bright horizontal band often observed in vertical cross-section (RHI) **weather radar** displays. It is due to strong reflections from snowflakes which become covered with a film of water as they fall through the 0°C level and begin to melt.

meniscus telescope A compact instrument, developed by Marksutov in 1941, in which the spherical aberration of a concave

spherical mirror is corrected by a meniscus lens. It differs from the Schmidt type in having a correcting plate with 2 spherical surfaces.

Mercury (1) See panel on p. 112. (2) Name given to the project which resulted in the first US manned orbital flight by John Glenn, 20 February, 1962.

mercury barometer An instrument used for measuring the pressure of the atmosphere in terms of the height of a column of mercury which exerts an equal pressure. In its simplest form, it consists of a vertical glass tube about 80 cm long, closed at the top and having its lower open end immersed in mercury in a dish. The tube contains no air, the space above the mercury column being known as a *Torricellian vacuum*.

meridian That great circle passing through the poles of the celestial sphere which cuts the observer's horizon in the north and south points, and also passes through his zenith. Also *meridian of longitude*.

meridian circle A telescope mounted on a horizontal axis lying due east and west, so that the instrument itself moves in the meridian plane. It is used to determine the times at which stars cross the meridian, and is equipped with a graduated circle for deducing declinations. Also *transit circle*.

Mercury

The nearest major planet to the Sun and the smallest of the terrestrial planets. Telescope observation of Mercury from the Earth is very difficult, partly because of its small size and partly because it can never be more than 28° from the Sun on the celestial sphere. Its proximity to the Sun, as viewed from the Earth, arises from the fact that its orbit lies inside the Earth's. For the same reason, Mercury (like Venus, the other inferior planet) exhibits the same cycle of phases, similar to those of the Moon. Hardly any surface detail can be discerned and very little was known about the planet until the fly-by of Mariner 10 in 1974 and 1975. The space probe was put in an orbit around the Sun such that it encountered Mercury three times before it suffered permanent failure. The images returned have allowed mapping of about 35% of the surface of Mercury.

Ancient, heavily cratered terrain accounts for 70% of the area surveyed. The most significant single feature in the *Caloris basin*, a huge impact crater with a diameter of 1300 kilometres, one quarter of the diameter of the planet. The basin has been filled by a relatively smooth plain and terrain of the same type covers parts of the ejecta blanket. The impact took place 3800 million years ago and produced a temporary revival of the volcanic activity that had mostly ceased 100 million years earlier, creating the smoother areas inside and around the basin. On the point on Mercury diametrically opposite the impact site, there is curious chaotic terrain that must have been created by the shock wave. Characteristic features found on Mercury are lobate *scarpes rupes*, which take the form of cliffs between 50 and 3000 kilometres high. These are believed to be the result of shrinkage of the planetary crust as it cooled. In places they cut across craters. The average density of Mercury is only slightly less than that of the Earth. Its smaller size and lower interior pressure leads to the conclusion that Mercury has a substantial iron core accounting for 70% of its mass and 75% of its total diameter. There is also a magnetic field of about 1% of the Earth's as further evidence for the metallic core.

The planet's rotation is such that a 'day' on Mercury lasts two 'years'. This leads to immense temperature contrasts. At perihelion, the subsolar point reaches 430° C. The night-time temperature plunges to −170°C.

The high daytime temperatures and the small mass of the planet make it impossible to retain an atmosphere. The small amounts of helium detected may be the product of radioactive decay of surface rocks or captured from the **solar wind**.

meson A hadron with a baryon number of 0. Mesons generally have masses intemediate between those of electrons and nucleons and can have negative, zero or positive charges. Mesons are *bosons* and may be created or annihilated freely. There are three groups of mesons: π-mesons (pions), K-mesons (kaons) and η-mesons.

mesopause The top of the mesosphere at about 80 to 85 km.

mesosphere Region of the atmosphere lying between the **stratopause** and **mesopause** (50–85 km), in which temperature generally decreases with height.

Messier catalogue A listing of 108 galaxies, star clusters and nebulae drawn up by the French comet hunter, Charles Messier, in 1770. Objects are designated M1, M2 and so on. Many of these alphanumeric names are still widely used in astronomy.

metal matrix composite Usually a refractory metal reinforced by a different fibre, e.g. silica reinforced aluminium or boron-titanium.

meteor A 'shooting star'. A small body which enters the Earth's atmosphere from interplanetary space and becomes incandescent by friction, flashing across the sky and generally ceasing to be visible before it falls to Earth. See **bolide**.

meteor craters Circular unnatural craters of which Meteor Crater in Arizona is best known; believed to be caused by the impact of meteorites.

microgravity

The condition of near weightlessness which is induced by free fall or un-powered space flight; it is characterized by the virtual absence of gravity-induced convection, hydrostatic pressure and sedimentation. Typical environments in the range 10^{-4} to 10^{-5} g are experienced in manned space systems but can be 10^{-6} g or less in unmanned free-flying systems. The effects of small forces, like surface tension, which would be swamped on Earth, become evident in microgravity. Experiments can be performed whose results will enhance Earth-bound techniques. Also, some procedures can be carried out which are impossible in a one-g environment. An example is the manipu-lation of matter in the liquid or gaseous form, possibly leading to the manufac-ture of new materials, better crystals or purer vaccines.

The phenomenon of *space-sickness* is experienced by one in two astronauts during the first few days of near-zero g exposure because the body's balance mechanisms do not always adapt rapidly to the new environment. The import-ance of this topic to working in space has led to a wide range of orientation and vestibular-related investigations in microgravity.

The term microgravity is also used when referrring to the scientific discipline that is concerned with the evaluation of processes in a near-zero g environ-ment. This includes certain aspects of Fluid Physics, and the Life and Material Sciences.

meteoric shower A display of meteors in which the number seen per hour greatly ex-ceeds the average. It occurs when Earth crosses the orbit of a meteor swarm and the swarm itself is in the neighbourhood of the point of section of the two orbits.

meteorite Mineral aggregates of cosmic origin which reach the Earth from inter-planetary space. Cf. **meteor** and **bolide**.

meteorological satellite An artificial Earth satellite (usually in a **geosynchronous** orbit) used for weather observation and weather forecasting. See **Earth obser-vation**.

meteor stream Streams of dust revolving about the Sun, whose intersection by the Earth causes meteor shows. Some night-time showers have orbits similar to those of known comets; daytime showers, detected by radio-echo methods, have smaller orbits, similar to those of minor planets.

Metis A tiny natural satellite of **Jupiter**, dis-covered in 1979 by Voyager 2 mission.

Metonic cycle A period of 19 years, which is very nearly equal to 235 synodic months, this relationship having been introduced in Greece in 433 BC by the astronomer Meton; its effect is that after a full cycle the phases of the Moon recur on the same days of the year.

metre The Système Internationale (SI) fun-damental unit of length. The metre is defined (1983) in terms of the velocity of light. The metre is the length of path travelled by light in vacuum during a time interval of 1/299 792 458 of a second. Originally intended to represent 10^{-7} of the distance on the Earth's surface between the North Pole and the Equator, formerly it has been defined in terms of a line on a platinum bar and later (1960) in terms of the wavelength from ^{86}Kr.

metre-kilogram(me)-second-ampere See **MKSA**.

MF Abbrev. for *Medium Frequency*. Fre-quencies from 300 to 3000 kHz.

microburst Dangerous vertical gust of wind having a core of about 1.5 miles (2.5 km) diameter in which downward velocities of 4000 ft/m (20 m/s) can occur down to low altitude.

microgravity See panel on this page.

microlight Aircraft whose empty weight does not exceed 330 lb (150 kg). In US *ultralight* is used for weights up to 254 lb (115 kg).

microlux A unit for very weak illumi-nations, equal to one-millionth of a lux.

micrometeorites Extremely small par-ticles, typically of mass less than 10^{-6} gm and diameter less than 10^{-4} m, which are present in space; these hyper-velocity par-ticles represent a hazard for space-flight and must be protected against. They do not burn up in the Earth's atmosphere, but drift down to the surface. **Comets** are probably abun-dant sources of new micrometeorites.

micrometer Measures small angular separ-ations in the telescope. It consists of three frameworks carrying spider-webs close to the image plane; one is fixed and the others are each adjustable by micrometer heads, by which the separation is read, with a graduated circle giving the angular relation of a double star. A *micrometer gauge* is a mechanical device for measuring thickness.

micrometre One-millionth of a metre. Symbol μm. Formerly *micron*.

microwave background Discovered in 1963, a weak radio signal which is detectable in every direction with almost identical intensity. It has a temperature of 2.7 K. The slight asymmetry is due to the motion of our Galaxy relative to this radiation. The radiation is the relic of the early hot phase in the **Big Bang** universe.

microwave spectrometer An instrument designed to separate a complex microwave signal into its various components and to measure the frequency of each; analogous to an *optical spectrometer*. See **spectrometer**.

microwave spectroscopy The study of atomic and/or molecular resonances in the microwave spectrum.

microwave spectrum The part of the electromagnetic spectrum corresponding to microwave frequencies.

middle marker beacon A marker beacon associated with the **ILS**, used to define the second predetermined point during a beam approach.

mid-wing monoplane A monoplane wherein the main planes are located approximately midway between the top and bottom of the fuselage.

MIG Abbrev. for *Miniature Integrating Gyro*. See **floated rate integrating gyro, tuned rotor gyro**.

MIL-1553 B Standard requirement for airborne digital databus. Originally US but now international.

Milankovitch theory of climatic change The theory that large oscillations in climate are related to changes in solar radiation received by the Earth as a result of (1) the variations in eccentricity of the Earth's orbit (periods of 10^5 and 4.10^5 years); (2) variations in the obliquity of tilt of the Earth's axis (period 4.10^4 years); (3) the precession of the equinoxes (period 2.10^4 years). There is evidence for the two shorter cycles in data obtained from analysis of deep-sea bottom cores.

Milky Way See **galaxy**.

millibar See **bar**.

millilambert A unit of brightness equal to 0.001 lambert; more convenient magnitude than the lambert.

millilux Unit of illumination equal to one-thousandth of a lux.

millimass unit Equal to 0.001 of atomic mass unit. Abbrev. *mu*.

Mil-Spec *Military Specification* issued in the US, which lays down basic requirements to be observed by design teams in the development of aircraft. Abbrev. also *MS*.

Mimas The natural satellite orbiting closest to **Saturn**, 400 km in diameter.

minimum burner pressure valve A device which maintains a safe minimum pressure at the burners of a gas turbine when it is idling.

minimum flying speed The minimum speed at which an aircraft has sufficient lift to support itself in level flight in standard atmosphere. There is a close relationship with the weight, which affects the **wing loading** and the term must be stated with the weight (and altitude if the **International Standard Atmosphere** at sea level is not implied) and true airspeed specified.

Minkowski diagram A spacetime diagram used to represent the positions and times of events relative to an inertial reference frame.

minor planet Term used generally in professional astronomy for **asteroid**. Also *planetoid*.

Mir Advanced USSR space station, developed from experience gained with **Salyut**; it has the capability for considerable growth by adding further modules. See **space station**.

mirage An effect caused by total reflection of light at the upper surface of shallow layers of hot air in contact with the ground, the appearance being that of pools of water in which are seen inverted images of more distant objects. Other types of mirage are seen in polar region, where there is a dense, cold layer of air near the ground. See **fata morgana**.

Mira stars Long-period variable stars named after Mira Ceti; more than 3000 are known, with periods from 2 months to 2 years, and all are red giant stars.

missile There are two basic types of missile, in the current sense: *guided* and *ballistic*. The former is controlled from its launch until it hits its target; the latter, always of long-range, surface-to-surface type, is controlled into a precision ballistic path so that its course cannot be deflected by countermeasures. See **guided atmospheric flight**.

mission Succession of events which must happen to achieve the objectives stipulated; it includes everything which must be done from conception to the delivery of the results (more loosely, the actual flight of the spacecraft).

mission adaptive wing A wing whose

section profile is automatically adjusted to suit different flight conditions, e.g. Mach number, lift and altitude.

mission control centre Room or building in which are assembled the means necessary to visualize and control a space system so that its mission objectives can be achieved.

mission specialist Member of the crew of the **Space Shuttle** whose responsibilities are concerned with mission aspects, such as the control of the Orbiter's resources to a payload, the handling of payload equipment and the performance of the experiments in orbit.

mist A suspension of water droplets (radii less than 1 μm) reducing the visibility to not less than 1 km. See **fog**.

mixing length The average distance travelled by an eddy which is transporting heat, momentum or water vapour in the atmosphere.

mixture control An auxiliary control fitted to a carburettor to allow of variation of mixture strength with altitude. May be manually operated or automatic.

MKSA *Metre-Kilogram(me)-Second-Ampere* system of units, adopted by the International Electrotechnical Commission, in place of all other systems of units. See **SI units**.

M$_{ne}$ Abbrev. for the maximum permissible indicated **Mach number**: a safety limitation, the suffix means 'never exceed' because of strength or handling considerations. The symbol is used mainly in operational instructions.

M$_{no}$ Abbrev. for *Normal Operating Mach number*, usually of a jet airliner, the term being used mainly in flight operations instructions for flight levels above 7600 m.

mock moons Lunar images similar to **mock suns**. Also *paraselenae*.

mock suns Images of the Sun, not usually very well defined, seen towards sunset at the same altitude as the Sun and 22° from it on each side. They are portions of the 22° ice *halo* formed by ice crystals which, for some reasons, are arranged with their axes vertical. Also *parhelia*.

mode Situation or method of performing a specified task.

modified refractive index Sum of the refractive index of the atmosphere at a given height and the ratio of the height to the radius of the Earth.

modulation (1) Change of amplitude or frequency of a carrier signal of given frequency. (2) Changing from one key to another in music. The continual change from one fundamental frequency to another in speech.

module A separate, and separable, compartment of a space vehicle.

mogas Abbrev. for *MOtor GASoline*, 91 to 93 octane.

moment (1) See **hinge-, pitching-, rolling-, yawing-**. (2) Of a force or vector about a point, the product of the force or vector and the perpendicular distance of the point from its line of action. In vector notation, **r n F**, where **r** is the position vector of the point, and **F** is the force or vector. (3) Of a force or vector about a line, the product of the component of the force or vector parallel to the line and its perpendicular distance from the line. (4) Of a couple. See **couple**.

moment of inertia Of a body about an axis: the sum Σmr^2 taken over all particles of the body where m is the mass of a particle and r its perpendicular distance from the specified axis. When expressed in the form Mk^2, where M is the total mass of the body ($M = \Sigma m$), k is called the *radius of gyration* about the specified axis. Also used erroneously for *second moment of area*.

momentum A dynamical quantity, conserved within a closed system. A body of mass M and whose centre of gravity G has a velocity v has a *linear momentum* of Mv. It has an *angular momentum* about a point O defined as the moment of the linear momentum about O. About G this reduces to $I\omega$ where I is the moment of inertia about G and ω the angular velocity of the body.

momentum wheel A flywheel, part of an attitude control system, which stores momentum by spinning; three wheels with their axes at right angles can serve to stabilize a satellite's attitude. Also *inertia wheel*.

monochromatic By extension from *monochromatic light*, any form of oscillation or radiation characterized by a unique or very narrow band of frequency.

monochromatic light Light containing radiation of a single wavelength only. No source emits truly monochromatic light, but a very narrow band of wavelengths can be obtained, e.g. the cadmium red spectral line, wavelength 643.8 nm with a *half-width* of 0.0013 nm. Light from some lasers have extremely narrow line widths.

monochromatic radiation Electromagnetic radiation (originally *visible* radiation) of one single frequency component. By extension, a beam of particulate radiation comprising particles all of the same type and energy. *Homogeneous* or *mono-energic* is preferable in this sense.

monochromator Device for converting heterogeneous radiation (electromagnetic or particulate) into a homogeneous beam by absorption, refraction or diffraction processes.

monocoque A fuselage or nacelle in which

Moon

The Moon is the Earth's natural satellite and its major properties are shown in the table.

The surface has heavily cratered highlands and smooth dark **maria**. The highland rocks were formed 3.9–4.4 billion years ago, whereas the maria are relatively younger basalts aged 3.1–3.8 billion years. The Moon formed about 4.6 billion years ago and initially the surface was molten, perhaps on account of heating from intense **meteorite** bombardments. As the surface solidified, meteorites continued to rain on the Moon, and their impacts were responsible for most of the craters still visible today. By 3.8 billion years ago the interior had heated sufficiently from the decay of radioactive elements to start volcanic activity. Lava flowing on to the surface flooded the basins created in earlier bombardments to form the maria. When this activity ceased 3.1 billion years ago the geological evolution of the Moon came to an end. Almost all craters were formed by meteorite impacts, although there are a small number of volcanic caldera as well. *Rilles* are conspicuous features of the maria: they are sinuous valleys stretching hundreds of kilometres through which molten rock once flowed.

The lunar crust is about 65 km thick, and is made of relatively light materials. The mantle is silica-rich, and constitutes most of the interior. There may be an iron-rich core at the centre and this may still be liquid or plastic. Mass concentrations or *mascons* are present beneath the larger basins, possibly because the lava is denser in those locations.

It is not possible to state with certainty how the Moon was formed. Dynamical arguments show that naive theories such as capture from elsewhere or fission of a single proto Earth-Moon are wrong. In their place are condensation models, which state that the Earth and Moon formed near to each other at about the same time, models that invoke the formation and condensation of a ring of matter round the proto-Earth and models that suggest the Moon is a fragment of a larger body that broke up.

The Moon rotates on its axis at the same rate as it orbits the Earth. This resonance has been caused by tidal interactions between the Earth and Moon. The Moon bulges towards the Earth.

The first spacecraft to orbit the Moon was Luna 3 in 1959, which gave the first views of the far side. The Apollo missions (1968-72), arguably the most expensive programme ever attempted, resulted in 382 kg of rock being brought to Earth. No spacecraft has visited the Moon since 1976 and no future missions have been adopted by the major funding agencies. Orbiting space stations, planned for the late 1990s, will make the Moon an easier goal and it is possible a base will be built as the intermediate stage of a mission to Mars.

Lunar data:

Radius	1732.8 km
Mass	7.348×10^{22} kg (1.23% Earth mass)
Mean density	3.342×10^3 kg m^{-3}
Surface gravity	1.618 m s^{-1}
Escape velocity	2.38 km s^{-1}
Mean distance from Earth	384 400 km
Axial revolution period	27.321 666 days
Synodic revolution period	29.530 588 days
Apparent magnitude	-12.7 mag (max)

all structural loads are carried by the skin. In a *semimonocoque*, loads are shared between skin and framework, which provides local reinforcement for openings, mountings etc. See **stressed skin construction**.

monoplane A heavier-than-air aircraft, either an aircraft or glider, having one main supporting surface.

monopole Spherical radiator whose surface moves inwards and outwards with the same phase and amplitude everywhere. Any sound source which produces an equivalent sound field is also called a monopole, e.g. any small source generating volume flow.

monopropellant Single propellant which produces propulsive energy as the result of a chemical reaction, usually induced by the presence of a catalyst. Cf. **bipropellant**.

monopulse A radar system with an antenna system with two or more overlapping lobes in its radiation pattern. From the transmission of a single pulse and analysis of error signals due to the target being off-axis in one or more lobes, detailed information about direction can be obtained. Used in many gun-control and missile guidance systems.

monsoon Originally winds prevailing in the Indian Ocean, which blow S.W. from April to October and N.E. from October to April; now generally winds which blow in opposite directions at different seasons of the year. Similar in origin to land and sea breezes, but on a much larger scale, both in space and time. Particularly well developed over southern and eastern Asia, where the wet summer monsoon from the S.W. is the outstanding feature of the climate.

month See **anomalistic-**, **sidereal-**, **synodic-**, **tropical-**.

Moon See panel on p. 116.

mooring tower A permanent tower or mast for the mooring of airships. Provided with facilities for the transference of passengers and freight, and arrangements for replenishing ballast, gas and fuel.

morning star Popularly, a planet, generally Venus or Mercury, seen in the eastern sky at or about sunrise; also, loosely, any planet which transits after midnight.

most economical range The range obtainable when the aircraft is flown at the height, airspeed and engine conditions which give the lowest fuel consumption for the aircraft weight and the wind conditions prevailing.

MOU Abbrev. for *Memorandum Of Understanding*.

movement area That part of an airport

reserved for the take-off, landing and movement of aircraft.

moving-target indicator A device for restricting the display of information to moving targets. Abbrev. *MTI*.

Mpc Abbrev. for **Megaparsec**.

MS See **Mil-Spec**.

MSW Abbrev. for **Magnetic Surface Wave**.

MTBF Abbrev. for **Mean Time Between Failures**.

MTI Abbrev. for *Moving-Target Indicator*.

multiple echo Perception of a number of distinct repetitions of a signal, because of reflections with different delays of separate waves following various paths between the source and observer.

multiple star A system in which three or more stars, united by gravitational forces revolve about their common centre of gravity.

multiplex Use of one channel for several messages by **time-division multiplex** or **frequency division**.

multirow radial engine A radial *aero-engine* with two or more rows of cylinders.

multisensor The use of more than one sensor to obtain information.

multispeed supercharger A gear-driven **supercharger** in which a clutch system allows engagement of different ratios to suit changes in altitude.

multistage Said of a space-vehicle having successive rocket-firing stages, each capable of being jettisoned after use.

multistage compressor A gas turbine compressor with more than one stage; each row of blades in an **axial-flow compressor** is a stage, each **impeller** is a stage in a **centrifugal-flow compressor**. In practice, all axial compressors are multistage, while almost all centrifugal compressors are single stage. Occasionally an initial axial stage is combined with a centrifugal delivery stage.

multistage supercharger A supercharger with more than one impeller in series.

multistage turbine A turbine with two or more disks joined and driving one shaft.

muon Fundamental particle with a rest mass equivalent to 106 MeV; it is one of the *leptons* and has a negative charge and a half-life of about 2 μs. Decays to electron, neutrino and antineutrino. It participates only in *weak* interactions. The *antimuon* has a positive charge and decays to positron, neutrino and antineutrino.

mush Condition of flight at the stall when the aircraft tends to maintain **angle of attack** while losing height rather than the sharp nose down pitch which is more common.

NO

n Symbol for neutron.

ν Symbol for (1) frequency, (2) neutrino, (3) Poisson's ratio, (4) kinematic viscosity.

N Symbol for **newton**.

N Symbol for neutron number.

*N*_A Symbol for Avogadro number.

NA Abbrev. for **Numerical Aperture**.

nacelle A small streamlined body on an aircraft, distinct from the fuselage, housing engine(s), special equipment or crew.

nacreous clouds Clouds composed of ice crystals in 'mother of pearl' formations, found at a height of 25 to 30 km. They may be **wave clouds**.

nadir That pole of the horizon vertically below the observer; hence the point on the celestial sphere diametrically opposite the **zenith**.

NASA Abbrev. for *National Aeronautics and Space Administration,* responsible for civil space activities in the US, both research and development.

NASDA Abbrev. for *National Space Development Agency,* the Japanese space agency mainly responsible for applications of space activities and launch systems.

natural abundance Same as *abundance* or *abundance ratio* which is, for a specified element, the proportion or percentage of one isotope to the total, as occurring in nature.

natural evaporation The evaporation that takes place at the surface of ponds, rivers etc. which are exposed to the weather; it depends on solar radiation, strength of wind and relative humidity.

natural radioactivity That which is found in nature. Such radioactivity indicates that the isotopes involved have a half-life comparable with the age of the Earth or result from the decay of such isotopes. Most such nuclides can be grouped in one of three **radio-active series**.

Nautical Almanac An astronomical ephemeris published annually in advance, for navigators and astronomers. First published in 1767, it is now called the *Astronomical Ephemeris.* An abridged version, for the use of navigators, is given the original title *The Nautical Almanac.*

nautical twilight The interval of time during which the Sun is between 6° and 12° below the horizon, morning and evening. See **astronomical twilight, civil twilight.**

navaid Abbrev. for *navigational aid.*

navigable semicircle The left hand half of the storm field in the northern hemisphere, the right hand half in the southern hemisphere, when looking along the path in the direction a *tropical revolving storm* is travelling. Cf. **dangerous semicircle.**

navigation flame float A pyrotechnic device, dropped from an aircraft, which burns with a flame while floating on the water. Used to determine the drift of the aircraft at night.

navigation lights Aircraft navigation lights consist of red, green and white lamps located in the port wing tip, starboard wing tip and tail respectively.

navigation smoke float A pyrotechnic device, dropped from an aircraft, which emits smoke while floating on the water. Used for ascertaining the direction of the wind or the drift of the aircraft.

navigation systems See panel on p. 119.

NAVSTAR Name for the **Global Positioning System** using satellites spaced around the world.

NDB Abbrev. for *Non-Directional Beacon.* See **beacon.**

N-display A radar **K-display** in which the target produces two breaks on the horizontal time base. Direction is proportional to the relative amplitude of the breaks, and range is indicated by a calibrated control which moves a pedestal signal to coincide with the breaks.

neap tides High tides occurring at the Moon's first or third quarter, when the Sun's tidal influence is working against the Moon's, so that the height of the tide is below the maximum in the approximate ratio 3:8.

near field See **far field.**

nebula A term applied to any celestial object which appears as a hazy smudge of light in an optical telscope, its usage predating photographic astronomy. It is now more properly restricted to true clouds of **interstellar medium.** Nebulae may be either bright or dark. Galaxies are sometimes referred to as *extragalactic nebulae,* the most famous of which is the **Andromeda nebula.**

nebular hypothesis One of the earliest scientific theories of the origin of the solar system, stated by Laplace. It supposed a flattened mass of gas extending beyond Neptune's orbit to have cooled and shrunk, throwing off in the process successive rings which in time coalesced to form the several planets.

NEF See **noise exposure forecast value.**

negative *g* (1) In a manoeuvring aircraft,

navigation systems

Navigation is concerned with two factors: firstly determining the position of an aircraft or spacecraft at a particular time and secondly determining a course to be followed in order to arrive at a destination.

Historically, ship navigation involved Sun and star sightings, accurate time-keeping and reference to known landmarks, when available. The navigator evaluated position and courses by manual trigonometric calculations. In modern automatic systems in *aircraft*, celestial navigation or *astronavigation* is used, supplemented by radio or radar fixes and increasingly sophisticated systems which perform all calculations automatically. By connecting the latter to the flight control system aircraft can fly themselves from airport to airport. The same fundamental methods apply to *spacecraft* but a higher degree of automation is required. Position and velocity are usually determined by tracking using radio ranging (frequency shift), doppler or radar techniques. The control actions to adjust these parameters can then be computed as part of the guidance process. Normally, a spacecraft's trajectory requires initial guidance just after launch, mid-course guidance for trajectory correction and terminal guidance for assuring that the target position will be encountered. Two *frames of reference* are used: *absolute* in which three mutually perpendicular axes are fixed relative to external space (i.e. the stars) and *relative* in which the axes are fixed to arbitrary directions, e.g. local map orientations on the Earth's surface or from moving aerospacecraft.

Dead reckoning is still used by aircraft and involves plotting position by calculations involving speed, course, time, wind effect and previous known position but more accurate methods are general. **Radio communication** is used to determine location, and to obtain heading information and warnings of obstructions or hazards. *OMEGA* is an accurate long-range radio *navaid* of very low frequency and hyperbolic type which covers the whole Earth from eight ground stations and is usable down to sea level. **Radar** is also used. **See radar mapping**. See **Global Positioning System** (GPS, Navstar), **Joint Tactical Information Distribution System** (JTIDS).

There is also a range of devices which are completely contained within the aircraft and do not rely on ground beacons or transmissions, and are therefore not susceptible to external interference. The basic method is an **inertial navigation system** (*INS*), using highly accurate gyros and accelerometers to give the present position in relation to the start with a typical accuracy of 1 in 10^9. INS is particularly used in launch systems and manned spacecraft. In a *strapped down INS* no gyroscopes are employed and the accelerometers are fixed to the airframe. Computers are used to transform data from body axes to inertial axes. These are supplemented by methods which compare the terrain below to stored information and so prevent the degradation with time of the accuracy of the INS system. These are called **tercom** (terrain comparison matching) or **terprom** (terrain profile matching). Infrared radiation detected as heat is used to depict ground features over-flown by military aircraft (using FLIR, *Forward Looking Infra Red*) and for terminal guidance of anti-tank and air-air missiles.

Specialized instruments which are used include **fibre optic gyro, floated rate integrating gyro, laser gyro, rate gyro, tuned rotor gyro, Doppler** and **LORAN** navigators.

Neptune

The eighth planet in order of distance, 30 astronomical units from the Sun, and the outermost of four gaseous giants. Its maximum magnitude is +7.7 so that it is never visible to the naked eye. The discovery, on 23 September 1843 at the Berlin observatory was the result of a prediction made by Leverrier of Paris. He and John Adams of Cambridge University had independently analysed the motion of Uranus and come to the conclusion that a more remote massive planet was perturbing the orbit. Neptune orbits the Sun in 164.79 years, has a diameter of 50 000 km and a mass 17.23 that of Earth.

On account of its distance, little was known of this planet until the spectacular mission in August 1989 of the Voyager 2 spacecraft which returned over 20 000 photographs. The atmosphere is dynamic, possibly sustained in motion by internal sources of heat. There is a huge dark spot, which is a hurricane similar to Jupiter's Great Red Spot. Bright cirrus clouds of methane are 50–75 km above the main cloud layer. Windspeeds reach 1000 km per hour. The magnetic field is tilted at 50° to the rotation axis and its strength is similar to Earth. Aurora occur at the magnetic poles.

Neptune has three complete rings, one of which is 4000 km wide. There are seven small natural satellites or moons with diameters 50–420 km, and one large moon, Triton.

Triton has a radius of 1360 km. The surface is covered in an icy slush of methane. The relatively smooth surface is evidence that the moon had extensive volcanic activity in the first billion years of its existence.

any force acting opposite to the normal force of gravity. (2) The force exerted on the human body in a gravitational field or during acceleration so that the force of inertia acts in a foot-to-head direction, causing considerable blood pressure on the brain. Also *minus g*. *Negative g tolerance*, in practice, is the degree of tolerance 3 *g* for 10–15 sec.

negative stagger See **stagger**.

nephoscope An instrument for observing the direction of movement of a cloud and its angular velocity about the point on the Earth's surface vertically beneath it. If the cloud height is also known its linear speed may be calculated.

Neptune See panel on this page.

Nereid The second substantial natural satellite of **Neptune**, diameter 300 km.

net pyrradiometer *Radiation balance meter.* Instrument for measuring the difference of the total radiations falling on both sides of a plane surface from the solid angle 2π respectively.

net wing area The **gross wing area** minus that part covered by the fuselage and any nacelles.

neutral equilibrium The state of equilibrium of a body when a slight displacement does not alter its potential energy.

neutral point (1) A small region of the daylight sky from which scattered sunlight is unpolarized; such points were discovered by Arago, Babinet and Brewster. (2) A point in the field of a magnet where the Earth's magnetic field (usually the horizontal component) is exactly neutralized. (3) That c.g. position in an aircraft at which longitudinal stability is neutral. *Stick-fixed neutral point* is the c.g. position at which control column movement to trim a change in speed is zero. *Stick-free neutral point* is the c.g. position at which the stick force needed to trim a change in speed is zero. (4) See **gravipause**.

neutrino A fundamental particle, a **lepton**, with zero charge and zero mass. A different type of neutrino is associated with each of the four charged leptons. Its existence was predicted by Pauli in 1931 to avoid β-decay infringing the laws of conservation of energy and angular momentum. As they have very weak interactions with matter, neutrinos were not observed experimentally until 1956. See **antineutrino**.

neutrino astronomy Term applied to attempts to detect **neutrinos** from the Sun, with the aim of discovering more exactly the conditions in the solar core.

neutron Uncharged subatomic particle, mass approximately equal to that of the proton, which enters into the structure of atomic nuclei. Interacts with matter primarily by collisions. Spin quantum number of neutron = +D , rest mass = 1.008 665 a.m.u., the charge is zero and the magnetic moment −1.9125 nuclear Bohr magnetons. Although stable in nuclei, isolated neutrons decay by β-emission into protons, with a half-life of 11.6 minutes.

neutron star A small body of very high density (ca 10^{12} kg/dm^3) resulting from a supernova explosion in which a massive star collapses under its own gravitational forces, the electrons and protons combining to form neutrons.

new moon The instant when Sun and Moon have the same celestial longitude; the illuminated hemisphere of the Moon is then invisible.

new star See **nova.**

New Style A name given to the system of date-reckoning which was established by the Gregorian calendar. Abbrev. *NS.*

newton Symbol *N*. The unit of force in the SI system, being the force required to impart, to a mass of 1 kg, an acceleration of 1 m/sec^2. 1 newton =0.2248 pounds force.

Newtonian telescope A form of reflecting telescope due to Newton, in which the object is viewed through an eyepiece in the side of the tube, the light reflected from the main mirror being deflected into it by a small plane mirror inclined at 45° to the axis of the telescope and situated just inside the principal focus.

Newton's laws of motion (1) Every body continues in a state of rest or uniform motion in a straight line unless acted upon by an external impressed force. (2) The rate of change of momentum is proportional to the impressed force and takes place in the direction of the force. (3) Action and reaction are equal and opposite, i.e. when two bodies interact the force exerted by the first body on the second body is equal and opposite to the force exerted by the second body on the first. These laws were first stated by Newton in his *Principia*, 1687. Classical mechanics consists of the applicaton of these laws.

Newton's rings Circular concentric interference fringes seen surrounding the point of contact of a convex lens and a plane surface. Interference occurs in the air film between the two surfaces. If r_n is the radius of the nth ring, R is the radius of curvature of the lens surface and λ the wavelength. $r_n = \sqrt{nR\lambda}$. See **contour fringes.**

NGC Abbrev. for the *New General Catalogue* of all nebulous objects known in 1888. Together with the supplementary Index Catalogue (IC) it lists 13 000 galaxies, clusters and nebulae.

NGTE rigid rotor A helicopter self propelling rotor system evolved at the *National Gas Turbine Establishment* which uses the principle of the **jet flap** to obtain very high lift co-efficients.

nimbostratus Grey cloud layer, often dark, the appearance of which is rendered diffuse by more or less continuously falling rain or snow, which in most cases reaches the ground. It is thick enough throughout to blot out the Sun. Low ragged clouds frequently occur below the layer, with which they may or may not merge. Abbrev. *Ns.*

NOAA Abbrev. for *National Oceanic and Atmospheric Administration*, a US body which manages and operates environmental satellites, and provides data to users worldwide.

noctilucent clouds Thin but sometimes brilliant and beautifully coloured clouds of dust or ice particles at a height of from 75 to 90 km. They are visible about midnight in latitudes greater than about 50° when they reflect light from the Sun below the horizon.

node (1) One of the two points at which the orbit of a celestial object intersects a reference plane, such as the **ecliptic** or **equator.** The path crossing from south to north is the *ascending node*; the *descending node* has the opposite sense. (2) The location of a minimum in the sound pressure or particle velocity when waves superimpose and result in standing waves.

noise (1) Socially unwanted sounds. (2) Interference in a communication channel.

noise abatement climb procedure Means of flying a civil aircraft from an airport so as to climb rapidly until the built-up area is reached and thereafter reducing power to just maintain a positive rate of climb until the area is overflown or 5000 ft is reached.

noise control Reduction of unwanted noise by various methods, e.g. absorption, isolation, antisound.

noise exposure forecast value Used in evaluating the annoyance caused by fluctuating noises. Abbrev. *NEF.*

noise factor, figure Ratio of noise in a linear amplification system to thermal noise over the same frequency band and at the same temperature.

noise footprint The contour beneath an aircraft of constant noise level, measured in dB or derived units.

noise level The **loudness level** of a noise signal.

noise meter See **objective noise meter.**

noise ratio See **signal/noise ratio.**

noise suppressor A turbojet propelling nozzle fitted with fluted members which induct air to slow and break up the jet efflux, thereby reducing the noise level.

non-destructive testing Use of probing systems to test structures for integrity without impairing their quality instead of removing samples for conventional analysis.

non-hydrostatic model Numerical forecasting model in which the **hydrostatic approximation** is not used so that the effects of vertical accelerations can be accounted for.

non-relativistic Said of any procedure in which effects arising from relativity theory are absent or can be disregarded, e.g. properties of particles moving with low velocity, e.g. 1/20th that of light propagation.

non-return flow wind tunnel A straight-through wind tunnel in which the air flow is not recirculated.

noon The instant of the Sun's upper culmination at any place. See also **mean noon**.

normal axis See **axis**.

normal flight (1) All flying other than aerobatics, including straight and level, climbing, gliding, turns and *sideslips* for the loss of height or to counteract drift. (2) A licensing category for certifying whether *airworthy*.

Northern Lights The *Aurora Borealis.* See **aurora**.

nosedive See **dive**.

nose heaviness The state in which the combination of the forces acting upon an aircraft in flight is such that it tends to pitch downwards by the nose.

nose ribs Small intermediate ribs, usually from the front spar to the leading edge only, of planes and control surfaces. They maintain the correct wing contour under the exceptionally heavy air load at that part of the aerofoil.

nose-wheel landing gear See **tricycle landing gear**.

notch aerial A radio aerial, usually for *HF*, formed by cutting a notch out of the aircraft's skin and covering it with a dielectric material to its original profile.

nova Classically, any new star which suddenly becomes visible to the unaided eye. In modern astronomy, a star late in its evolutionary track which suddenly brightens by a factor of 10^4 or more. The rise in brightness is so rapid (days) that the star is seldom noticed until it is indeed of naked eye visibility; after a few weeks it returns to the pre-nova magnitude. Astronomers consider that most novae (pl. also *novas*) are members of close **binary stars**. One component is a **white dwarf**, and this is the source of the explosion: matter is transferred from the other, highly evolved, companion and triggers a new burst of nuclear reactions.

nowcasting A system of rapid and very short range (1 to 2 h) forecasting of phenomena such as heavy rain and thunderstorms, based on real-time processing of simultaneous observations from a network of remote-sensing devices (including **weather radars** and **meteorological satellites**) combined with simple extrapolation techniques.

noy Unit of perceived noisiness by which equal-noisiness contours, e.g. for 10 noys, 20 noys etc., replace equal-loudness contours.

nozzle See **propelling nozzle**.

nozzle guide vanes In a gas turbine, a ring of radially positioned aerofoils which accelerate the gases from the combustion chamber and direct them on to the first rotating turbine stage.

NPL type wind tunnel The *closed-jet, return-flow* type is often called the original NPL type, and the *closed-jet, non-return flow* the standard NPL type, as they were first used by the UK *National Physical Laboratory.*

NTI Abbrev. for *Noise Transmission Impairment.*

NTSB Abbrev. for *National Transportation Safety Board*, US.

nuclear energy In principle, the binding energy of a system of particles forming an atomic nucleus. More usually, the energy released during nuclear reactions involving regrouping of such particles (e.g. fission or fusion processes). The term *atomic energy* is deprecated as it implies rearrangement of atoms rather than of nuclear particles.

nuclear fission The spontaneous or induced disintegration of the nucleus of a heavy atom into two lighter atoms. The process involves a loss of mass which is converted into nuclear energy.

nuclear fusion The process of forming atoms of new elements by the fusion of atoms of lighter ones. Usually the formation of helium by the fusion of hydrogen and its isotopes. The process involves a loss of mass which is converted into nuclear energy. The basis of possible fusion reactors.

nuclear propulsion The use of the energy released by a nuclear reaction to provide propulsive thrust through heating the working fluid or providing electric power for an ion or similar propulsion system.

nucleating agent Substance used for seeding clouds to control rainfall and fog formation. See **rainmaking**.

nucleosynthesis The synthesis of elements other than hydrogen and helium by means of nuclear fusion reactions in stellar interiors and **supernova** explosions.

nucleus

nucleus (1) The central core of a comet, about 1–10 km across, and consisting of icy substances and dust. (2) The central part of a **galaxy** or **quasar**, possibly the seat of unusually energetic activity within the galaxy. (3) Term used generally in astronomy to indicate any concentration of stars or gas in the central part of a **nebula**.

numerical aperture Product of the refractive index of the object space and the sine of the semiaperture of the cone of rays entering the entrance pupil of the objective lens from the object point. The resolving power is proportional to the numerical aperture. Abbrev. *NA*.

numerical forecast A forecast of the future state of the atmosphere made by solving the equations of a **numerical forecast model** on a digital computer. The initial data are obtained from an **objective analysis**.

numerical forecast(ing) model A set of differential equations, with suitable boundary conditions, for the production of a **numerical forecast** and usually describing an artificially simplified atmosphere.

numerical weather prediction See panel on p. 124.

nutating feed That to a radar transmitter which produces an oscillation of the beam without change in the plane of polarization. The resulting radiation field is a *nutation field*.

nutation (1) The periodic variation of the inclination of the axis of a spinning top (or gyroscope) to the vertical and, therefore, (2) an oscillation of the Earth's pole about the mean position. The latter has a period of about 19 years, and is superimposed on the precessional movement.

nutation field See **nutating feed**.

N-wave See **supersonic boom**.

NWP Abbrev. for **Numerical Weather Prediction**.

O and C building Abbrev. for the *Operations and Check-out Building* at Kennedy Space Center, Florida, used for the integration and check-out of payload elements destined for the Space Shuttle.

objective analysis A method of processing original weather observations by computer to give all values of the atmospheric variables needed to produce a numerical forecast, in contrast to a subjective analysis made by scrutinizing observations plotted on charts. The computer programs, after checking for transmission and other errors, produce interpolated grid-point values dynamically consistent with the equations of the forecast model. There are often options

Ohm's law of hearing

to allow manual incorporation of late observations and other data not easily processed automatically.

objective noise meter Sound level meter in which noise level to be measured operates a microphone, amplifier and detector, the last named indicating noise level on the phon scale. The apparatus is previously calibrated with known intensities of the *reference tone*, 1 kHz, suitable weighting networks and an integrating circuit being incorporated in the amplifier to simulate relevant properties of the ear in appreciating noise.

objective prism A narrow angle (ca 1°) prism placed in front of the objective lens or primary mirror of a telescope. Causes the image of each star to give a small spectrum. Those of many objects within a small field can thus be recorded simultaneously, either photographically or (by **optical fibres**) photo-electrically.

obliquity of the ecliptic The angle at which the **celestial equator** intersects the **ecliptic**. At present this angle is slowly decreasing by 0.47 arc seconds a year, due to **precession** and **nutation**. It varies between 21°55′ and 24°180′. Its value in 2000 will be about 23°26′34″.

obstruction lights Lights fixed to all structures near airports which constitute a hazard to aircraft in flight.

obstruction markers See **airport markers**.

occlusion The coming together of the **warm** and **cold fronts** in a depression so that the warm air is no longer in contact with the Earth's surface. If the *warm frontal zone* is cut off from contact with the surface it is a *cold occlusion*. If the *cold frontal zone* is cut off from contact with the surface it is a *warm occlusion*.

occultation The hiding of one celestial body by another interposed between it and the observer, as the hiding of the stars and planets by the Moon, or the satellites of a planet by the planet itself.

ogee wing A wing of ogee plan form and very low aspect ratio which combines low *wave drag* in supersonic flight with high lift at high incidence through separation vortices at low speed.

ohm SI unit of electrical resistance, such that 1 ampere through it produces a potential difference across it of 1 volt. Symbol Ω.

Ohm's law of hearing Law of psychoacoustics. A simple harmonic motion of the air is appreciated as a simple tone by the human ear. All other motions of the air are analysed into their harmonic components which the ear appreciates as such separately.

123

numerical weather prediction

Abbrev. *NWP*. A method of predicting the future state of the atmosphere and the associated weather, not by traditional methods whereby a human forecaster studies and analyses charts of meteorological observations and uses his physical understanding and experience to forecast developments, but by mathematical calculation using a digital computer. For NWP it is necessary: (1) to devise a conceptual model of the atmosphere and its mode of behaviour according to the laws of physics, a model which is inevitably simpler than the full complexity of Nature but not oversimple; (2) to write down the mathematical equations describing how the model works using the laws of motion, of conservation of energy, momentum and matter, together with the gas laws and those of thermodynamics including the phase changes of water; (3) to solve these equations given certain initial and boundary conditions by methods of numerical approximation that tend towards the 'true' solutions of the unapproximated equations.

Certain physical processes, such as the turbulent exchange of heat and moisture at the Earth's surface or the formation of clouds and the fall-out of precipitation, are so complicated that they cannot be modelled in full detail but have to be *parametrized* i.e. represented in a quasi-empirical and statistical way as functions of directly modelled values of wind, pressure, temperature and humidity. (There is as much art and skill in NWP as there used to be in old-fashioned forecasting, but the relationship to fundamental physical processes is closer.)

In modern operational NWP the forecasts produced are continually updated by the introduction of new observations in a type of rolling process. Some years ago a separate process of 'objective computer analysis' was performed on the new observations using the appropriate forecast as a first guess, and this objective analysis would have to be *initialized*, or subtly modified, so that it could be used as a set of initial conditions for the model equations without producing spurious 'shock waves' in the forecast. More recently, the absorption and initialization of new data have been carried out much more as a unified process.

The *primitive equations* now used for NWP describe, in theory, all scales of motion from source waves to meteorological waves thousands of kilometres long. In the 1950s, however, when computers were less powerful, transformations of the equations were used which automatically filtered out unwanted non-meteorological solutions albeit with some loss of detail and accuracy in the forecast.

At various standard intervals after the commencement of a forecast, the resulting data are obtained from the computer in whatever forms are operationally convenient for the large variety of uses to which they are put. For example, data relevant to the operation of commercial aircraft are directly transmitted in numerical form to the airlines' own computers; for guidance to forecasters at official meteorological service outstations, information is supplied in a variety of graphical ways as well as numerically. Methods are also being developed whereby staff at remote locations can interrogate a central forecast data store in order to obtain data suited to their individual needs.

oil cooler See **fuel-cooled-**.

oil-dilution system In a reciprocating aero-engine, a device for diluting the lubricant with fuel as the engine is stopped so that there is less resistance when starting in cold weather.

okta, octa One-eighth of the sky area used in specifying cloud cover for airfield weather condition reports.

Olbers' Paradox A paradox expressed in 1826 by Heinrich Olbers: 'why is the sky dark at night?' In an infinitely large, unchanging, universe populated uniformly with stars and galaxies, the sky would be dazzling bright, which is not the case. This simple observation therefore implies that the universe is not an infinite static arrangement of stars. In fact, modern cosmology postulates a finite expanding universe.

Old Style A name given to the system of date-reckoning superseded by the adoption of the **Gregorian calendar**.

oleo Main structural member of the support assemblies of an aircraft's *landing gear*. Of telescopic construction and containing oil so that on landing the oil is passed under pressure through chambers at a controlled rate thereby absorbing the shock. Also *oleo leg, shock strut, shock absorber*.

oleo-pneumatic Means of absorbing shock loads by a combination of air compression and oil pressure created by forcing the latter through an orifice.

Omega Long range radio navigation aid of very low frequency covering the whole Earth from 8 ground transmitters. It can be received down to sea level.

omega equation A weather diagnostic equation for the vertical velocity in **pressure co-ordinates** dP/dt conventionally denoted by ω. It is obtained by eliminating the time derivatives from the thermodynamic and **vorticity equations**, and applying the **quasi-geostrophic approximation**. With the omission of some of the smaller terms the co-equation may be written

$$\nabla^2(\sigma\omega) - \frac{f(\zeta + f)}{g}\frac{\partial^2 w}{\partial p^2} =$$

$$\frac{1}{f}\nabla^2 J\left(\phi, \frac{\partial\phi}{\partial p}\right) - \frac{1}{g}J(\phi, g + f)$$

where σ is a measure of atmospheric stability, f is the **Coriolis parameter**, ζ the vertical component of relative vorticity, γ the geopotential, g the acceleration of gravity, and J indicates an operator such that

$$J(u, v) \equiv \frac{\partial u}{\partial x}\frac{\partial v}{\partial y} - \frac{\partial u}{\partial y}\frac{\partial v}{\partial x}.$$

omnidirectional Simple aerial, mounted on a spacecraft, radiating energy equally in all directions.

omnidirectional radio beacon A VHF radio beacon radiating through 360° upon which an aircraft can obtain a bearing. Used for navigation by **VOR** with **DME**. Abbrev. *ORB*.

ONERA Abbrev. for *Office National d'Etudes et de Recherches Aerospatiales*, Fr.

on-top altitude clearance   Air-traffic control clearance for **visual flight rules** flying above cloud, haze, smoke or fog.

open clusters Galactic clusters of stars of a loose type containing at most a few hundred stars; the stars of a cluster have a common motion through space, and are associated with dust and gas clouds, e.g. *Hyades, Pleiades, Praesepe, Ursa Major cluster*.

open-jet wind tunnel A windtunnel in which the working section is not enclosed by a duct.

OPF Abbrev. for *Orbiter Processing Facility* at Kennedy Space Center, Florida, used for refurbishment activities and the loading of payloads into the Space Shuttle Orbiter.

opposition The instant when the geocentric longitude of the Moon or of a planet differs from that of the Sun by 180°.

optical-electronic devices Used to locate weakly radiating sources by the detection of their infrared emission. The radiation is collected by optical mirrors or lenses and concentrated on a sensitive infrared detector. Detailed maps of the Earth's surface, weather mapping and non-destructive testing of materials and components are some of the non-military applications.

optical fibre Fibres of ultra-pure glass, having a central core of higher refractive-index glass than the outer *cladding*, capable of conducting modulated light signals by total internal reflection. Optical-fibre cables consist of such cores either singly or several per cable, with further cladding and armouring for mechanical protection. Benefits include small diameters, high potential bandwidth and lower cost than copper. *Monomode* and *multimode* fibres are different classes.

OR Abbrev. for (1) *Operational Requirement*, (2) *Operational Research*.

ORB Abbrev. for **Omnidirectional Radio Beacon**.

orbit See panel on p. 126.

orbit decay The change in orbit parameters of a space vehicle caused by air drag which becomes more rapid as the surface of a planet is approached due to increasing atmospheric density, eventually resulting in entry or re-entry of the vehicle.

ornithopter Any flying machine that derives its principal support in flight from the air reactions caused by flapping motions of the wings, this motion having been imparted to the wings from the source of power being carried.

orographic ascent The upward displacement of air blowing over a mountain.

orographic rain Rain caused by moisture-laden winds impinging on the rising slopes of hills and mountains. Precipitation is

orbit

The path of a heavenly body and, by extension, an artificial satellite, space vehicle etc., moving about another under gravitational attraction. It is the shape of a conic section with a focus at the centre of mass. When in free orbit with no engine firing, a space vehicle is subject only to gravity and its motion obeys **Kepler's Laws of planetary motion**, and its motion is characterized by its orbital velocity and its inclination to the equator of the primary body. The time taken to complete one orbit is called the *period* and if the vehicle moves in the opposite sense to the primary body, the orbit is termed *retrograde*.

Depending on the eccentricity, e, different types of orbit as in diagram (a) occur. For an elliptical orbit, the space vehicle's velocity, V, can be deduced as:

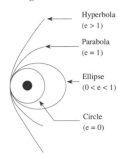

$$V^2 = G(M + m)\left(\frac{2}{r} - \frac{1}{S}\right)$$

where G is the gravitational constant, M and m are the masses of the primary body and space vehicle respectively, r is the distance of the vehicle from the focus and S is the semi-major axis of the ellipse. The closest point of approach to the primary body is called the

(a) Types of orbit.

pericentre-perigee for Earth and *perihelion* for the Sun. The furthest point from the body is the *apocentre-apogee* for Earth and *aphelion* for the Sun. The geometry of orbital motion of an Earth satellite is illustrated in diagram (b). The *inclination* of the orbit is the angle between the orbital and equatorial planes. Perturbations of an orbit may occur if the gravitational field is assymmetrical (e.g. the influence of Earth's equatorial bulge) and effects such as the rotation of the line of apsides or regression of the nodes may result. A *transfer orbit* must be used for sending a space vehicle to another planet; first, an Earth orbit is established and then the transfer orbit achieved by the injection of energy. For the orbital transfer between co-planar circular orbits a **Hohmann orbit** involves the minimum energy.

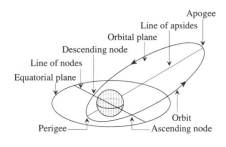

(b) Geometry of orbital motion.

Since the Earth moves under a satellite in low-Earth orbit (*LEO*), its surface may be observed only within the limits of the orbit inclination. A polar orbit provides a complete view of the Earth and, if it is Sunsynchronous, ensures continued passages over points of interest at the same local time. The geosynchronous orbit (*GEO*), in which a satellite moves in an equatorial orbit at the same speed as the Earth rotates beneath it (i.e. *geostationary*), is particularly interesting and is used by communications satellites because they can be 'seen' continuously from a ground station. Transfer from LEO to GEO is effected by means of a geosynchronous transfer orbit (*GTO*).

An *aircraft* circling a given point is said to orbit that point and **air-traffic control** instructions incorporate the term. See **holding pattern**.

caused by the cooling of the moist air consequent upon its being forced upwards.

orrery A mechanical model of the solar system showing the relative motions of the planets by means of clockwork; much in vogue in the 18th century. (Named after Charles Boyle, Earl of Orrery.)

oscillation See **longitudinal-**, **phugoid-**.

oscillations Any motion that repeats itself in equal intervals of time is a *periodic motion*. A particle in periodic motion moves back and forward over the same path and is said to be *oscillating* or *vibrating*. If the oscillations are not precisely repeated due to frictional forces which dissipate the energy of motion, the oscillations are said to be *damped*.

osculating orbit The name given to the instantaneous ellipse whose elements represent the actual position and velocity of a comet or planet at a given instant (*epoch of osculation*).

outer marker beacon A marker beacon, associated with the **ILS** or with the **standard beam approach system**, which defines the first predetermined point during a beam approach.

outer section See **main plane**.

out-gassing Spontaneous liberation of gas from a material in a space environment, sometimes termed *off-gassing*. To avoid contamination the material is left for some time before nearby instruments are used.

outside loop See **inverted loop**.

overhang (1) In multiplanes, the distance by which the tip of one of the planes projects beyond the tip of another. (2) In a wing structure, the distance from the outermost supporting point to the extremity of the wing tip.

overhaul period See **time between overhauls**.

oxidant The oxygen-bearing component in a

bipropellant rocket, usually liquid oxygen, high-test hydrogen peroxide, or nitric acid.

oxygen-isotope determinations A method of using ^{16}O and ^{18}O isotope ratio measurements from cores taken from Greenland and Antarctic ice-sheets or from marine fossils in cores from the sea bottom. The results may be used to estimate (1) the temperature at which the original snow fell before turning to ice, (2) the sea temperature at the time of deposition of marine fossils and (3) the global ice volume, thus giving a chronology of the ice ages during the Pleistocene.

ozone layer That region of the **stratosphere**, between about 20 and 40 km above the Earth's surface, where ozone makes up a greater proportion of the air than at any other height. Although this proportion is only a few parts per million, it nevertheless exerts a vital influence by absorbing much of the ultraviolet radiation in sunlight and preventing it from reaching the Earth's surface where it has considerable biological effect; in addition it changes the thermal structure of the middle atmosphere owing to the heating caused by such absorption and is also a contributor to the **greenhouse effect**. Ozone is formed, mainly in the tropical stratosphere, by complex photochemical reactions and is distributed over the whole world by large-scale *eddy diffusion*. See **general circulation of the atmosphere**. It is removed predominantly by catalytic chain reactions involving molecules such as H, OH, NO, NO_2, Cl and Br which initiate thousands of ozone-destroying cycles before they are themselves removed; the supply of such active molecules has been much increased since 1950 by man-made pollutants including chlorofluorocarbons and this is affecting the subtle dynamic equilibrium of the ozone layer. See **atmospheric pollution**.

PQ

Pa Abbrev. for **Pascal**.

paddle plane See **cyclogyro**.

pancaking The alighting of an aircraft at a relatively steep angle, with low forward speed.

panel absorber Light panels mounted at some distance in front of a rigid wall in order to absorb sound waves incident on them.

panspermia Hypothesis (S. Arrhenius, 1907) that life on Earth was introduced billions of years ago by **extraterrestrials** in space.

pantobase The fitment of a landplane with **hydroskis**, enabling it to taxi, take off from and alight on water or snow.

PAR Abbrev. for **Precision-Approach Radar**.

parabolic flight The flying of a special parabolic trajectory by a suitably fitted aeroplane to reproduce the conditions of **freefall** over a period of minutes.

parabolic mirror, reflector One shaped as a paraboloid of revolution. Theoretically produces a perfectly parallel beam of radiation if a source is placed at the focus (or vice versa). Such mirrors are used in reflecting telescopes, car headlamps etc. See **dish** for *parabolic reflector*.

parabolic velocity The velocity which a body at a given point would require to describe a parabola about the centre of attraction; smaller values give an ellipse, larger values a hyperbola. Also *escape velocity*, since it is the upper limit of velocity in a closed curve.

parabrake See **brake parachute**.

parachute An umbrella-shaped fabric device of high drag (1) to retard the descent of a falling body or (2) to reduce the speed of an aircraft or item jettisoned therefrom. Commonly made of silk or nylon, sometimes of cotton or rayon where personnel are not concerned. See **antispin-, automatic-, brake-, ribbon-, ring slot-, pilot chute**. See **parasheet**.

parachute flare A pyrotechnic flare, attached to a parachute released from an aircraft to illuminate a region.

paraglider Inflatable hypersonic re-entry kite of highly swept back wing shape with rounded leading edge, proposed for Gemini spacecraft, but not used.

parallactic angle The name given to that angle in the astronomical triangle formed at the heavenly body by the intersection of the arcs drawn to the zenith and to the celestial pole. See **astronomical triangle**.

parallax (1) Generally, the apparent change in the position of an object seen against a more distant background when the viewpoint is changed. Absence of parallax is often used to adjust two objects, or two images, at equal distances from the observer. (2) In astonomy, the apparent displacement in the position of any celestial object caused by a change in the position of the observer. Specifically, the change due to the motion of the Earth through space. The observer's position changes with the daily rotation of the Earth (*diurnal parallax*), the yearly orbit (*annual parallax*) and motion through space generally (*secular parallax*). The term 'parallax' is often loosely used by astronomers to be synonymous with 'distance' because the annual parallax is inversely proportional to distance. See **spectroscopic parallax**.

paraselenae See **mock moons**.

parasheet A simplified form of parachute for dropping supplies, made from one or more pieces of fabric with parallel warp in the form of a polygon, to the apices of which the rigging lines are attached.

parhelia See **mock suns**. Sing. *parhelion*.

parking orbit The waiting orbit of a spacecraft between two phases of a mission.

parsec The unit of length used for distances beyond the solar system. It is the distance at which the **astronomical unit** subtends one second of arc, and is therefore 206 265 AU, 3.086×10^{13} km, 3.26 light-years. From *parallax second*. Abbrev. *pc*. Kiloparsec (kpc) and megaparsec (mpc) for 1000 pc and 1 000 000 pc are widely used in galactic and extragalactic contexts.

partial pressure suit A laced airtight overall for aircrew members in very high-flying aircraft. It has inflatable cells to provide the wearer with an atmosphere and external body pressure in the event of cabin pressure failure. Essential for survival above 50 000 ft (15 000 m).

particle velocity In a progressive or standing sound wave, the alternating velocity of the particles of the medium, taken as either the maximum or rms velocity.

pascal The SI derived unit of pressure or stress, equals 1 newton per square metre. Abbrev. *Pa*.

Pasiphae The eighth natural satellite of **Jupiter**.

passive homing guidance A missile guidance system which homes on to radiation (e.g. infrared) from the target.

passive radar That using microwaves or infrared radiation emitted from source, and hence not revealing the presence or position of the detecting system. Military use.

passive satellite See **active satellite**.

pax Airline passengers.

payload (1) That part of an aircraft's load from which revenue is obtained. (2) That part of a military aircraft which is devoted to offensive or defensive actions (bullets, bombs, chaff etc). (3) That part of a spacecraft additional to that used for structure and for maintaining its essential functions; usually this implies the instruments and supporting hardware for performing experiments. (4) That part of a launch system which is the 'useful' mass placed on the desired trajectory and orbit.

payload integration Process of bringing together individual experiments, their support equipment and **software** into a payload entity, in which all interfaces are compatible and whose operation has been fully checked-out.

payload specialist A highly qualified scientific member of the crew of the Space Shuttle, whose sole responsibility is the operation of the experiments of a payload. He or she is not necessarily a professional astronaut.

PCM Abbrev. for *Pulse Coded Modulation*.

P-display That of a **PPI** unit. Map display produced by intensity modulation of a rotating radial sweep.

pendulum damper A short heavy pendulum, in the form of pivoted balance weights, attached to the crank of a radial aero-engine in order to neutralize the fundamental torque impulses and so eliminate the associated critical speed.

penetrating shower Cosmic-ray shower containing mesons and/or other penetrating particles.

pentad Period of 5 days; being an exact fraction of a normal year, it is useful for meteorological records.

penumbra See **umbra**.

percentage tachometer An instrument indicating the rev/min of a turbojet engine as a percentage, 100 per cent corresponding to a pre-set optimum engine speed. Provides better readability and greater accuracy and also enables various types of engine to be operated on the same basis of comparison.

periastron That point in an orbit about a star in which the body describing the orbit is nearest to the star; applied to the relative orbit of a double star.

perigee The nearest point to the Earth in the

orbit of a spacecraft, missile or planetary body. See **orbit**.

perihelion That point in the orbit of any heavenly body moving about the Sun at which it is nearest to the Sun; applied to all the planets and also to comets, meteors, spacecraft etc. Pl. *perihelia*. See **orbit**.

perimeter track A taxi track round the edge of an airport.

period-luminosity law A relationship between the period and absolute magnitude, discovered by Miss Leavitt to hold for all Cepheid variables; it enables the distance of any observable Cepheid to be found from observation of its light curve and apparent magnitude, this indirectly deduced distance being called the *Cepheid parallax*.

period of revolution The mean value, derived from observations, of one complete revolution of a planet or comet about the Sun, or of a satellite about a planet.

Perseids A major **meteor** shower visible for up to two weeks before and after peaking on 12 August each year, the date on which the Earth crosses the orbit. The maximum hourly rate is about 70 meteors. This show is associated with a comet of 1862.

perturbation (1) In astronomy, any small deviation in the equilibrium motion of any celestial object caused by any change in the gravitational field acting on it. (2) In astronautics, any disturbance to a planned trajectory or orbit, caused by the effects of drag, gravitation, solar pressure etc.

Pfund series A series of lines in the far infrared spectrum of hydrogen. Their wave numbers are given by the same expression as that for the Brackett series but with $n_1 = 5$.

phase Of the Moon, the name given to the changing shape of the visible illuminated surface of the Moon due to the varying relative positions at the Earth, Sun, and Moon during the synodic month. Starting from *new moon*, the phase increases through *crescent, first quarter, gibbous*, to *full moon*, and then decreases through *gibbous, third quarter*, waning to *new moon* again. The inferior planets show the same phases, but in the reverse order; the superior planets can show a gibbous phase, but not a crescent.

phased array An antenna consisting of an array of identical radiators (waveguides, horns, slots, dipoles etc.) with electronic means of altering the phase of power fed to each of them. This allows the shape and direction of the radiation pattern to be altered without mechanical movement and with sufficient rapidity to be made on a pulse-to-pulse basis.

Phobos One of the two natural satellites of

Mars. The other is *Deimos*.

Phoebe The ninth natural satellite of Saturn.

phon Unit of the objective loudness on sound-level scale which is used for deciding the apparent loudness of an unknown sound or noise, when a measure of loudness is required. This is effected either by subjective comparison by the ear, or by objective comparison with a microphone-amplifier and a weighting network. The reference sound pressure level at 1 kHz is 0.0002 microbar $= 20$ μN/m^2.

phonmeter Apparatus for the estimation of loudness level of a sound on the phon scale by subjective comparison. Also *phonometer*.

photoelectric photometry The determination of stellar magnitude and colour index by means of a photoelectric device used at the focus of a large telescope.

photographic zenith tube Instrument for the exact determination of time; it consists of a fixed vertical telescope which photographs stars as they cross the zenith; instrumental and observational errors are thus eliminated, the instrument being entirely automatic. Abbrev. *PZT*.

photometry The accurate quantitative measurement of the amount of electromagnetic energy (most usually, light) received from a celestial object. Techniques are visual, photographic and **photoelectric**.

photon Quantum of light of electromagnetic radiation of energy $E = h\nu$ where h is Planck's constant and ν is the frequency. The photon has zero rest mass, but carries momentum $h\nu/c$, where c is the velocity of light. The introduction of this 'particle' is necessary to explain the photoelectron effect, the Compton effect, atomic line spectra, and other properties of electromagnetic radiation.

photonics Computing and data transmission using photons in place of electrons with, as a consequence, a greatly increased data transfer rate.

photosphere The name given to the visible surface of the Sun on which sunspots and other physical markings appear; it is the limit of the distance into the Sun that we can see.

phugoid oscillation A longitudinal fluctuation in speed of long periodicity, i.e. a velocity-modulation, in the motion of an aircraft, accompanied by rising and falling of the nose.

picture noise See **grass**.

pilot balloon A small rubber balloon, filled with hydrogen, used for determining the direction and speed of air currents at high altitudes, the balloon is observed by means of a theodolite after being released from the ground.

pilot chute A small parachute which extracts the main canopy from its pack.

pinpoint An aircraft's ground position as fixed by direct observation. Cf. **fix**.

pion See **meson**.

pip Significant deflection or intensification of the spot on a CRT, giving a display for identification or calibration. Also *blip*.

pitch (1) The distance forward in a straight line travelled by a propeller in one revolution at zero slip; often colloq. though wrongly applied to the blade incidence. (2) Angular displacement along the lateral axis. See **pitching**. (3) Spacing between evenly spaced items, e.g. rivets.

pitch control The *collective* and *cyclic pitch* (i.e. blade incidence) *controls* of a helicopter's mainrotor(s).

pitching The angular motion of a ship or aircraft in a vertical plane about a lateral axis.

pitching moment The component of the couple about the lateral axis, acting on an aircraft in flight.

pitch setting The blade angle of adjustable- or variable-pitch **propellers**.

Pitot-static tube Tube inserted nearly parallel to flow stream. It has two orifices, one facing flow and hence receiving total pressure, and the other registering the static pressure at the side. The pressure difference between the two orifices registers dynamic air pressure ($\frac{1}{2} \rho \varpi^2$, where ρ is the air density and v the velocity). This is displayed on the airspeed indicator.

pitot tube See **Pitot-static tube**.

plage Dark or bright areas on calcium or hydrogen spectroheliograms, identified as areas of cool gas or heated gas respectively on the Sun's surface; cf. *flocculi*, which refers to small patches only.

plain flap A wing flap in which the whole trailing edge (apart from the ailerons) is lowered so as to increase the camber. Also, occasionally, *camber flap*.

Planck's law Basis of quantum theory, that the energy of electromagnetic waves is confined in indivisible packets or quanta, each of which has to be radiated or absorbed as a whole, the magnitude being proportional to frequency. If E is the value of the quantum expressed in energy units and ν is the frequency of the radiation, then $E = h\nu$, where h is known as *Planck's constant* and has dimensions of energy $\times$ time, i.e. action. Present accepted value is 6.626×10^{-34} J s. See **photon**.

Planck's radiation law

Planck's radiation law An expression for the distribution of energy in the spectrum of a black-body radiator:

$$E_\nu d\nu = \frac{8\pi h\nu^3}{c^3(e^{h\nu/kT}-1)}d\nu$$

where E_ν is the energy density radiated at a temperature T within the narrow frequency range from ν to $\nu+d\nu$, h is Planck's constant, c the velocity of light, e the base of the natural logarithms and k Boltzmann's constant.

planet The name given in antiquity to the seven heavenly bodies, including the Sun and Moon, which were thought to travel among the fixed stars. The term in now restricted to those bodies, including the Earth, which revolve in elliptic orbits about the Sun; in the order of distance they are: Mercury, Venus, Earth, Mars, Jupiter, Saturn, Uranus, Neptune and Pluto. The two planets, Mercury and Venus, which revolve within the Earth's orbit are designated *inferior planets*, the planets Mars to Pluto are *superior planets*. Planets reflect the Sun's light and do not generate light and heat.

planetarium A building in which an optical device displays the apparent motions of the heavenly bodies on the interior of a dome which forms the ceiling of the auditorium.

planetary nebula A shell of glowing gas surrounding an evolved star, from which it is ejected. There is no connection with planets: the name derives from the visual similarity at the telescope between the disk of such a nebula and that of a planet. They represent late stages in the evolution of stars 1–4 times more massive than the Sun. Some thousands are known in our Galaxy.

planetoid See **minor planet**.

planing bottom The part of the under surface of a flying-boat hull which provides hydrodynamic lift.

plan-position indicator Screen of a CRT with an intensity-modulated and persistent radial display, which rotates in synchronism with a highly directional antenna. The surrounding terrain is thus painted with relevant reflecting objects, such as ships, aircraft, and physical features. Abbrev. *PPI*. See **azimuth stabilized PPI**.

plasma Synonym for the positive column in a gas discharge.

platform Term sometimes applied to a spacecraft used as a base for experiments in space research, usually unmanned.

Pleiades, The The name given to the open cluster in the constellation *Taurus*, of which the seven principal stars, forming a well-known group visible to the naked eye, each

have a separate name.

plenum chamber A sealed chamber pressurized from an air intake. Centrifugal flow turbojets having double-entry impellers (see **double-entry compressor**) have to be mounted in plenum chambers to ensure even air pressure on both impeller faces.

plume Snow blown over the ridge of a mountain.

Pluto Ninth planet of the solar system discovered by Clyde Tombaugh on 18 February 1930 as a result of a systematic search. The orbit is very elliptical and inclined at 17° to the ecliptic. For a proportion of its 248-year orbit it is actually closer to the Sun than **Neptune**. Its diameter is about 3000 km (less than our Moon) with a relative density similar to water. Pluto has one moon, Charon, found in 1978. Possibly the two objects are escaped satellites of Neptune.

pluviometry The study of precipitation, including its nature, distribution and techniques of measurement.

POCC Abbrev. for *Payload Operations Control Centre*, a room or building where experimenters, suitably supported by their own specialized equipment, gather to monitor and control their experiments on-board a space vehicle. Data links are provided directly or via a communications satellite.

pod See **engine pod**.

pogo effect Unstable, longitudinal oscillations induced in launch system, mainly due to fuel **sloshing** and engine vibration.

Pointers Popular name for the two stars of the Great Bear, α and β Ursae Majoris; they are roughly in line with the Pole Star and so help to identify it.

polar axis (1) That diameter of a sphere which passes through the poles. (2) In an equatorial telescope, the axis, parallel to the Earth's axis, about which the whole instrument revolves in order to keep a celestial object in the field.

polar control See **twist and steer**.

Polaris The brightest star in the constellation Ursa Minor which currently lies (by chance) within 1° of the north celestial pole. Its altitude is approximately equal to the latitude of the observer. This star was much used for simple navigation in the northern hemisphere. The effects of **precession** are gradually reducing its usefulness, as the north pole is drifting away from this particular star. Also *pole star*.

polarizing monochromator A filter consisting of a succession of quartz crystals and calcite or Polaroid sheets; the light passing through is restricted to a narrow band, useful in observing the solar chromosphere.

polar platform An unpressurized platform in Sun-synchronous orbit, used particularly for remote sensing.

polar response curve The curve which indicates the distribution of the radiated energy from a sound reproducer for a specified frequency. Also the relative response curve of a microphone for various angles of incidence of a sound wave for a given frequency. Generally plotted on a radial decibel scale.

polar sequence An adopted scale for determining photographic stellar magnitudes. It consists of a number of stars near the North Pole which are used as a standard of comparison; they range from Polaris to the faintest observable.

poles See **celestial-**, **terrestrial-**.

pole star See **Polaris**.

polymorph The term applied by Barnes Wallis to his supersonic aircraft designs incorporating *variable sweep* wings.

polyplexer Device acting as duplexer and lobe switcher.

population types The two broad types of stellar population. Population I includes hot blue stars such as those in the Sun's neighbourhood; they are found in the arms of spiral galaxies, and share in the galactic rotation. Population II stars are found in the central regions of galaxies and in globular clusters, where dust and gas are absent; they are red stars, having high velocities and do not share in the regular rotation of the system.

porpoising Oscillating symmetrical movements of a seaplane, flying-boat or amphibian, when planing: pitching instability on the water, as distinct from instability under airborne conditions. Also for landplanes during take-off and landing due to undercarriage forces.

position angle A measure of the orientation of one point on the celestial sphere with respect to another. The position angle of any line with reference to a given point is the inclination of the line to the hour circle passing through the point; it is measured from 0° to 360° from the north point round through east.

position error That part of the difference between the *equivalent* and *indicated airspeeds* due to the location of the **pressure head** or **static vent**. Position error is not a constant factor, but varies with airspeed due to the variations in the airflow around an aircraft at different **angles of attack**.

positive coarse pitch An extreme blade angle which is reached and locked after engine failure to reduce the drag of a non-feathering propeller.

positive g The force exerted on the human body in a gravitational field or during acceleration so that the force of inertia acts in a head-to-foot direction. See **negative g**.

potential energy Universal concept of *energy* stored by virtue of position in a field, without any observable change, e.g. after a mass has been raised against the pull of gravity. A body of mass m at a height h above the ground possesses potential energy mgh, since this is the amount of work it would do in falling to the ground. In electricity, potential energy is stored in an electric charge when it is taken to a place of higher potential through any route. A body in a state of tension or compression (e.g. a coiled spring) also possesses potential energy.

potential evapotranspiration The theoretical maximum amount of water vapour conveyed to the atmosphere by the combined processes of evaporation and transpiration from a surface covered by green vegetation with no lack of available water in the soil.

potential instability The condition of a layer of the atmosphere which is in a state of **static stability** but in which instability would appear if it were lifted bodily until it became saturated. Also *convective instability* in the US.

potential temperature The temperature which a given sample of air would have if brought by an adiabatic process to a standard pressure, conventionally 1000 mb. If the pressure and absolute temperature of the sample are P and T, then

$$\theta = T\left(\frac{P_0}{P}\right)^{\frac{\gamma-1}{\gamma}}$$

where γ is the ratio of the specific heats of a perfect gas. θ is related to the entropy S by $S = c_p \log\theta + \text{constant}$, where c_p is the specific heat at constant pressure.

potential vorticity The vorticity which a column of air between two isentropic surfaces would have if it were brought by an adiabatic process to an arbitrary standard latitude and then stretched or shrunk to an arbitrary standard thickness. It is a conservative air mass property for adiabatic processes. If the original values of the vertical component of vorticity, the **Coriolis parameter**, and the *thickness* are ζ_0, f_0 and h_0, while those after the standardization are ζ_s, f_s and h_s, then

$$(\zeta_0 + f_0)/h_0 = (\zeta_s + f_s)/h_s$$

See **Ertel potential vorticity**.

power Rate of doing work. Measured in joules per second and expressed in *watts* (1 W = 1 J/s). The foot-pound-second unit of power is the *horsepower*, which is a rate of working equal to 550 ft-lbf per second. 1 horsepower is equivalent to 745.7 watts. 1 watt is equal to 10^7 CGS units, that is 10^7 erg/second.

power-assisted controls Primary flying controls wherein the pilot is aided by electric motors or double-acting hydraulic jacks.

power controls A primary flying control system where movement of the surfaces is done entirely by a power system, commonly hydraulic, but sometimes electrohydraulic or electric. The system is always at least duplicated (power source, supply lines and operating rams) and both circuits are usually running continuously. Reversion to manual control was common in early systems, but the trend is toward a multiplication of reserve circuits and supply sources. See **power-assisted controls, q-feel.**

power loading The gross weight of a propeller driven aircraft divided by the take-off power of its engine(s). For jet aircraft, *thrust loading.*

power rating The power, authorized by current regulations, of an aero-engine under specified conditions, e.g. maximum take-off rating, combat rating, maximum continuous rating, weak-mixture cruising rating etc. The conditions are specified by r.p.m. and, for piston engines **manifold pressure** and torque (in large engines), for turboprops **jet-pipe temperature** and torque, for turbojets exhaust **gas temperature**, for rocket motors **combustion-chamber** pressure.

power unit An engine (or assembly of engines) complete with any extension shafts, reduction gears or propellers.

Poynting-Robertson effect Small particles of dust in the solar system slowly fall into the Sun as a result of this effect. Solar radiation causes them to lose angular momentum, and as a result they drift closer to the Sun. For particles smaller than one micron, **radiation pressure** is great enough to counteract the effect and indeed blow the finest dust out of the solar system altogether.

PPI Abbrev. for **Plan-Position Indicator.**

Praesepe A well-known open cluster in Cancer.

precession A regular cyclic motion of a dynamical system in which, with suitably chosen co-ordinates, all except one remain constant, e.g. the regular motion of the inclined axis of a top around the vertical.

precession of the equinoxes The westward motion of the equinoxes caused mainly by the attraction of the sun and moon on the equatorial bulge of the earth. This *luni-solar precession* together with the smaller planetary precession combine to give the general precession amounting to 50.27″ per annum. The equinoxes thus make one complete revolution of the ecliptic in 25 800 years, and the earth's pole turns in a small circle of radius 23°27′ about the pole of the ecliptic, thus changing the coordinates of the stars.

precipitable water The total mass of water in a vertical atmospheric column of unit area, or its height if condensed in liquid form.

precipitation Moisture falling on the Earth's surface from clouds; it may be in the form of rain, hail or snow.

precision-approach radar A primary radar system which shows the exact position of an aircraft during its approach for landing. Abbrev. *PAR.*

prepreg Fibrous composite material consisting of unidirectional fibres embedded in a matrix of resin prepared in the form of sheet or strip ready for forming; then combined into the final product consisting of several plies arranged in different directions .

preset guidance The guidance of controlled missiles by a mechanism which is set before launching and is subsequently unalterable.

pressure altitude Apparent altitude of the local ambient pressure related to the **International Standard Atmosphere.**

pressure cabin An airtight cabin which is maintained at greater than atmospheric pressure for the comfort and safety of the occupants. Above 20 000 ft (6 000 m) a differential of 6.5 lbf/in^2 (45 kN/m^2) is usual, and above 40 000 ft (12 000 m) one of 8.25 lbf/in^2 (57 kN/m^2). Pressurization can be either by a shaft driven *cabin blower*, or by air bled from the compressor of turbine main engines.

pressure co-ordinates A system of co-ordinates used in **numerical forecasting** in which the vertical ordinate is pressure, p. In this system, the vertical velocity w is replaced by the total derivative, following motion, of the pressure, i.e. $\dfrac{dp}{dt}$.

pressure drag The summation of all aerodynamic forces normal to the surface, resolved parallel to free stream direction; sum of **form drag** and **induced drag.**

pressure gradient (1) The rate of change of the atmospheric pressure horizontally in a certain direction on the Earth's surface as shown by isobars on a weather chart. (2) The rate of change of pressure with distance over the ground, normal to the isobars. The force acting on the air is the *pressure-gradient force*.

pressure head A combination of a **static pressure** and a **Pitot tube** which is connected to opposite sides of a differential pressure gauge, for giving a visual reading corresponding to the speed of an airflow. Sometimes *pitot-static tube*.

pressure helmet A flying helmet for the crew of high-altitude aircraft or spacecraft for use with a *pressure*, or *partial pressure suit*. Usually of plastic, with a transparent face-piece, which may be in the form of a visor, the helmet incorporates headphones, microphone and oxygen supply, and there is usually a feeding trap near the mouth.

pressure jet A type of small jet-propulsion unit fitted to the tips of helicopter rotor blades, in which small size (to give low drag for **autorotation**) is of greater importance than the losses due to ejecting the efflux at pressures as high as 2 or 3 atmospheres.

pressure-pattern flying The use of barometric pressure altitude to obtain the most favourable winds for long distance, high altitude aerial navigation.

pressure ratio (1) The absolute air pressure, prior to combustion, in a gas turbine **aero-engine**, **pulse-jet** or **ramjet**, divided by the ambient pressure: analogous to the *compression ratio* of a reciprocating engine. (2) The ratio of air pressures between different compressor stages or stations in an aero-engine.

pressure suit An airtight fabric suit, similar to that of a diver, for very high altitude and space flight. It differs from the *partial pressure suit* in being loose-fitting, with bellows or other form of pressure-tight joint, to permit limited movement by the wearer.

pressure waistcoat A double-skinned garment, covering the thorax and abdomen, through which oxygen is passed under pressure on its way to the wearer's lungs to aid breathing at great heights, i.e. above 40 000 ft (12 000 m).

pressurized Fitted with a device that maintains nearly normal atmospheric pressure, e.g. in an aircraft.

pre-TR cell A gas-filled RF switching valve which protects the **transmit-receive tube** in a radar receiver from excessive power. Also acts as a block to receiver frequencies other than the fundamental.

primary bow See **rainbow**.

primary radar One in which the incident power from the transmitter is reflected from the target to form the return signal or *echo*.

Cf. **secondary radar**.

primary structure All components of an aircraft structure, the failure of which would seriously endanger safety, e.g. wing or tailplane spars, main fuselage frames, engine bearers, portions of the skin which are highly stressed.

prime contract That for the whole aircraft or weapon system from design and manufacture to test and supply, including the management of subcontractors for completion to time and cost.

priming pump A manual or electric fuel pump which supplies the engine during starting where an injection carburettor or fuel injection pump is fitted.

primitive equation model A numerical forecast model that uses the *primitive equations*, not the **filtered equations**.

primitive equations These are the fundamental equations of motion of a fluid modified only by the use of the **hydrostatic approximation** and the neglect of viscosity. The primitive equations comprise three prognostic equations (the x and y components of the momentum equation and the thermodynamic equation of energy) and three diagnostic equations (the continuity equation, the hydrostatic approximation and the equation of state). These equations form a closed set in the dependent variables of velocity, pressure, density and temperature. The solutions include gravity waves but not vertically propagating sound waves.

principal axes of a body At any given point O: three mutually perpendicular axes $Oxyz$ such that the three products of inertia about the co-ordinate planes are all zero. The principal axes at the centre of gravity are the axes of symmetry if any exist.

principle of equivalence A statement of the theory of relativity: an observer has no means of distinguishing whether his laboratory is in a uniform gravitational field or is in an accelerated frame of reference. See **relativity, general relativity**.

principle of relativity A universal law of nature which states that the laws of mechanics are not affected by a uniform rectilinear motion of the system of co-ordinates to which they are referred. Einstein's relativity theory is based on this principle, and on the postulate that the observed value of the velocity of light is constant and is independent of the motion of the observer. See **relativity**.

probe A space vehicle, esp. an unmanned one, sent to explore near and outer space and to collect and transmit data back to Earth. If a planet or its environment is explored, the vehicle is sometimes referred to as a *planetary probe*.

procurement Organizational procedure for obtaining equipment, supplies, services and personnel.

products of inertia Of a body about two planes, the sum Σmxy taken over all particles of the body where m is the mass of a particle and x and y the perpendicular distances of the particle from the specified planes.

profile drag The 2-dimensional drag of a body, excluding that due to lift; the sum of the surface-friction and form drag.

prognostic chart A chart of the **meteorological elements** which are expected to exist in the near future. A forecast weather chart.

prominence A streamer of glowing gas visible in the outer layers of the solar atmosphere. Several types are seen in the upper **chromosphere** and lower **corona**. All consist of regions of higher density and lower temperature than the surrounding gas, which is why they can be seen. Although they are best seen at the rim of the Sun during an **eclipse**, they are frequently detectable above the **photosphere** by using a **spectroheliograph**.

proof load The load which a structure must be able to withstand, while remaining serviceable.

propagation loss The transmission loss for radiated energy traversing a given path. Equal to the sum of the *spreading loss* (due to increase of the area of the wavefront) and the *attenuation loss* (due to absorption and scattering).

propagation of light Consists of transverse electromagnetic waves propagated through free space with a velocity of 2.9979 $\times 10^8$ m/s and wavelengths about 400 to 800 nm. The ratio (velocity of light in free space/velocity in medium) is the refractive index of the medium. According to the special theory of relativity, the velocity of light is absolute and no body can move at a greater speed.

propellant Comprehensive name for the combustibles for a chemical rocket motor. For liquid propellants it comprises the *fuel* (hydrocarbons, such as kerosine and hydrazine) and the *oxidant* (such as liquid oxygen and fluorine). With solid propellants, combustible materials (such as perchlorate and aluminium powder) are prepared prior to firing *in situ*.

propeller See panel on p. 136.

propeller brake A shaft brake to stop, or prevent windmilling of, a turboprop, principally to avoid inconvenience on the ground.

propeller efficiency The ratio of the actual thrust power of a propeller to the torque power supplied by the engine shaft; 80–85%

is a typical value.

propeller governor A mechanical means of controlling propeller speed.

propeller hub The detachable fitting by which a propeller is attached to the power-driven shaft, usually also containing the variable pitch control gear.

propeller solidity The proportion of the disk occupied by blades, measured at 70% radius as standard.

propeller turbine engine See **turboprop**.

propelling nozzle The constricting nozzle at the outlet of a turbojet **exhaust cone** or *jet pipe* which reduces the gases to slightly more than ambient atmospheric pressure and accelerates them to raise their kinetic energy, thereby increasing the thrust.

proper motion That component of a star's own motion in space which is at right angles to the line of sight, so that it constitutes a real change in the position of the star relative to its neighbouring stars.

propulsive duct Generic term for the simplest form of reaction-propulsion aero-engine having no compressor/turbine rotor. Thrust is generated by initial compression due to forward motion, the form of the duct converting kinetic energy into pressure, the addition and combustion of fuel, and subsequent ejection of the hot gases at high velocity. See **pulse-jet**, **ramjet**. Also *Athodyd* (*Aero THermODYnamic Duct*)

propulsive efficiency (1) The propulsive horsepower divided by the torque horsepower. (2) In a turbojet, the net thrust divided by the gross thrust.

propulsive lift See panel on p. 137.

proton-proton chain A series of thermonuclear reactions which are initiated by a reaction between two protons. It is thought that the proton-proton cycle is more important than the carbon cycle in the cooler stars.

Proxima Centauri The nearest star to the Sun; a faint companion to the double star *Alpha Centauri* in the constellation Centaurus, its distance being 4.3 light-years.

proximity fuse Miniature radar carried in guided missiles, shells or bombs so that they explode within a preset distance of the target.

psychrometer See **wet and dry bulb hygrometer**.

Ptolemaic system The final form of Greek planetary theory as described in Claudius Ptolemy's treatise. In this the Earth was the centre of the world, the planets, including the Sun and Moon, being supposed to revolve round it in motions compounded of eccentric circles and epicycles; the fixed stars were supposed to be attached to an

propeller

A rotating hub with helical blades which convert shaft power to aerodynamic thrust, power being supplied by an **aero-engine** or, rarely, human power. Left- and right-hand rotation is termed anticlockwise and clockwise respectively when seen from behind and a propeller can be a *pusher* or a *tractor*. The term *airscrew* is obsolete. A 'propeller' can be air-driven to provide electric power to a missile pod or act as a *ram air turbine* to provide emergency hydraulic power in the event of a main engine failure.

Four bladed propeller

The hub at the centre of the disk usually contains the variable pitch control gear. The inset shows a cross section of one blade.

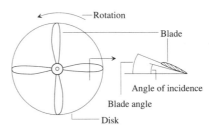

Important features: *diameter* determines power and thrust, and requires ground clearance; *pitch* is the distance advanced by a propeller during one complete revolution; *advance ratio* is the aircraft velociy divided by the product of revolutions per second and the diameter; *disk area* is the area swept out by the propeller disk; *blade angle* is that made by any section of the blade and the plane of rotation of the disk; *blade twist* is the variation in blade angle along the blade; *tip speed* is the circumferential speed of the blade tip in the plane of the disk; *helical tip speed* is the total speed of the blade tip relative to the airflow, i.e. the *vector sum* of tip speed and aircraft speed; *solidity* is the proportion of the disk occupied by blades, measured at 70% radius as standard.

Categories of propeller:

Fixed pitch which has a fixed shape as one made of wood.

Variable pitch in which blade angle is made to be altered.

Adjustable pitch in which blade angle can be changed between flights by maintainance action.

Controllable pitch in which blade angle can be changed in flight.

Constant speed in which blade angle is varied automatically by a propeller speed control to maintain a constant engine speed.

Contra-rotating refers to a double propeller, both of which are co-axial and rotating in opposite directions and driven either by one engine through a gear box or by two tandem engines, the rear driving through the hollow shaft of the front.

Ducted fan or *propeller* is one mounted within a duct.

Unducted fan or *UDF* is a multi-bladed, heavily loaded propeller driven by a gas turbine, possessing many of the features of a fan but without the surrounding duct.

The *slipstream* of a propeller is the region of airflow in the wake of the propeller in which the velocity is greater than the flight speed as the result of kinetic energy supplied by propulsion. The increase in drag of parts of the aircraft washed by the slipstream is called *propeller interference* and it may alter pitching moments and change the lift. Similarly, the presence of the aircraft can modify the flow through the propeller and change its thrust.

See **propeller-brake, -efficiency, -governor, -hub**.

propulsive lift

Propulsive lift is the means of providing a force on an aircraft in the lift direction by the use of engine power. Strictly, the helicopter employs propulsive lift but it is separately treated under **rotorcraft**. The means described here include all other kinds of propulsive lift.

The first mention in history was the vectoring (swivelling) contra-rotating propeller of Sir George Cayley (1837) designed for application to lighter-than-air craft. The first glimpse of today's reality appeared in a report to the UK government by the consultant Bramson in his assessment of the claims of Flt.Lt. F.Whittle for his experimental jet engine. It was stated that the thrust-weight ratio of the jet engine was so high that it could produce a thrust of the same magnitude as the weight of the whole aircraft. Prototypes of jet-lift aircraft were flown in the early 1950s, after which a great number of design solutions were built and flown.

The following classes are recognized:

VTO, vertical take-off.
VTOL, vertical take-off and landing.
STOL, short take-off and landing (not necessarily with propulsive lift, e.g. with low wing loading and high lift flaps).
V/STOL, vertical and short take-off and landing.
STOVL, short take-off and vertical landing.

Historical types include:

'Flying bedstead', the first Rolls-Royce flying demonstrator, 1953. Two Nene engines and compressed air reaction controls.
X-13, US Ryan Vertijet tailsitter, 1956.
SC-1, Short's VTO flat riser research aircraft with lift and cruise engines, 1958.
P 1127, Hawker prototype with Pegasus vectored thrust engine, 1960.
Balzac V-001, Dassault Mirage conversion with 8 lift and one cruise engine, 1963.
VJ 101, German supersonic prototype. Rolls-Royce-MAN RB 145 rotating engine pod on each wing tip, 1964.
US/FRG, US project, 1964, with cruise and retractable lift engines and variable sweep wing. Not built.

The *Harrier* (see diagram) derived from the P 1127, is the only successful jet-lift aircraft. The central Pegasus 4-nozzle vectored thrust engine delivers all the thrust under the control of simple throttle and nozzle vectoring levers. It can hover, take off and land vertically and fly off with greater loads in STOL using wing lift and partly deflected thrust. Balance in low speed flight is by reaction controls fed by engine bleed air.

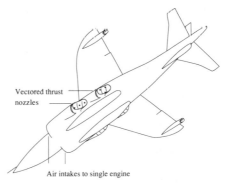

Vectored thrust nozzles

Air intakes to single engine

V/STOL permits military operations off roads and from forest clearings and revolutionizes naval aviation from large and small carriers. Variations include *Sea Harrier* and *AV-8B* for US marines. *VIFF* is Vectoring In Forward Flight, permitting in-flight manoeuvres to outfly adversaries in aerial combat. *Skijump* is the use of an inclined ramp at the end of an aircraft carrier deck to allow marine operations at high weights and in rougher sea conditions.

Future prospects. Supersonic STOVL are under experiment; *Skyhook* is a way of collecting a hovering Harrier by a 'smart crane' mounted on a small ship; civil VTOL aircraft with analogous characteristics have been designed.

pulsar

Pulsating star. Pulsars were discovered in 1967 by Cambridge radio astronomers A. Hewish and S.J. Bell. A pulsar is a radio source characterized by extremely regular bursts of radio waves. The periodicities of the known pulsars vary from a couple of milliseconds up to four seconds. These very short periods indicate that pulsars are magnetized neutron stars, just one km in size, that are rapidly rotating. Their distribution is concentrated towards the plane of the Milky Way, and this indicates that they are the relics of relatively young stars. The nearest is about 100 pc from the Sun, and the Galaxy probably contains about 10^5 pulsars altogether, some 1% of which are catalogued. A pulsar is formed as an endpoint in the evolution of a massive star. Although the details are uncertain, the general picture is that stars of more than a few solar masses end spectacularly as **supernova** explosions, in which a huge remnant of the outer layers is flung into space and the exhausted nuclear core implodes to form a **neutron star**. Conservation of angular momentum and magnetic flux during the collapse accounts for the rapid spin of the neutron star. One of the fastest pulsars is directly associated with the **Crab Nebula**, and is the central engine responsible for its continuing X-ray and radio emission.

The periods of pulsars gradually decline with time, in a highly regular manner, as their rotation slows. The rate of decline indicates that a pulsar will remain detectable for a few million years. The radio signals from pulsars act as important probes of the magnetic field in our Galaxy (20 nT in the solar neighbourhood) as well as electron densities.

Precise measurements of the pulse arrival times and the positions of pulsars have important applications in **astrometry**. The accuracy of these measurements is so great that all data have to be reduced to values at the centre of mass or *barycentre* of the solar system, with full relativistic correction for the motion of the Earth. It is partly for this reason that *dynamical time* (see **time**) replaced **ephemeris time** in 1984: a clock on Earth gains 1.6 ms per year relative to one at the barycentre. Accurate timing measurements enable the relationship between the **ecliptic** and **equatorial** co-ordinate systems to be more closely defined. **Proper motions** can be measured for some pulsars, and they indicate space velocites of 170 km/sec.

A few pulsars are members of **binary** systems, in which the pulsar is in orbit around a companion star. These pulsars have periods of just a few milliseconds. It appears that transfer of matter from the companion to the pulsar has, via angular momentum transfer, led to a speeding up of the pulsar. In one case the surface of the accelerated pulsar is now travelling at one-tenth the velocity of light. Timing observations of binary pulsars are a highly sensitive test of the **general theory of relativity**, and they may yet provide evidence for the radiation of **gravity waves**.

There is still uncertainty as to the emission mechanism. The phenomena observed are consistent with rotating lighthouse models, in which a narrow cone of radiation sweeps across the Earth once per rotation. The site of the emission could be the polar caps or the location in the **magnetosphere** where the rotation speed approaches the velocity of light.

outer sphere concentric with the Earth.

pull-out The transition from a dive or spin to substantially normal flight.

pull-out distance A naval term for the distance travelled by the hook of an aircraft while arresting on the deck of a carrier. Also *run-out distance*.

pulsar See panel on this page.

pulsating star Same as **Pulsar**. See panel.

pulse compression Techniques which permit high range resolution (for which short pulses are necessary) while transmitting relatively long pulses in order to increase transmitter power and thereby enhance detection capability. Commonly, the pulse is frequency- or phase-modulated; modulation information is fed to the pulse-compression circuits in the receiver in order that a matched-filter effect will enable the receiver to respond only to echoes bearing the same modulation. See **coding**.

pulsed Doppler radar One in which the Doppler shift of the signals received from a moving target is used to measure its velocity. The pulsed Doppler technique, but not a **CW radar** using Doppler measurement, can also give range and position information.

pulsed-radar system One transmitting short pulses at regular intervals and displaying the reflected signals from the target on a screen. See **CW radar**.

pulse-forming line An artificial line which generates short high-voltage pulses for radar.

pulse jet A propulsive duct with automatic air intake valves, or a frequency-tuned jet pipe, so that pressure builds up between 'firings', thus achieving thrust at reasonable economy at moderate airspeeds, e.g. 200–400 mph (350–650 km/h).

pulse repetition frequency, rate The average number of pulses in unit time.

purge To clean and flush out liquids (usually propellants) from tanks to prevent build-up of explosive mixture; dry nitrogen or helium is used.

push-broom sensor Term applied to a detecting instrument which employs a line of detectors in juxtaposition for recording a line of a scene without recourse to mechanical scanning.

pyranometer Instrument for measuring either the diffuse or the total global solar radiation. Also *solarimeter*.

pyrgeometer Instrument for measuring the longwave atmospheric radiation or the outward radiation from the Earth's surface.

pyrheliometer Instrument for measuring direct solar radiation, excluding the diffuse and reflected components.

QA Abbrev. for *Quality Assurance*.

Q-band Frequency band mostly in radar, 36 to 46 GHz; now superseded by Ka-band. See **K-band**.

Q-code Telecommunications code using three letter groups: QAA–QNZ for aeronautics; QOA–QQZ for Maritime uses; QRA–QUZ for all services. Examples: QAH = 'What is your height above ....?'; QAM = 'What is the latest met. report?'; QBA = 'What is the horizontal visibility at .....?'.

q-feel A term given (because of the use of $q = D \; \rho \varpi^2$, i.e. **dynamic pressure**) to a device which applies an artificial force on the

control column of a power-controlled aircraft proportional to the aerodynamic loads on the control surfaces, thereby simulating the natural 'feel' of the aircraft throughout its speed range.

quadrantal points Points of the compass which in moving from north correspond to the headings NE (45°), SE (135°), SW (225°) and NW (315°). Cf. **cardinal points**.

quadrature Position of the Moon or a superior planet in elongation 90° or 270°, i.e. when the lines drawn from the Earth to the Sun and the body in question are at right angles.

quadrupoles Radiator producing a sound field of two adjacent dipoles in antiphase. The eddies in a subsonic jet of gas are quadrupoles.

qualification test An evaluation of a flight article or its equivalent to verify that it functions correctly under the specified conditions of flight; normally the test conditions are more severe than those expected.

quantum (1) General term for the indivisible unit of any form of physical energy; in particular the *photon*, the discrete amount of electromagnetic radiation energy, its magnitude being $h\nu$ where ν is the frequency and h is Planck's constant. (2) An interval on a measuring scale, fractions of which are considered insignificant.

quantum field theory The overall theory of fundamental particles and their interactions. Each type of particle is represented by appropriate *operators* which obey certain commutation laws. Particles are the quanta of fields in the same way as photons are the quanta of the electromagnetic field. So *gluon* fields and *intermediate vector boson* fields can be related to *strong* and *weak* interactions.

quantum gravity The theory which would unify gravitational physics with modern **quantum field theory**.

quarter The term applied to the phase of the Moon at quadrature. The first quarter occurs when the longitude of the Moon exceeds that of the Sun by 90°, the last quarter when the excess is 270°. The two other quarters are the **new moon** and **full moon**.

quarter-chord point The point on the **chord line** at one quarter of the chord length behind the leading edge. **Sweepback** is usually quoted by the angle between the line of the quarter-chord points and the normal to the aircraft fore-and-aft centre-line.

quark A type of fundamental particle that forms the constituents of *hadrons*. There are currently believed to be six types of quarks (and their antiquarks). In the simple quark theory, the baryon is composed of three

quasar

When radio astronomers made catalogues in the 1950s, the positional accuracy was not high enough to enable optical astronomers to find visible counterparts. By 1960 improvements in radio interferometers had shown that a handful of sources seemed to be unusual stars in our Galaxy. They had novel emission line spectra seemingly associated with highly excited states of rare elements. In 1963, lunar **occultation** was used to pinpoint source 3C 273 with unprecedented accuracy, and this too matched a *quasi-stellar object*. Its optical spectrum had a familiar pattern of lines, the Balmer series of hydrogen; what was unusual was the **redshift** of $z = 0.16$, then thought of as a high value. Once the possibility of such high redshifts was indicated, other spectra could be explained: source 3C 48 has a redshift of 0.37, and its emission lines are due to neon and oxygen. The quasi-stellar objects, or *quasars*, were not galactic stars, but the remotest objects then known. About 20 quasars are known with redshifts exceeding 4.0, corresponding to distances of about 4000 Mpc.

The optical spectrum of a typical quasar is non-thermal. They have intense broad emission lines characteristic of highly ionized gas with a temperature of 10 000 K, the line widths corresponding to velocities of 10 000 km/sec or so. In addition there are narrow-width lines associated with forbidden transitions. Higher redshift quasars have complex systems of absorption lines also. These have multiple redshifts, associated with matter expelled from the quasars perhaps, or existing in intervening galaxies along the line of sight.

The active core of a quasar is only about 1 parsec across, as shown in high resolution maps made with interferometers. Some objects are highly variable on timescales of a few days, which would indicate an active region only the size of our solar system, but with the energy output hundreds of times larger than a normal galaxy. In objects where it is possible to measure the motion of individual components within the core, the separation sometimes changes so fast that the apparent velocity of separation is up to ten times the speed of light. This *superluminal motion* is partly illusory: matter travelling at close to the velocity of light along a direction close to the line of sight will have an apparent transverse velocity greater than light. However, when motion is analyzed completely relativistically, with proper allowance for the redshift of the object, the paradox is resolved.

The optical and radio continuum spectra are typical of **synchroton** radiation. The power-law spectrum indicates that most emission is coming from relativistic electrons travelling through magnetic fields of about 10 nT. An important goal of theoretical research is to find mechanisms for accelerating the electrons up to the requisite energy, and to replenish the supply over millions of years. The rather simple physics of the synchroton process shows that the energy requirement in relativistic electrons is about 10^{53}J. It appears unlikely that conventional nuclear reactions in stars can account for this energy. More exotic mechanisms, such as matter falling on to compact objects or **black holes** are currently favoured.

Although radio astronomy was instrumental in finding quasars, only about 10 per cent are strong radio sources. Quasars are now discovered through optical surveys using objective prisms. Automatic plate measuring machines can scan photographs obtained with *Schmidt telescopes* to find new quasars very efficiently, so that a few thousand are now catalogued.

Quasars, together with radio galaxies, are important in **cosmology**. To express this simply, because they are the most distant objects, they act as probes of the universe at early times, as well as standard candles with which the geometry of the universe can be calibrated. In practice it is difficult to disentangle effects due to the expansion and ageing of the universe from effects intrinsic to an evolving population of radio sources.

quarks, an antibaryon is composed of three antiquarks, and a meson is composed of a quark and an antiquark. No quark has been observed in isolation. See **flavour**, **colour**.

quasar See panel on p. 140.

quasi-biennial oscillation Alternation of easterly and westerly wind regimes in the equatorial stratosphere with an interval between successive corresponding maxima of from 24 to 30 months. A new regime starts above 30 km and propagates downwards at about 1 km per month. Abbrev. *QBO*.

quasi-geostrophic approximation An approximation to the dynamical equations governing atmospheric flow, esp. the **vorticity equation**, whereby the horizontal wind is replaced by the **geostrophic wind** in the term representing the vorticity, but not in the term representing the **divergence**.

quasi-stationary front A **front** which is moving slowly and irregularly so that it cannot be described as either a **cold front** or a **warm front**.

quasi-stellar object See **quasar**.

QSO Abbreviation for *Quasi-Stellar Object*. See **quasar**.

R

RA Abbrev. for **right ascension**.

radar See panel on p. 143.

radar absorbent material   See panel on p. 143.

radar astronomy   See panel on p. 143.

radar beacon   A fixed radio transmitter whose radiations enable a craft to determine its own direction or position relative to the beacon by means of its own radar equipment. See **transponder**.

radar indicator   See panel on p. 143 .

radar mapping   Cartography using radar data, especially from Sideways Looking Radar (SLAR). May also be used as a navigation system.

radar performance figure   See panel on p. 143.

radar range   See panel on p. 143.

radar range equation   See panel on p. 143.

RADARSAT   Canadian satellite remote sensing programme.

radar scan   See panel on p. 143.

radar screen   See **radar indicator** in the panel.

radial engine   An aircraft or other engine having the cylinders arranged radially at equal angular intervals round the crankshaft. See **double-row-**, **master connecting-rod**.

radial velocity   See **line-of-sight velocity**.

radiant   Point on the celestial sphere from which a series of parallel tracks in space, such as those followed by the individual meteors in a shower, appear to originate.

radiant heat   Heat transmitted through space by *infrared radiation*.

radiant intensity   The energy emitted per second per unit solid angle about a given direction.

radiation   The dissemination of energy from a source. The energy falls off as the inverse square of the distance from the source in the absence of absorption. The term is applied to electromagnetic waves (radio waves, infrared, light, X-rays, γ-rays etc.) and to acoustic waves. It is also applied to emitted particles (α, β, protons, neutrons etc.). See types of radiation: *black-body, electromagnetic, infrared, ultraviolet, visible* etc. See **Planck's radiation law, Stefan-Boltzmann law, Wien's laws for radiation from a black body**.

radiation pressure   Minute pressure exerted on a surface normal to the direction of propagation of a wave. It is due to the rate of transfer of momentum by the wave. For electromagnetic waves incident on a perfect reflector this pressure is equal to the energy density in the medium. In quantum physics the radiation consists of *photons* and the radiation pressure is due to the transfer of the momentum of the photons as they strike the surface. Radiation pressures are very small, 10^{-5} N m^{-2} for sunlight at the Earth's surface. For sound waves in a fluid the pressure gives rise to 'streaming', i.e. a flow of the fluid medium.

radiative equilibrium   The normal state of matter inside stars in which the temperature in every part generates a gas pressure which exactly balances the pressure due to the self-gravity of the star. There is, therefore, no convection in this idealized situation.

radiator flaps   See **gills**.

radioactive series   Most naturally occurring radioactive isotopes belong to one of three series that show how they are related through radiation and decay. Each series involves the emission of an α-particle, which decreases the *mass number* by 4, and β- and γ-decay which do not change the mass number. The natural series have members having mass number: (a) $4n$ (thorium series); (b) $4n + 2$ (*uranium-radium series*); (c) $4n + 3$ (*actinium series*). Members of the $4n + 1$ (*plutonium series*) can be produced artificially. Also *radioactive chain*.

radioactivity   Spontaneous disintegration of certain natural heavy elements (e.g. radium, actinium, uranium, thorium) accompanied by the emission of α-rays, which are positively charged helium nuclei; β-rays, which are fast electrons; and γ-rays, which are short X-rays. The ultimate end-product of radioactive disintegration is an isotope of lead.

radio altimeter   Device for determining height by electronic means, generally by detecting the delay in reception of reflected signals, or change in frequency. Also *radar altimeter*.

radio approach aids   Those which assist landing in bad visibility, notably ILS and MLS. Also *radio* or *electronic landing aids*.

radio astronomy   The exploration of the universe by detecting radio emission from a variety of celestial objects. The frequency spans a vast range from 10 MHz to 300 GHz. A variety of antennas are used, from single dishes to elaborate networks of telescopes forming intercontinental radio **interferometers**. The principle sources of radio emission are: the Sun, Jupiter, interstellar hydrogen, emission nebulae, pulsars, supernova remnants, radio galaxies, quasars, and the cosmic background radiation of the

---

radar

RAdio Detection And Ranging is the use of reflected electromagnetic radiation, normally with a wavelength in the radio frequency spectrum between 30 m and 3 mm to give information on distant objects. Information obtainable includes; range, range rate, bearing, height, relative velocity and imaging. If the transmitted wave is reflected back from the object's surface this is termed *primary radar* but if the target responds with a transmitted pulse from its transponder this is termed *regenerated* or *secondary radar*.

There are a number of terms commonly used. ***Radar absorbent material*** is that which is attached or built in to an aircraft skin which attenuates the return echo, thus reducing the radar signal. A ***radar altimeter*** is used in aircraft and spacecraft for determining a vehicle's height above a surface and for measuring the height of small objects, e.g. hills and waves, on a planetary surface. ***Radar astronomy*** uses pulses of radio waves to detect the distances and map the surface morphology of objects in the Solar System. It has been applied with great success to the mapping of Venus. ***Radar bombing*** is level bombing using a radar bombsight. ***Radar clutter*** is the unwanted indications on a display due to atmospheric interference, lightning, natural static, ground/sea returns or hostile **electronic countermeasures.** ***Radar cross section*** (RCS) is the apparent size of an object as judged by its displayed echo, determined (in the absence of counter measure activity) by true size, range, aspect geometric shape, materials, surface texture and treatment, and other factors including intervening dust and precipitation. Normally defined by the ratio Pr/Ps where Pr is radar power received at the object and Ps is the power reflected, plotted as polar (1) in the horizontal plane. ***Radar indicator*** is the display on a CRT of a radar system output, either as a radial line or a co-ordinate system for range and direction. The echo signal gives a brightening of the luminous spot, which remains on for some seconds because of afterglow. See **plan-position indicator.** ***Radar performance figure*** is the ratio of peak transmitter power to the minimum signal detectable by the receiver. ***Radar range*** is usually given as that at which a specified object can be detected with 50% reliability. It is used in the *radar range equation* which is a mathematical expression for primary radar, which relates transmitter power, antenna gain, wavelength, effective area of the target, receiver sensitivity and *radar range*. The ***radar scan*** is (1) the circular, rectangular or other motion of a radar as it searches for a target, (2) the physical movement of a radar antenna or of the radial line on plan position indicator. The ***radar signature*** is the characteristic pattern of an object as displayed by detection and identification equipment (DoD, NATO). A signature analysed with sufficient accuracy can identify the source type and even a specific example of an emitter. Analogous signatures can be obtained by infrared, optical, acoustic (e.g. sonic boom) and Doppler (velocity) means.

The frequency needs for radar must take into account both the bandwidth needed and the operationally desirable band. The principle radar frequencies cover the range 200 MHz – 4 GHz. Examples are the band 3 – 35 GHz employed for navigation and the frequency 13.5 GHz used for space altimetry.

The appendix gives a table and other information on the letter designations of the relevant frequency bands.

---

universe itself.

radio beacon Stationary radio transmitter which transmits steady beams of radiation along certain directions for guidance of ships or aircraft, or one which transmits from an omnidirectional antenna and is used for the taking of bearings, using an identifying code. Also *aerophare*.

radio beam Concentration of electromagnetic radiation within narrow angular limits, such as is emitted from a highly directional antenna.

radio bearing Direction of arrival of a radio signal, as indicated by a loop, goniometer, interferometer or any directional receiving system as used for navigational purposes. Often giving angle between the fixed station and a reference direction, e.g. true or magnetic North, hence *true radio bearing, magnetic radio bearing*.

radio command Command guidance using a radio link.

radio communications See panel on p. 145.

radio compass Originally a rotating loop, later rendered more sensitive by a goniometer system and by display on a cathode-ray tube. Any device, depending on radio, which gives a bearing. Now superseded by *automatic direction finding* and other *navaids*.

radio control Control of vehicle trajectory by commands tansmitted over a radio link.

radio countermeasures Those activities of **electronic countermeasures** which are concerned with telecommunications.

radio direction-finding Passive reception of direction-finding signals from radio beacons or navigational transmitters, as distinct from active radar. Abbrev. *RDF*.

radio duct See **radiotropospheric duct**.

radio fix The position of a space vehicle, aircraft or other vehicle obtained by a radio navigation aid by traditional crossing of position lines; the same applied to a fixed radio emitter.

radio galaxy About one galaxy in a million is an intense source of cosmic radio waves, caused by synchrotron emission of relativistic electrons. See **galaxy**.

radio horizon In the propagation of electromagnetic waves over the Earth, the line which includes the part of the Earth's surface which is reached by direct rays.

radiolocation Former term for **radar**.

radiometer Instrument devised for the detection and measurement of electromagnetic radiant energy and acoustic energy, e.g. thermopile, bolometer, microradiometer.

radio range (1) Specific system of radio homing for aircraft, in which crossed loops are separately modulated with complementary signals, which coalesce on reception when the aircraft is on course. (2) For transmissions which are not affected by ionospheric or tropospheric peneomena, the optical or *line of sight* path between two points.

radiosonde Instrument for measuring temperature, pressure and humidity at successive levels in the atmosphere which is

carried aloft on a balloon and transmits the measurements by radio. The balloon also carries a radar target so that upper winds may be derived from ground measurements.

radio sonobuoy Sonar devices immersed or dropped into water; can be active (emitting) or passive, directional or non-directional, providing a radio read-out, usually on command.

radio telegraphy Use of a radio channel for telegraph purposes, e.g. by interrupted carrier, change of frequency, or modulation with interrupted audio tone.

radio telephony Use of a radio channel for transmission of speech. Methods include simple modulation, suppressed carrier and one sideband, inverted sidebands and carrier, scrambling before modulation, one in a group modulation or pulse modulation.

radio telescope An instrument for the collection, detection and analysis of radio waves from any cosmic source. All such telescopes consist of a radio antenna, detector and amplifier. Antenna systems may be arrays of **dipoles**, single dishes or **interferometers**. The purpose of a radio telescope is to measure the intensity of radio emission and establish its spectrum. Image analysis can be used to give a picture similar to a photograph. See **astronomical telescope**.

radiotropospheric duct Stratum in which, because of a negative gradient of refractive modulus, there is an abnormal concentration of radiated energy.

radius of action Half the range in still air of a military aircraft, taking safety and operational requirements into account; the total range is out and home again.

radius vector The line joining the focus to the body which moves about it in an orbit, as the line from the Sun to any of the planets, comets etc.

radome Housing for radar equipment, transparent to the signals, e.g. a plastic shell on aircraft or a balloon on the ground.

RAE Abbrev. for *Royal Aircraft Establishment*, Farnborough, UK.

RAeS Abbrev. for *Royal Aeronautical Society*, UK.

rain Result of condensation of excess water vapour when moist air is cooled below its dew point. Rain falls when droplets increase in size until they form drops whose weight is equivalent to the frictional air resistance. The greater proportion of raindrops have a diameter of 0.2 cm or less; in torrential rain a small proportion may reach 0.4 cm. Rain effects important geological work by assisting in the mechanical disintegration of

radio communications

Any form of communication involving the transmission and reception of electromagnetic waves, from a frequency of 10 kHz up to more than 300 GHz. Information is conveyed by **modulation** of the information it is desired to impart on to a *carrier*. The information may be letters represented by code (e.g. Morse), speech, telemetry, pictures (either facsimile or television), digital signals or computer data. In broadcasting, radio communication is a one way process serving many listeners or viewers, or it may be two-way as in tele-communication systems. In the latter, communication may be between two mobile users in different vehicles or from a mobile vehicle and a fixed station, from one microwave tower to another in terrestrial communication or via a **communications satellite**.

The wavebands used are as follows (but see also the appendix):

VLF, very low frequency: <30 kHz

LF, low frequency: 30–300 MHz

MF, medium frequency: 300 kHz–3 MHz

HF, high frequency: 3–30 MHz

VHF, very high frequency: 30–300 MHz

The following are generally used for radar as well as communication.

SHF, super high frequency: 3–30 GHz

EHF, extremely high frequency: 30–300 GHz

Un-named: 300 GHz–3 THz

See **radio-altimeter, -approach aids, -beacon, -bearing, -compass, -control, -countermeasures, -fix, -sonde, -sonobuoy, -tropospheric duct**, for other terms.

Telemetry is the real-time transmission of data by a radio link, e.g. from missile or spacecraft to ground station; nowadays invariably in digital form and vital to unmanned reconnaissance systems and space probes. With the latter, data has been received from the outer parts of the solar system from transmitters radiating less than 1 w. Data can be pressure, velocity, surface angular position or any other instrument output as well as digitized picture information. A single data channel can be shared by using multiplex techniques of various kinds. v. *telemeter*; adjs. *telemetry* as in *telemetry system* and *telemetering*.

Spacecraft communications are mandatory for the delivery of commands and the acquisition of housekeeping and payload data. It is particularly important when telescience operations are involved. If experiments are operated by an astronaut for an investigator on the ground, the communication of information (real time TV of experiments and voice links) to and from the spacecraft is an essential element of the technique. Spacecraft communication is provided mainly by the S and K radio bands.

Radio tracking is used to follow spacecraft by radio interferometry including Doppler determination of speed.

rocks; also chemically, in bringing about solution of carbonates etc.; and, through the agency of running water, in redistributing the products of erosion and disintegration.

rain band An absorption band in the solar spectrum on the red side of the D lines, produced by water vapour in the Earth's atmosphere.

rainbow A rainbow is formed by sunlight which is refracted and internally reflected by raindrops, the concentration of light in the bow corresponding to the position of minimum deviation of the light. The angular radius of the primary bow is 42°, this being equal to 360° minus the angle of minimum deviation for a spherical drop. The colours, ranging from red outside to violet inside, are due to dispersion in the water. See **secondary bow**.

rain gauge An instrument for measuring the amount of rainfall over a given period, usually 24 hr. The usual form consists of a sharp-trimmed funnel, 5 in diameter, leading into a narrow-necked graduated collecting vessel. The *Dines tilting siphon* is of the continuously recording type, noting both the time and the amount of rainfall.

rainmaking Artificial stimulation of precipitation by scattering solid carbon dioxide on supercooled clouds, or by silver iodide nucleation.

rain stage That part of the condensation process taking place at temperatures above 0°C so that water vapour condenses to water liquid.

RALS Abbrev. for **Remote Augmented Lift System**.

RAM Abbrev. for *Radar Absorbent Material*. See **radar**.

ram-air turbine A small turbine motivated by ram (i.e. free stream) air; used (1) to drive fuel pumps, hydraulic pumps, or electrical generators in guided weapons because of the absence of shaft drives in rockets and ramjets; (2) as an emergency power source for driving hydraulic pumps or electrical generators for high-speed aircraft, particularly those with power controls. Abbrev. *RAT*.

ram intake A forward-facing engine (or accessory) air intake which taps the kinetic energy in the airflow and converts it into pressure energy by diffusion; in supersonic flight very high pressure ratios can be obtained.

ramjet The simplest **propulsive duct** deriving its thrust by the addition and combustion of fuel with air compressed solely as a result of forward speed. In subsonic flight kinetic energy is converted into pressure by **diffuser**, or widening duct, which also slows it sufficiently to permit combustion to be maintained; about Mach 1 the shock wave generated by the air-intake lip improves the compression when it decelerates the air to subsonic velocity prior to diffusion. At high Mach numbers, 1.5 and upward, two shock waves are required for the dual purpose of raising the pressure and slowing the air for combustion – pressure ratios of 6 : 1 are attainable at Mach 2, 36 : 1 at Mach 3. In supersonic flight the jet efflux of a ramjet has to be accelerated to high velocity by a

venturi or **convergent-divergent nozzle**. See **aero-engine**.

ramp (1) Parking area for an aircraft at an airport. (2) Inner wall of supersonic intake creating shock waves and improving pressure recovery. Frequently movable as on Concorde. (3) Inclined launcher for missile or RPV.

ramwing Special kind of aircraft designed to fly very low over water thereby gaining advantage in lift-drag ratio by capturing the ground effect. May one day rival the low cost ocean freight ship.

random noise Noise due to the aggregate of a large number of elementary disturbances with random occurrence in time.

range Maximum horizontal distance covered by a projectile.

range-height indicator A display used in conjunction with a **plan-position indicator** for airport control. It displays a vertical plane on which the elevation and bearing of the target may be seen.

range tracking The process of continuously monitoring the delay between the transmission of a pulse and reception of an echo. Tracking requires that the time elapsed between pulse and echo be measured, that the echo be identified as the target rather than random noise, and that the range-time history of the target be maintained.

RAT Abbrev. for **Ram-Air Turbine**.

rated altitude The height measured in the **International Standard Atmosphere**, at which a piston aero-engine delivers its maximum power. Cf. **power rating**.

rate gyro A single degree-of-freedom gyro which measures an angular rate by precession of the gyro against a spring restraint.

rate of climb Generally rate of ascent from Earth. In performance testing, the vertical component of the air path of an ascending aircraft, corrected for standard atmosphere. See **International Standard Atmosphere**.

rate of climb indicator See **vertical speed indicator**.

rating The engine power permitted by regulations for a specified use, e.g. maximum, take-off, combat, maximum continuous.

RATOG Abbrev. for *Rocket Assisted Take-Off Gear*. See **take-off rocket**.

Rayleigh criterion Criterion for the resolution of interference fringes, spectral lines and images. The limit of resolution occurs when the maximum of intensity of one fringe or line falls over the first minimum of an adjacent fringe or line. For a telescope with a circular aperture of diameter D this criterion gives the smallest angular separation of the two images of point objects as 1.22 λ/D.

(λ is the wavelength of the light.)
Rayleigh-Jeans law An expression for the distribution of energy in the spectrum of a black-body radiator:

$$E_\nu d\nu = \frac{8\pi\nu^2 kT}{c^3} d\nu$$

where E_ν is the energy density radiated at a temperature T within a narrow range of frequencies from ν to $\nu+d\nu$, k is Boltzmann's constant and c the velocity of light. The formula holds only for low frequencies. See **Planck's radiation law**.
Rayleigh limit One-quarter of a wavelength, the maximum difference in optical paths between rays from an object point to the corresponding image point for perfect definition in a lens system.
RCS See **reaction control system**.
R-display An expanded A-display in which an echo on a radar screen can be *magnified* by an expanded sweep for close examination.
RDT & E Abbrev. for *Research, Development, Test and Evaluation*, US.
reaction chamber The chamber in which the reaction or combustion of a rocket's fuel and oxidant occur.
reaction control system *RCS*. A set of small thrusters, suitably placed on a spacecraft to control its attitude in pitch, roll and yaw.
reaction propulsion The scientifically correct expression for all forms of jet and rocket propulsion; they act by ejection of a high-velocity mass of gas, from which the vehicle reacts with an equal and opposite momentum, according to Newton's Third Law of Motion.
real-time analyser Analyser which calculates the spectrum (**Fourier analysis**) of a signal so fast that no input data are excluded from the analysis. Uses *Fast Fourier Transform* (FFT) (a particularly fast numerical method for calculating Fourier transforms).
rebecca-eureka Radar system on aircraft carrying low-power interrogator transmitters (rebecca), working with fixed beacon responders (eureka), sending coded signals when triggered by interrogator pulses.
recording altimeter A barographic type of instrument which traces height against time.
recovery time That required by a *transmit-receive tube* in a radar system to operate (usually measured to the point where receiving sensitivity is 6 dB below maximum).
rectified airspeed See **calibrated airspeed**.
recurrent novae A small group of novae

which have shown more than one outburst of light, as T Coronae Borealis in 1866, 1898 and 1933; they show smaller ranges of brightness than most novae.
red giant See **Hertzsprung-Russell diagram**.
redshift-distance relation See panel on p. 148.
red spot See **Jupiter**.
Redux bonding A proprietary method of joining primary sheet metal aircraft structures with a two-component adhesive under controlled heat and pressure. It is widely used for the making sandwiches of **honeycomb structure**, for doubling sheet metal and for attaching **stringers** or skin stiffeners.
red variables See **Mira stars**.
re-entry See panel on p. 149.
re-entry corridor See panel on p. 149.
re-entry thermal protection See panel on p. 149.
reference climatological station A meteorological station where a homogeneous series of observations of weather elements over a period of at least 30 years have been, or are expected to be, made under approved conditions.
reflecting telescope A telescope using a mirror to bring light rays to a focus, it was first applied to astronomy by Isaac Newton who recognized its merits in overcoming the **chromatic aberration** of lenses. William Herschel developed techniques for casting large primary mirrors from 1783, and from that time the largest telescopes have been reflectors. Several configurations are used under different circumstances: *Cassegrain, Newtonian, Gregorian, Maksutov, Schmidt* and *coudé*. All of the world's largest telescopes are reflectors.
reflection coefficient Complex ratio of reflected pressure to incident pressure when a plane sound wave is incident on a discontinuity. The complex ratio includes changes in amplitude and in phase during the reflection.
reflector sight Mirror gunsight which projects the aiming reticule and computed correction information for speed and deflection on to a transparent glass screen. Used first in World War II. Modern sights (HUD) incorporate elaborate radar information and firing instructions which allow an attack and breakaway under completely blind conditions.
refracting telescope A telescope which uses lenses to bring light rays to a focus; it was first applied to astronomy by G. Galileo who resolved the stars of the Milky Way and discovered the satellites of Jupiter. In its modern form, using lenses corrected for

redshift-distance relation

Distances to galaxies are found by a variety of techniques. For nearby galaxies the regular variations of the **Cepheid variable stars** give absolute magnitudes, which yield a distance when compared to observed magnitudes. Other candidate objects where brightness comparisons give a distance indication are the brightnesses of O-type stars, **novae, supernovae** and clouds of ionized hydrogen. These methods work to a distance of about 65 million light-years from Earth.

The spectrum of a galaxy tells us its velocity relative to Earth. When V.M.Slipher obtained the first galaxy spectra in 1921-5 there were two surprising results: all galaxies had spectra which were shifted towards the red, indicating that they were going away from us, and the velocities were large, up to 2000 km/sec in the original sample. The amount of shifting to the red, or size of the *redshift*, is directly proportional to velocity, unless the velocity is a fair fraction of the speed of light. So, if the wavelength of the **Balmer series** of hydrogen is found to be, say, 1% longer for some galaxy, we speak of a redshift of z = 0.01, corresponding to a velocity of 3000 km/sec.

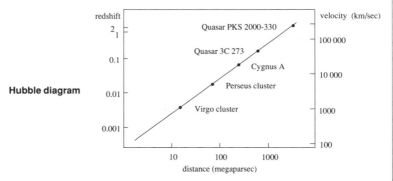

Edwin Hubble at Mount Wilson built on this discovery of redshifts. He obtained velocities from the redshifts and distances for faint and distant galaxies. In 1929 he found a simple linear relationship between the distance (d) to a galaxy and the velocity (V): $V=Hd$. This relation came to be known as the *Hubble law*, now more properly called the *Hubble relation*. A graph showing the correlation between velocity (or redshift) and distance (or absolute luminosity) is termed the *Hubble diagram* (see above). The scale factor H was historically termed the *Hubble constant*. However, it might vary with time as the universe evolves, and is therefore now called the *Hubble parameter*. Determining this parameter is a major goal of extragalactic astronomy: its value is 50–100 km/sec per **megaparsec** of distance. The reciprocal of this quantity has the dimension of time and is known as the *Hubble time*. A rough indication of the age of the universe, its value is 10–20 billion years.

Within the general framework of the **Big Bang** universe it is possible for the ultimate fate of the universe to be one of infinite expansion or ultimate collapse back to a singularity. By plotting the Hubble diagram out to large distances one can, in principle, determine whether the universe is open or closed, according to how the relation curves at high redshifts. This is one part of the mission of the *Hubble space telescope*.

Although the Hubble parameter has an uncertainty of perhaps a factor of two, this has not prevented astronomers from turning the Hubble relation round and using it to estimate distances to remote galaxies and **quasars**. If the redshift z is determined, the distance is obtained directly from $d = cz/H$, where c is the velocity of light. About 20 quasars are known with redshifts of about 4.0 or larger. These are the most distant observable objects.

re-entry

That period of return to Earth (or any other planet) when a spacecraft passes through the atmosphere to land on the surface. During re-entry (sometimes *entry*) the spacecraft decelerates and is subject to intense heating generated by atmospheric friction. The magnitude of the problem is indicated by the deceleration which may be greater than 10 g and the stagnation temperature which may reach many thousands of degrees. Furthermore, a sheath of ionized air around the spacecraft can black-out radio communication for several minutes.

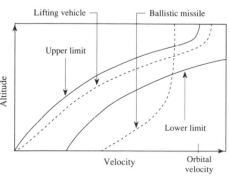

For a safe passage through the atmosphere a manned spacecraft must keep within a narrow *re-entry corridor* which widens as the velocity is reduced and is also dependent on atmospheric density. It can be as low as about 10 km, giving rise to stringent guidance requirements. The lower side of the corridor or *undershoot boundary* is fixed by the excessive heating experienced below it, and the upper side or *overshoot boundary* marks that region above which the atmospheric density is so low that the spacecraft can not be slowed down but skips back into space.

Re-entry corridor.

Re-entering spacecraft must be designed to withstand the hazardous re-entry conditions, and their trajectory and angle of entry optimized. Re-entry may be *direct*, as for ballistic missiles, or *lifting* when use is made of a spacecraft's lifting capabilities to reduce the high heating rates but, thereby, increasing the total heat transferred. In both cases, large values of the drag parameter, $m/C_D A$ (where m is the mass, C_D the drag coefficient and A the drag reference area), imply large stagnation heating rates that are inversely proportional to the square root of the nose radius. In addition, lifting re-entry provides lower g loads and more manoeuvrability, and is preferable for manned flight. The re-entry vehicle can be shielded from the intense heat by the following techniques, alone or in combination: (1) an ablation shield which vaporizes and carries the heat away; a material of large latent heat (e.g. glass resin) is used; (2) a heat sink of high thermal capacity (e.g. copper); (3) a good thermal insulator (e.g. fibreglass); (4) radiative cooling using a high thermal emissivity surface.

As an example, the US Space Shuttle Orbiter, re-enters in about 55 minutes over a distance of 8000 km. Its velocity is reduced from 7.8 km/sec to 0.1 km/sec and a maximum deceleration of only 3 g is encountered. The vehicle's angle of attack is gradually reduced and its aerodynamic surfaces utilized. A *cross-range* of 1600 km on each side of its nominal re-entry track is possible. Temperatures of 1600° C at the stagnation points and 300° C over leeward surfaces are encountered. To protect the aluminium structure, reinforced carbon-carbon inserts are used at the leading edges and silica tiles over the main body.

chromatic aberration, this telescope is still used by visual observers and amateurs.
refractive index The absolute refractive index of a transparent medium is the ratio of

the phase velocity of electromagnetic waves in free space to that in the medium. It is given by the square root of the product of the complex relative permittivity and complex relative permeability. Symbol n.

refractive modulus One million times the excess of **modified refractive index** above unity in M units.

Regnault's hygrometer A type of hygrometer in which the silvered bottom of a vessel contains ethoxyethane, through which air is bubbled to cool it, its temperature being indicated by a thermometer.

regolith The layer of fine powdery material on the **Moon** produced by the repeated impact of **meteorites**.

reheat Injection of fuel into the jet pipe of a turbojet for the purpose of obtaining supplementary thrust by combustion with the unburnt air in the turbine efflux. Reheat is the British, and original, term, but is gradually being superseded by the American term *afterburning*, with *afterburner* for the device itself.

relative humidity The ratio of vapour pressure in a sample of moist air to the **saturation vapour pressure with respect to water** at the same temperature.

relative visibility factor Ratio of apparent brightness of a monochromatic source to that of a source of wavelength 550 nm having the same energy.

relativistic mass equation When a particle is accelerated up to a velocity (v) which is more than a small fraction of the phase velocity of propagation of light in vacuum (*c*), it is said to be *relativistic* with mass increased according to the formula

$$m = m_0/\sqrt{1 - v^2/c^2},$$

where m_0 = rest mass (at low velocities). Required to be considered in cyclotron, betatron and linear accelerator design.

relativity Theory based on the equivalence of observation of the same event from different frames of reference having different velocities and accelerations. Introduced in 1905 and generalized a decade later, instein's *special relativity* was verified by observations and the precession of the perihelion of the planet Mercury. Important results of this restricted theory include the **relativistic mass equation**, the **mass-energy equation** and **time dilation**. See **principle of relativity**.

relight Term used for igniting an aircraft gas turbine in flight after shut down.

remote augmented lift system Engine designed for **STOVL** aircraft employing nozzles, remote from the engine, powered by compressed air ducted from the compressor and heated by fuel burnt in the nozzle in a manner similar to an *afterburner*.

remote mass-balance weight A mass balance weight which, usually because of limitations of space, is mounted away from the control surface, to which it is connected by a mechanical linkage.

remote sounding of the atmosphere See panel on p. 151.

rendezvous The meeting and bringing together of two spacecraft in orbit at a planned place and time; also the type of mission where a spacecraft encounters a target body with zero relative velocity at a preplanned time and place.

requirements The demands placed on any element of a space system (such as spacecraft, payload, subsystem, communications network, ground organization etc.) which have to be satisfied by that element.

réseau A reference grid of fine fiducial lines or points used in image analysis and measurement.

reserve buoyancy Potential buoyancy of a seaplane or amphibian which is in excess of that required for normal floating. The downward force required for complete immersion.

reserve factor Ratio of actual strength of an aircraft structure to estimated minimum strength for a specified load condition.

resolving power of a telescope Ability of an astronomical telescope to measure the angular separation of two images which are close together.

resonance See **ground-**.

resonance test A test in which an aircraft, while suspended by cables or supported on inflated bags, is excited by forced oscillations over a range of frequencies, so as to establish the natural frequencies and modes of oscillation of the structure.

resonant gap The interior volume of the resonant structure of a transmit-receive tube in which the electric field is concentrated.

responser Receiver of secondary radar signal from **transponder**.

rest-mass energy Rest-mass energy is c^2 times the *rest mass* of the particle, where c is the speed of light.

restoring moment A moment which, after any rotational displacement, tends to restore an aircraft to its normal attitude.

retrace See **flyback**.

retractable landing gear An alighting gear which can be withdrawn completely or nearly so from its operative position to reduce drag.

retractable radiator A liquid cooler for an aero-engine, capable of being withdrawn out of the airstream, for reducing drag and controlling the temperature of the cooling liquid.

remote sounding of the atmosphere

Remote atmospheric sounding methods have become increasingly important since the 1950s. They may be divided into *active*, in which a beam of electromagnetic or acoustic radiation (usually pulsed) is emitted from a transmitter and its reflection collected by an associated receiver, and *passive*, in which naturally occurring radiation is scanned and analysed. Active methods include radars on the ground or carried by satellites and aircraft, **lidars** and acoustic sounders. Passive methods include the reception of radiation emitted from the atmosphere and the surface of the Earth (both land and sea) by high resolution radiometers carried on satellites. Included in the atmospheric radiation are emissions from water droplets constituting fog and clouds as well as solar radiation reflected from the clouds' upper surfaces. Satellite radiometers can be arranged to look vertically downwards, thereby receiving radiation from all lower levels including the surface, or quasi-horizontally so that radiation is collected only from a vertically thin upper layer but over a long atmospheric path. This is called *limb sounding*, by means of which constituents present in very small concentrations can be studied.

Satellite radiometers are sensitive to visible and infrared radiation and more recently have been developed to measure microwave radiation which, unlike that at shorter wavelengths, can penetrate clouds. Measurements of radiation in different wave-bands emitted by CO_2 may be used to derive a good approximation to the vertical distribution of temperature. This is because different wavebands have different absorption co-efficients so that radiation from a strongly absorbed waveband will come predominantly from a level much nearer the satellite, i.e. at a much greater height, than that from a weakly absorbed band. Measurements of both visible and infrared radiation may be used to produce pictorial displays of cloud cover, and successive cloud images from a geostationary satellite can yield estimates of the wind field.

Radars, including Doppler radars which can provide measurements of air movements, are used chiefly at ground-based installations, for the detection of rain. However, satellite-borne radars or 'scatterometers' can measure back-scattered signals from capillary waves on the sea surface from which the surface wind may be estimated. Three Doppler radars, one pointing vertically and two at small angles to the vertical, can provide vertical profiles of the wind to great heights by measuring reflections from tiny variations in refractive index caused by small-scale turbulence which moves with the general wind-field; this is known as *wind profiling*.

Lidars, which are sensitive to light in the range from ultraviolet to infrared, are used to detect clouds and atmospheric pollution consisting of aerosols.

Acoustic sounding uses short bursts of high-frequency sound, the reflected and back-scattered signals being collected at synchronized receivers; it can provide information through most of the troposphere on temperature inversions and turbulent fluctuations of wind and temperature. Measurements of Doppler shifts in frequency can provide information on the general wind field.

retraction lock A device preventing inadvertent retraction of the landing gear while an aircraft is on the ground. Also *ground safety lock*.

retrograde motion (1) Motion of a comet (or satellite) whose orbit is inclined more than 90° to the ecliptic (or to the planet's equatorial plane). (2) Apparent motion of a planet from east to west among the stars, caused by a combination of its true motion with that of the Earth.

retrorocket A small rocket motor used for reducing the velocity of a space vehicle in landing, or in any manoeuvre calling for a thrust in the direction opposite to the motion.

return Refers to radar reflections, e.g. land (or ground) return, sea return.

return-flow system A gas-turbine combustion system in which the air is turned through 180° so that it emerges in the opposite direction to that in which it entered. Also sometimes *reverse-flow system*.

return-flow wind tunnel One in which the air is circulated round a closed loop to preserve its momentum and so reduce the power requirement.

reusability Refers to space hardware which may be used more than once. A system may be partially- or fully-reusable but, in either case, it implies recovery and refurbishment before reuse. Ant. *expendability*.

reversal of control Reversal of a control moment (or couple) which occurs when displacement of the control surface results in such high forces that distortion of the main structure counteracts the effect of the surface. This overloading is a function of airspeed, since control forces increase proportionally to the square of the velocity, and *reversal speed* is the lowest **EAS** at which reversal occurs.

reversible saturation-adiabatic process An idealized, alternating condensation-evaporation process occurring in the atmosphere, and which assumes that none of the condensation products are removed by precipitation.

reversing layer The name given to the lower part of the Sun's chromosphere where the absorption lines of the solar system are formed by 'reversal' from bright emission lines to dark absorption lines.

revolution The term generally reserved for orbital motion, as of the Earth about the Sun, as distinct from **rotation** about an axis.

Reynolds number The dimensionless group

$$\frac{\text{density} \times \text{velocity} \times \text{a linear measure}}{\text{viscosity}}$$

all values being in the same system of units, e.g. ft lb s or SI. If used in pipe or tube flow linear dimension is internal diameter and this has special application in heat exchange calculations. In other applications, e.g. mixers in vessels, the linear dimension may be, for example, the diameter of a moving part.

Rhea The fifth natural satellite of **Saturn**, and the second largest moon in its system, with a diameter of 1530 km.

rib A fore-and-aft structural member of an *aerofoil* which has the primary purpose of maintaining the correct contour of the covering, but is usually a stress-bearing component of the main structure. Ribs are usually set either parallel with the longitidinal axis or at right angles to the front spar. Cf. **nose ribs**.

ribbon parachute A parachute in which the canopy is made from light webbing

with spaces between, instead of conjoined fabric gores, so as to give greater strength against ripping for deployment at high speed. Commonly used for **brake parachutes**.

Richardson number A non-dimensional number Ri arising in the study of shearing flow in the atmosphere. If g is the acceleration of gravity, β is a measure of vertical stability (commonly $\frac{1}{\theta}\frac{\partial \theta}{\partial z}$ where θ is the *potential temperature*) and $\frac{\partial u}{\partial z}$ is a characteristic vertical wind shear, then

$$Ri = g\beta \Big/ \left(\frac{\partial u}{\partial z}\right)^2.$$

Turbulence is likely to be suppressed if Ri approaches 1.

ridge An outward v-shaped extension of the isobars from a centre of high pressure.

riding lamps Lamps displayed at night by a float plane or flying boat when moored or at anchor. Colours and positions as in the Maritime code.

RIG Abbrev. for *Rate Integrating Gyro*. See **floated rate-integrating gyro**.

rigging (1) The operation of adjusting and aligning the various components, notably flight and engine control, of an aircraft. (2) In airships and balloons, the system of wires by which the weight to be lifted is distributed over the envelope or gas-bag.

rigging angle of incidence See **angle of incidence**.

rigging diagram The drawing giving the manufacturer's instructions as to the positioning and aligning of the components and control systems of an aircraft.

rigging line See **shroud line**.

rigging position The position in which an aircraft is set up in order to effect the adjustment and alignment of the various parts, i.e. with the lateral axis and an arbitrarily chosen longitudinal datum line horizontal.

right ascension One of the two co-ordinates, used with **declination** for specifying position on the **celestial sphere** in the *equatorial co-ordinate system*. It is the angular distance measured eastwards along the **celestial equator** from the vernal **equinox** to the intersection of the hour circle passing through the body. (It is the celestial equivalent of longitude.) Its units are hours, minutes and seconds, and one hour of right ascension is 15°; the Earth's daily rotation takes the celestial sphere through one hour of right ascension in one hour of **sidereal time**. Abbrev. *RA*.

rocket engine

A device that provides thrust from the reaction of propellants in a combustion chamber to propel a vehicle to which it is attached. The reaction normally takes place between a fuel and an oxidizer. Two forms of liquid propellant are available. A *monopropellant*, e.g. H_2O_2, relies on the chemical reaction or decomposition of a single liquid, usually in the presence of a catalyst. A *bipropellant*, e.g. liquid O_2 and liquid H_2, consists of two liquids, the fuel and the oxidizer, which are stored separately. Solid propellants can also be used and are then contained within the combustion chamber, making the system inherently simple. Ignition is usually effected by a spark, after which the heat produced maintains the burning process. Sometimes the fuel and oxidizer are designed to ignite spontaneously on mixing; such propellants are called *hypergolic*.

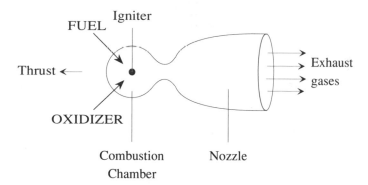

In a liquid rocket engine, such as that in the diagram, the propellants are introduced into a combustion chamber where they are burnt with the products of combustion being ejected through the nozzle at high velocity. Optimum design of the propellant injectors, combustion chamber, ignition system, nozzle and of the oxidizer-to-fuel ratio are all essential for efficient operation. Since a rocket engine works most efficiently in a vacuum, it is ideally suited for propulsion in space.

right-handed engine An aero-engine in which the propellor shaft rotates clockwise with the engine between the observer and the propeller.

rille A winding valley, with a U-shaped cross-section, found in the lunar **maria**.

ring slot parachute A parachute the canopy of which has slots all round its circumference, to give it greater stability.

ripcord (1) A cable used for opening the pack of a personal parachute. (2) An emergency release for gas in an aerostat envelope.

R & M Abbrev. for *Reliability and Maintainability*, US.

Roche limit The lowest orbit at which a satellite can withstand the tides raised within it by the primary body.

rocket engine See panel on this page.

rocket equation See **launch system**.

rocket propulsion See panel on p. 154.

Rogallo wing Delta shaped wing formed by three spars which meet at the apex and are covered with fabric. In flight the fabric becomes convex due to reduced air pressure over the top surface. Can be folded compactly and, although originally intended for re-entering spacecraft as a *paraglider*, has become very popular for **microlight** aircraft.

roger Radio code for 'I have received and understood all of your last transmission'.

roll Aerobatic manoeuvre consisting of a complete revolution about the longitudinal axis. In a *slow roll* the centre line of the aircraft follows closely along a horizontal straight line; an *upward roll* is similar, but considerable height is gained; a *hesitation*

---

rocket propulsion

A means of propelling objects into and in space, and based on the rocket or *reaction* principle. A system or vehicle powered by rocket propulsion carries its own source of energy in the form of propellants which are made to react and produce a stream of gas (or other) particles which are ejected with a high exhaust velocity. In accordance with Newton's third law of motion, a force is produced on the rocket system which is in the opposite direction to that of the ejected gases. This force is termed the thrust (T) and can be deduced from the rate of momentum change. It is related to the exhaust velocity (V_e) by the equation:

$$T = mV_e \ (T \text{ in newtons}) \ or \ T = \ mV_e \ / \ g_o \ (T \text{ in kgf})$$

where m is the mass of propellant expended per second and g_o is the acceleration due to gravity at the Earth's surface. It follows from these considerations that a high exhaust velocity is desirable. The ratio T/m, thrust divided by propellant consumption, is simply V_e using MKS units. If T is measured in kgf, then $T/m = V_e/g_o$, referred to as the specific impulse (I_{sp}) which has the units of seconds. Specific impulse can then be defined as the thrust in kgf obtained from one kgm of propellant burnt in one second. Again the total impulse is Tt, where t is the burning time, and $I_{sp} = \ Tt/mt$, so specific impulse can be regarded as the total impulse per unit mass of propellant consumed.

---

roll is one where the pilot brings his aircraft momentarily to rest in its rolling motion. A vertical upward or downward roll is usually called an *aileron turn* because these are the only control surfaces involved. A *flick roll* is an entirely different, very rapid and violent manoeuvre in which the aircraft makes its revolution along a helical path; high structural stresses are imposed and many countries ban this aerobatic. A *half-roll* is lateral rotation through 180°.

roll damper See **damper**.

rolling The angular motion of an aircraft tending to set up a rotation about a longitudinal axis. One complete revolution is called a **roll**.

rolling instability See **lateral instability**.

rolling moment The component of the couple about the longitudinal axis acting on an aircraft in flight.

rope See **window**.

Rossby number A non-dimensional number Ro defined as the ratio, for a particular class of motions in a rotating fluid such as the Earth's atmosphere, of inertial forces to **Coriolis forces**. $Ro = U/fL$ where U is a characteristic velocity, L a characteristic length, and f the **Coriolis parameter**. If Ro is large, then the effect of the Earth's rotation may be neglected.

Rossby wave A wave in the general atmospheric circulation, in one of the principal zones of westerly winds, characterized by large wavelength, (ca 6000 km), significant amplitude (ca 3000 km), and slow movement. First described by C.G. Rossby.

rotachute A 'parachute', usually for stores or the recovery of missiles, in which the normal retarding canopy is replaced by rotor blades, which are freely-revolving and act like a **rotor**.

rotaplane A heavier-than-air aircraft which derives its lift or support from the aerodynamical reaction of freely rotating rotors. See **rotorcraft**.

rotary engine An early type of aero-engine in which the crankcase and radially disposed cylinders revolved round a fixed crankshaft; not to be confused with the modern radial engine.

rotation The term generally confined to the turning of a body about an axis passing through itself, e.g. *rotation* of the Earth about its polar axis in 1 sidereal day.

rotation speed The speed during take-off at which the nosewheel of an aircraft is raised from the ground prior to **lift-off**.

rotor A system of revolving aerofoils producing lift, acting on a plane at right angles to the driving shaft.

rotor A large closed eddy which may form under **lee waves** of large amplitude; often associated with severe turbulence.

rotor cloud A whirling quasi-stationary cloud that forms in the upper part of a rotor

rotorcraft

A rotorcraft derives both lift and control from one or more rotors (rotating high-aspect wings) which are engine driven and usually rotate about a nearly vertical axis. The blade angle relative to the rotor disk is varied in two ways: *collective pitch*, in which the blades alter their angle in unison and thereby provide more or less lift force and *cyclic pitch* which varies throughout each revolution, decreasing incidence as the blade moves forward into the relative wind of the craft's motion and increasing it on the downwind portion. By these two means, which are under the control of the pilot, the rotorcraft can be manoeuvred as in an aircraft but can also hover, rise and descend vertically, and even fly backwards.

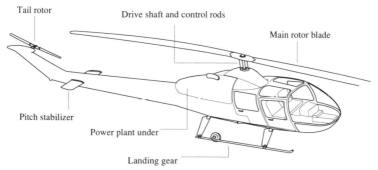

Tail rotor

Drive shaft and control rods

Main rotor blade

Pitch stabilizer

Power plant under

Landing gear

Elements of a helicopter with a single main rotor

Because the engine provides torque to rotate the rotor(s) the rotorcraft itself tends to rotate in the opposite direction. This is overcome by several means, usually by arranging two rotors to balance torque in some way. A tail rotor with a horizontal axis is common (see drawing). Two main rotors can also be used to cancel the disturbing torque.

These factors give the designer many possible configurations:

Tandem rotors, as in the Boeing-Vertol CH-47b, Chinook.

Co-axial rotors, as in the Soviet KA-26.

Intermeshing main rotors, as in the US Kaman UH-2C with 'synchropter' rotors.

Compound helicopter, which also has a fixed wing on the fuselage for lift at high forward speed, sometimes together with direct thrust from engine or rotor.

Rotor-tip jet propulsion, as in the Rotodyne in which fuel is fed to jet pods at the blade tip so avoiding shaft drive at the hub; but the method was exceptionally noisy.

Autogyro is a rotorcraft in which forward thrust is provided by a conventional engine driven propeller and lift comes from the rotor which is unpowered and free to rotate under the action of air flowing from below to above the disk. Usually autogyros perform *Short Take-Off and Landing* (STOL) but some can spin up the rotor for *Vertical Take-Off* (VTO) by clutching in the engine drive.

X-wing, a high-speed helicopter with a large chord rotor which can be stopped in high-speed flight so that the rotor blades behave as a pair of swept forward and a pair of swept aft wings. At lower speeds the X-wing uses circulation control by blown air to maintain high lift, as originally developed by the National Gas Turbine Establishment (UK).

Tilt rotor has engine-rotor units at each wing tip which can be rotated through 90°, providing vertical force for VTO but horizontal force for forward flight when it is airborne and the rotor blades have moved into the vertical plane. This gives a far higher flight speed than a conventional helicopter. The US XV-22 'Osprey' is an example.

to the lee of a range of hills. The *helm bar* near Cross Fell in Cumbria is a well-known example.

rotorcraft See panel on p. 155.

rotor head The structure at the top of the rotor pylon, including the hub member to which the blades of a rotorcraft are attached.

rotor hinge A hinge for the blades of a rotorcraft. See **drag hinge, feathering hinge, flapping hinge.**

rotor hub The rotating portion of the rotor head of a rotorcraft to which the rotor blades are attached.

rotor-tip jets Propulsive jets in the tips of a rotorcraft's blades that are used to obtain a drive with minimum torque reaction; they may be **pulse jets, ramjets**, combustion units fed with air and fuel from the fuselage, **pressure jets**, or small **rocket** units.

RPV Abbrev. for *Recoverable Pilotless Aircraft.*

RR Lyrae variable Variable star with a period of less than 1 day; common in globular clusters and used, like the cepheids, to measure galactic distances.

rudder A movable surface in a vertical plane for control of an aircraft in angles of yaw (i.e. movement in a horizontal plane about a vertical axis). Usually located at the rear end of the body and controlled by the pilot through a system of rods and/or cables or electric signals in **fly-by-wire.**

rudder bar, pedals A mechanism consisting of differential foot-operated levers by which the pilot actuates the rudder of a glider or an aircraft, or controls the pitch of a helicopter tail rotor through mechanical, hydraulic or electrical relaying devices.

runway threshold The usable limit of a runway; in practice it is usually the current downwind end which is intended.

runway visual markers See **airport markers.**

runway visual range In bad weather, the horizontal distance at which black-and-white markers of standard size are visible, the figure being transmitted to pilots approaching by **air-traffic control.** Abbrev. *RVR.*

S

sabin Unit of acoustic absorption; equal to the absorption, considered complete, offered by 1 ft^2 of open window to low-frequency reverberant sound waves in an enclosure. Obsolescent.

Sabine reverberation formula Earliest formula (named after investigator) for connecting the reverberation of an enclosure, T seconds, with the volume, V in cubic metres, and the total acoustic absorption in the enclosure, ΣaS, where a is the absorption coefficient of a surface of S square metres. The formula is $T = 0.16V/\Sigma a\Sigma$.

SAE Abbrev. for *Society of Automotive Engineers* (US). Gives name to a widely-used viscosity scale for classifying motor oils.

safety barrier A net which is erected on the forward part of the deck of an aircraft carrier to stop any aircraft which misses the *arrester gear*. A cable and/or nylon net which can be quickly raised to prevent an aircraft from over-running the end of a runway. The barrier is held by friction brakes or weights so that it imposes a 1 or 2 g deceleration on the aircraft.

safety height The height below which it is unsafe to fly on instruments because of high ground.

safety speed The lowest speed above stalling at which the pilot can maintain full control about all three axes. It is particularly applicable to multi-engined aircraft, where it is taken to be the minimum speed at which control can be maintained after complete failure of the engine most critical to directional control.

SAGE Abbrev. for *Semi-Automatic Ground Environment.* Air defence system whereby information is received from radar and other sources and is processed at a central station to give an evaluation of a situation.

sailplane A glider designed for sustained motorless flight by the use of air currents. The most advanced methods of streamlining and very high *aspect ratio* are used to reduce *drag* to the barest minimum.

SALR See **saturated adiabatic lapse rate**.

Salyut A modular USSR space station which has been improved steadily since its first use in 1971; it has demonstrated that astronauts can remain in orbit for periods up to a year.

sampled data tracking That used with high speed electronic beam switching, using **phased arrays**, allowing data to be gathered on each of several tracks virtually simultaneously.

sandwich construction Structural material, mainly used for skin or flooring, possessing exceptionally good stiffness for weight characteristics. It consists of two approximately parallel thin skins with a thick core having different mechanical properties, so that the tensile and compressive stresses develop in the skin, and the core both stabilizes these surfaces and gives great strength in bending; core materials range from balsa wood through metal-foil honeycomb (light-alloy or steel) to corrugated sheet.

SAR Abbrev. for **Synthetic Aperture Radar**.

SARAH Abbrev. for *Search and Rescue Homing*, a system for facilitating rescue when aircraft go down at sea, consisting of a small beacon transmitter which sends a coded pulse to rescue craft. Also used to guide support vessels to spacecraft after a sea landing.

saros cycle A cycle of 18 years 11 days, which is equal to 223 synodic months, 19 eclipse years and 239 anomalistic months. After this period the centres of the Sun and Moon return to their same relative positions and the same pattern of eclipses is repeated; known to the ancient Babylonians, it was used to predict eclipses. Adj. *saronic*.

satellite (1) Any small body orbiting under gravitational forces in a closed path around a much more massive body. The planets are, in this sense satellites of the Sun; the Moon is the natural satellite of the Earth; and the *Magellanic clouds* are satellite galaxies of the *Milky Way*. (2) A man-made device launched into orbit around the Earth, the Moon or other planets. They serve a variety of functions: relaying globally telephone and television signals; covert intelligence gathering of all kinds; remote sensing of the Earth and its environment; weather forecasting; and as platforms for astronomical telescopes.

saturated adiabatic A curve on an **aerological diagram** representing the temperature changes of a parcel of saturated air subjected to an adiabatic process, the state of saturation being maintained.

saturated adiabatic lapse rate The temperature lapse rate of air which is undergoing a *reversible natural adiabatic process* as shown by the *saturated adiabatic* lines on an **aerological diagram**. Abbrev. *SALR*.

saturated humidity mixing ratio The *humidity mixing ratio* of air which is saturated at a specified temperature and pressure. Saturation may be defined with reference

157

Saturn

The sixth major planet of the solar system in order from the Sun. Saturn is one of the four gas giants, second in size only to **Jupiter**. The equatorial diameter is 9.4 times the Earth's and the mass is 95 times greater. However, its average density is only 0.7 times that of water because the lightest elements, hydrogen and helium, make up the bulk of the mass. About 1 per cent of the volume is a rocky core. In the high pressure region surrounding the core, the hydrogen takes on the form of a metal and the outer half of the planet consists of a deep atmosphere. The visible features of the planet are cloud bands at the top of this atmosphere which are much observed by amateur astronomers and can be seen with a modest telescope.

The cloud patterns on Saturn do not show much colour contrast. Computer processing of images obtained by Voyagers 1 and 2 during their encounters in 1980 and 1981 reveal complex circulation currents, similar to those observed on Jupiter. Saturn rotates very rapidly, spinning once every 10 hours 32 minutes on average, though the rate varies with latitude. The resulting flattening at the poles is significant; the polar and equatorial diameters differ by 11%.

The most striking feature, visually, is the spectacular ring system which is detectable with ordinary binoculars. The rings lie in the planet's equatorial plane, which is tilted at an angle of 27° to its orbit round the Sun. As the relative positions of the Earth and Saturn change, the rings are presented at differing angles, sometimes appearing open, at other times being seen edge-on so that they disappear from view. The rings have the appearance of a series of zones of differing brightness, separated by numerous dark divisions. The two most marked divisions are the *Cassini* division (visible in the smallest telescope) and the *Encke* division. The Voyager images of the rings showed that they consist of many thousands of narrow concentric ringlets. The rings are only one kilometre thick and are made up of a huge number of separate rocks and particles, perhaps ranging in size from a kilometre down to a micrometre.

Before 1980, ten satellites of Saturn were known. A further eleven small satellites have been discovered since, some telescopically in 1980 when the ring system was edge-on (thus removing the glare) and some by the Voyager 1 and 2 spacecraft in 1980 and 1981.

---

either to liquid water or (below 0°C) to ice.

saturated vapour A vapour which is sufficiently concentrated to exist in equilibrium with the liquid form of the same substance.

saturation The condition in which a further increase in one controlling variable produces no further increase in the resultant effect. Important in radar when the input signal drives the receiver to a point when its output can increase no further; characterized in such cases by a severe increase in non-linear distortion.

saturation of the air The air is said to be saturated when the relative humidity is 100%.

saturation vapour pressure with respect to water The maximum **water vapour pressure**, which can occur when the water vapour is in contact with a free water surface at a particular temperature. The water vapour pressure existing when

effective evaporation ceases.

Saturn See panel on this page.

SAW Abbrev. for **Surface Acoustic Wave**.

SBA Abbrev. for **Standard Beam Approach**.

SBAC Abbrev. for *Society of British Aerospace Companies*.

S-band Loose definition, due to international disagreement, of microwave band in the 2 to 3 GHz region.

scan Systematic variation of a radar beam direction for search or angle tracking. See **C-display, R-display**.

scanner Mechanical arrangement for covering a solid angle in space, for the transmission or reception of signals, usually by parallel lines or *scans*.

scanning Coverage of a prescribed area by a directional radar antenna or sonar beam.

scanning loss That which arises from relative motion of a scanning beam across a target, as compared with zero relative motion.

scatterer Object which causes scattering, e.g. curved hard boards or fish bladders in the sea. Used in reverberation chambers.

scatterometer An instrument carried in a **meteorological satellite** for measuring the light scattered from the surface of the sea, thus yielding information on the height and movement of waves which can be used to derive estimates of the surface wind.

schlieren photography Technique by which the flow of air or other gas may be photographed, the change of refractive index with density being made apparent under a special type of illumination. The method is used in studying the behaviour of models in transonic and supersonic wind tunnels.

Schmidt optical system An optical system, for telescopes and for projection work, which uses a spherical mirror instead of a parabolic mirror. The resulting spherical aberration is corrected by using a moulded transparent plastic plate in front of the mirror; a *catadioptric* system.

scintillation The twinkling of stars, a phenomenon due to the deflection, by the strata of the Earth's atmosphere, of the light rays from what are virtually point sources.

scintillation counter Counter consisting of a *phosphor* or *scintillator*, e.g. NaI(Tl), which, when radiation falls on it, emits light which is detected and amplified by a photomultiplier, the height of the pulses from which are proportional to the energy of the event. These pulses are further amplified and passed to a single- or multi-channel pulse height analyser, to measure the energy and intensity of the radiation.

screeching A cacophonous form of unstable combustion which can occur with rockets, and occasionally in turbine engines, causing very rapid damage due to resonance stresses on the jet pipe or nozzle. Also *howl*.

screened horn balance A **horn balance** which is screened from the airflow by the fixed surface in front of it.

SDI Abbrev. for **Strategic Defence Initiative**.

sea breeze See **land and sea breezes**.

sea clutter Clutter generated by rough sea surfaces; potential source of difficulty with sea-skimming guided missiles.

sealed pressure balance An aerodynamic balance, used mainly on ailerons, consisting of a continuous projection forward of the hinge line within a cavity formed by close fitting **shrouds** projecting rearward from the main surface, the gap between the balance and main surface being sealed to

prevent communication of pressure between lower and upper surfaces. Sometimes called a *Westland-Irving balance* after its inventor and the company which developed it.

sea-level pressure Atmospheric pressure at mean sea level deduced from the pressure measured at the level of the observing station by taking into account the theoretical effect of a column between the two levels. Use of sea-level pressure on synoptic charts reveals the true meteorological patterns which would otherwise be totally obscured by the effect of altitude on the observations.

sea-level static thrust See **static thrust**.

sea marker Any device dropped from an aircraft on to water to make an observable patch from which the drift of the aircraft may be determined. Usually filled with a fluorescent substance for use during the day, and with a flame-producing device for night use.

seaplane An aircraft fitted with means for taking off from and alighting on water. See **float seaplane, flying boat, hydroskis**.

seaplane tank A long, narrow water tank with a powered carriage carrying equipment by which the water performance of a seaplane model can be observed and precisely measured.

search radar Radar designed to cover a large volume of space and to give a rapid indication of any target which enters it.

secondary bow A *rainbow* having an angular radius of 52°, the red being inside and the blue outside, usually fainter than the primary bow. It is produced in a manner similar to the primary bow except that two internal reflections occur in the raindrops.

secondary depression A **depression** embedded in the circulation of a larger primary depression.

secondary radar A radar which involves transmission of a second signal when the incident signal triggers a **transponder** beacon.

secondary surveillance radar See **ATCRBS**. Abbrev. *SSR*.

sector display A form of radar display used with continuously rotating antennae, *not* with sector scanning. (So termed because the display uses a long persistence CRT excited only when the antenna is directed into a sector from which a reflected signal is received.) Cf. **sector scan**.

sector scan Scan in which the antenna moves through only a limited sector. (Not to be confused with a **sector display**.)

secular acceleration A non-periodic term in the mathematical expression for the Moon's motion by which the mean motion increases ca 11″ per century; caused by perturbations and by tidal friction in shallow seas.

seeing A term used by telescopic observers to describe the quality of observing conditions as influenced by turbulence in the Earth's atmosphere.

seiche An apparent tide in a lake (originally observed on Lake Geneva) due to the pendulous motion of the water when excited by wind, Earth tremors or atmospheric oscillations.

selcal An automatic signalling system used to notify the pilot that his aircraft is receiving a call. It makes constant monitoring of the receiving equipment unnecessary. A contraction of *selective calling*.

selective absorption Absorption of light, limited to certain definite wavelengths, which produces so-called absorption lines or bands in the spectrum of an incandescent source, seen through the absorbing medium. See **Kirchhoff 's law, Fraunhofer lines**.

selector valve A valve used to direct the flow of the hydraulic fluid or compressed air in a system into the desired actuating current.

semi-diameter Half the angular diameter of a celestial body.

semi-monocoque construction See **monocoque** and **stressed-skin construction**.

semi-rigid airship See **airship**.

sensation unit Original name of the *decibel*; so called because it was erroneously thought that the subjective loudness scale of the ear is approximately logarithmic.

sensing Removal of 180° ambiguity in bearing as given by simple vertical loop antenna, by adding signal from open aerial.

separated lift Lift generated by very low aspect ratio wings (usually of *delta* or *gothic* plan form) at high angles of incidence (ca 20°) through separated vortices causing large suction forces. The **jet flap** is also a separated lift system.

separation The spacing of aircraft arranged by **air traffic control** to ensure safety, which may be vertical, lateral, longitudinal, or a combination of the three.

sequence valve A type of automatic selector valve in a hydraulic or pneumatic system, much used in aircraft, whereby the action of one component is dependent upon that of another.

serein The rare phenomenon of rainfall out of an apparently clear sky.

service ceiling The height at which the rate of climb of an aircraft has fallen to a certain agreed amount (in British practice, originally, 100 ft/min, 30 m/min, but for jet

aircraft 500 ft/min, 150 m/min).

service tanks See **fuel tanks**.

servocontrol A reinforcing mechanism for the pilot's effort. It may consist of **servo tabs**.

servo tab A control surface **tab** moved directly by the pilot, the moment from which operates the main surface, the latter having no direct connection with the pilot.

SETI Abbrev. for the *Search for Extra-Terrestial Intelligence*, i.e. investigating the possibility of intelligent life in the universe other than on Earth.

Seyfert galaxy Member of a small class of galaxies with brilliant nuclei and inconspicuous spiral arms. The intensely bright nuclei possess many of the properties of **quasars**. They are strong emitters in the infrared, and are also detectable as radio and X-ray sources. Carl Seyfert discussed this morphological type in 1943.

sferics Lightning flashes or other natural electrical impulses, esp. in relation to the determination of their location by simultaneous radio direction-finding using a number of aerials. The word is derived from atmo*spherics*.

shaft turbine Any gas turbine aero-engine wherein the major part of the energy in the combustion gases is extracted by a turbine and delivered, through appropriate gearing by a shaft. See **aero-engine**.

shear wave A transverse wave without compression of the medium.

sheet The general term for aircraft structural material under 0.25 in (6 mm) thick; above that it is usually called plate.

sheet lightning Diffuse illumination of clouds by distant lightning of which the actual path of the discharge is not seen.

Shelldyne TN for synthetic fuel having a higher than normal density for expendable turbojets. See **aviation fuels**.

shell star One of a number of stars of spectral type O, B or A, which is surrounded by a shell of luminous gas giving bright emission lines.

shimmy The violent oscillation of a castoring wheel (in practice nose or tail wheel of an aircraft) about its castor axis, which occurs when the co-efficient of friction between the surface and the tyre exceeds a critical value. It is usually suppressed by a friction, spring or hydraulic device called a *shimmy damper* (see **damper**) or by a twin-tread tyre.

shipboard aircraft Any aircraft designed or adapted for operating from an aircraft carrier; special modifications are strengthened landing gear, strong points for catapulting, arrester hook and, if large, folding wings.

shock absorber See **oleo**.

shock tube A device which generates high speed flows of air over short periods of time by the passage of a shock wave down a tube; used to simulate re-entry conditions in a shock tunnel.

shock wave A surface of discontinuity in which the airflow changes abruptly from subsonic to supersonic, i.e. from viscous to compressible fluid conditions, thus causing an increase in entropy and an abrupt rise in pressure and temperature. When a shock wave is caused by the passage of a supersonic body the airflow will decelerate to subsonic conditions through another shock wave. A supersonic body normally sets up a conical shock wave with its nose, the angle (= cosec $1/M$, where M is the Mach number) becoming increasingly acute the higher the speed, together with subsidiary shock waves from projections on the body and a decelerating shock wave from its tail. The nose and tail shock waves, either attached or travelling on after the passage of an aircraft, are the source of the pressure waves causing sonic booms. Abbrev. *shock*. See aerodynamics.

shooting star See **meteor**.

shoran Abbrev. for *SHOrt RANge navigation*. A precision position-fixing system using a pulse transmitter in an aircraft or other vehicle and **transponders** at two known fixed points.

short-period comet Comet with a period of less than 150 years, the orbit of which lies entirely within the solar system, e.g. *Halley's Comet*. They have been captured into the solar system through gravitational interaction with Jupiter or Saturn.

short period stability The rapid-incidence adjustment of a stable vehicle to a disturbance.

short-period variable See **variable star**.

short take-off and vertical landing Class of *V/STOL* aircraft, usually with supersonic capability, whose downward vectored, reheated, engine efflux is too energetic to permit regular vertical take-off operations from conventional surfaces. Abbrev. *STOVL*. See **propulsive lift**.

shower Result of impact of a high-energy cosmic-ray and photons so that a very large number of ionizing particles and photons are produced, directed downwards in a narrow cone.

shower unit The mean path length for reduction of 50% of the energy of cosmic rays as they pass through matter.

shroud (1) Rearward extension of the skin of a fixed aerofoil surface to cover the whole or part of the leading edge of a movable surface, e.g. flap, elevator, hinged to it. (2) See **jet pipe-**.

shroud Streamlined covering, part of a launch system, to protect the payload during launch and to reduce aerodynamic drag; it is ejected when a sufficiently high altitude is reached. Also *fairing*.

shrouded balance An aerodynamic balance in which the area of the hinge line moves within a space formed by shrouds projecting aft from the upper and lower fixed surfaces.

shroud line Any one of the cords attaching a parachute's load to the canopy. Also *rigging line*.

sidereal day The interval of time between successive passages of the vernal **equinox** across the same **meridian**. It is 23 h 56 m 4.091 s of mean solar time. In order to prevent the *sidereal day* changing in the middle of the night, when observations are taking place, the sidereal day begins at sidereal noon, when the vernal equinox crosses the local meridian. The time for the Earth to rotate once relative to the distant stars is longer than the sidereal day by about 0.008 s, due to the **precession of the equinoxes**.

sidereal month The interval (27.321 66 sidereal days) for the Moon to complete one orbit of the Earth relative to the distant stars.

sidereal period The interval between two successive positions of a celestial body in the same point with reference to the fixed stars; applied to the Moon and planets to indicate their complete revolution relative to the line joining the Earth and Sun.

sidereal time Time measured by considering the rotation of the Earth relative to the distant stars (rather than the Sun, which is the basis of civil time). The sidereal time at any instant is the same as the **right ascension** of objects exactly on the **meridian**. See **time**.

sidereal year The interval between two successive passages of the Sun in its apparent annual motion through the same point relative to the fixed stars; it amounts to 365.256 36 days, slightly longer than the tropical year, owing to the annual precessional motion of the equinox.

siderostat An instrument designed on the same principle as the coelostat to reflect a portion of the sky in a fixed direction; applied specially to a form of telescope called the *polar siderostat*, in which the observer looks down the polar axis on to a mirror.

sideslip The component of the motion of an

aircraft in the plane of its lateral axis; generally a piloting or stability error, but also used intentionally to obtain a steep glide descent without gaining speed or to improve weapon aiming. *Angle of sideslip* is that between the plane of symmetry and the direction of motion.

sigma co-ordinates A co-ordinate system used in numerical forecasting in which the vertical ordinate σ is pressure p divided by surface pressure p_s, i.e. $\sigma = p/p_s$.

SIGMET A SIGMET message is a warning issued by an aviation meteorological watch and forecast office of the occurrence or expected occurrence of one or more meteorological hazards to aircraft including thunderstorms, severe **clear air turbulence**, marked **lee waves** and severe icing.

signal/noise ratio Difference in dB between the wanted signal and the unwanted background noise.

silo Underground chamber housing a guided missile which is ready to be fired.

simultaneity Basic consequence of *relativity*. Two events that are simultaneous according to one observer if a clock with him records the same time, may occur at different times according to another observer in another reference frame moving relative to the first.

single-entry compressor A centrifugal compressor which has vanes on one face only.

single-stage-to-orbit Space system which can launch a payload into orbit without staging, i.e. using one engine which can provide the necessary thrust throughout all the flight regimes of the complete ascent.

Sinope The ninth natural satellite of Jupiter.

Six's thermometer A form of **maximum and minimum thermometer** consisting of a bulb containing alcohol joined to a capillary stem bent twice through 180°C. A long thread of mercury is in contact with the alcohol in the stem, and this mercury moves as the alcohol in the bulb expands and contracts. Each end of the mercury thread pushes a small steel index in front of it, one of which registers the maximum temperature and the other the minimum.

ski jump ramp Curved ramp fitted at the forward flight deck of an aircraft carrier to give improved take-off performance to vectored-thrust V/STOL aircraft. The end slope is typically 8 to 15° and allows (a) shorter take-off at given weight, (b) take-off at higher weight and (c) operations in higher wave states.

skin The outer surface other than fabric of an

aircraft structure; the outer surface in a **sandwich construction**.

skin friction See **surface-friction drag**.

Skylab A manned laboratory placed in orbit in May 1973; the largest part (the workshop) consisting of a refurbished, empty S-IVb tank, originally foreseen as a stage of the Saturn rocket. Launched unmanned, it was visited by a crew of 3 astronauts on 3 occasions for stay-times of 28, 56 and 84 days. It re-entered the Earth's atmosphere in July 1979.

slab tail A one-piece horizontal tail surface, pivoted and power operated so as to serve as a stabilizing tailplane, elevator and, through a lower gearing, trimming tab.

SLAET Abbrev. for *Society of Licensed Aircraft Engineers and Technologists*.

slat An auxiliary aerofoil which constitutes the forward portion of a slotted aerofoil, the space between it and the main portion of the structure forming the slot.

sleet A mixture of rain and snow, or partially melted snow. US *ice pellets*.

slip flow The molecular shearing which replaces normal gas flow conditions at **hypersonic** velocities above *Mach 10* where the mean free path is of the same order of dimensions as the body. See **Mach number**.

slipper tank An **auxiliary tank** mounted externally, close up under wing or fuselage.

slipstream The helical airflow from a propeller, faster than the aircraft.

slip tank Same as **drop tank**.

sloshing Bulk motion of liquid propellants in their tanks when subject to accelerations, particularly at launch. This must be accounted for in the structural design of a rocket vehicle.

slotted aerofoil Any aerofoil having an air passage (or slot) directing the air from the lower to the upper surface in a rearward direction. Slots may be permanently open, closable, automatic or manually operated.

slotted flap A trailing-edge *flap* which opens a slot between itself and the main aerofoil as it is lowered or extended.

slow Measuring mode of a **sound-level meter** with a time constant of one sec.

slow-running cut-out See **fuel cut-off**.

smart Originally applied to guided and self-homing bombs for attacking point targets, now used for any device showing 'artificial intelligence' capability.

SMC Abbrev. for **Standard Mean Chord**.

snaking An uncontrolled oscillation in yaw, usually at high speed, of approximately constant, but small, amplitude (ca 1°).

Snell's law A wave refracted at a surface makes angles relative to the normal to the surface, which are related by the law $n_1 \sin\theta_1 = n_2 \sin\theta_2$. n_1 and n_2 are the refractive indices on each side of the surface, and θ_1 and θ_2 are the corresponding angles. The two rays and the normal at the point of incidence on the boundary lie in the same plane.

snow Precipitation in the form of small ice crystals, which may fall singly or in flakes, i.e. tangled masses of snow crystals. The crystals are formed in the cloud from water vapour.

snow stage That part of the condensation process taking place at temperatures below 0°C so that water vapour condenses directly to ice.

S/N ratio Abbrev. for **Signal/Noise ratio**.

soaring The art of sustained motorless flight by the use of thermal upcurrents and other favourable air streams.

sodar Acoustic method for finding atmospheric layers. In principle like sonar, used in meteorology.

software A general term for all types of computer *program* and their associated *documentation*. Cf. **hardware**.

solar antapex The point on the celestial sphere diametrically opposite to the *solar apex*.

solar apex The point on the celestial sphere towards which the solar system as a whole is moving at the rate of 20 km/s. It is located in the constellation Hercules in equatorial coordinates, RA 271° and declination +30° approximately.

solar array *Solar panel.* Bank of solar cells, mounted on a panel structure, and extended from a satellite, which converts solar energy into electrical energy using *photovoltaic* conversion. Also *solar battery, solar cell array*.

solar cell Photoelectric cell using silicon, which collects photons from the Sun's radiation and converts the radiant energy into electrical energy with reasonable efficiency. Used in satellites and for remote locations lacking power supplies, e.g. for radiotelephony in the desert.

solar constant The total electromagnetic energy radiated by the Sun at all wavelengths per unit time through a given area normal to the solar beam at the mean distance of the Earth. Its current value is 1.37 kW per m^2. It is not, in fact, truly constant, and variations of order 0.1% are detectable.

solar corona The outermost layer of the Sun's atmosphere, visible as a halo of light during a total **eclipse**. Its temperature is 0.5–2 million K, and it is a strong source of

X-rays. Its visible radiation can be studied at any time with a *coronagraph*.

solar day See **apparent-, mean-**.

solar eclipse See **eclipse**.

solar energy utilization Energy from the Sun can be converted (1) to thermal energy using a working fluid, (2) to electrical energy using a photovoltaic cell, (3) to mechanical energy using the radiation pressure (solar sailing in space) and (4) to chemical energy by photosynthesis.

solar flare Bright eruption of the Sun's surface associated with sunspots, causing intense radio and particle emission.

solar flocculi The name given to small bright and dark markings on the solar chromosphere as seen on calcium or hydrogen spectroheliograms.

solar granulation The mottled appearance of the Sun's photosphere, small bright granules (about 1000 km in diameter) being seen in rapid change against the darker background.

solar panel See **solar array**.

solar parallax The mean value of the angle subtended at the Sun by the Earth's equatorial radius. It cannot be measured directly, but is derived from Kepler's laws when the distance between the Earth and any other planet is known. The value of 8.80″ for the solar parallax has been in use for half a century, but recent measures of the distance of Venus by radar in 1961 gave the value 8.794 05″, corresponding to a value of 149.6×10^6km (92 957 130 miles) for the astronomical unit.

solar radio noise Radio emission from the atmosphere of the Sun, which is investigated by many techniques in **radio astronomy**. Sudden bursts are associated with solar **flares** in particular. Sunspots, and the activity in their vicinity, are also strong sources of radio emission.

solar rotation The rotation of the Sun takes place in the same direction as the orbital motion of the planets. The rotation is nonuniform, taking 24.65 days at the Sun's equator, but increasing to about 34 days near the poles. Because of the Earth's motion, the equatorial solar rotation has a synodic period of 27 days, and this interval is apparent in the recurrence of magnetic storms, aurorae etc.

solar sailing Means of movement in space using the pressure of the **solar wind** on a large surface or solar sail, suitably deployed.

Solar System The term designating the Sun and the attendant bodies moving about it under gravitational attraction; comprises nine major planets, and a vast number of asteroids, comets and meteors.

solar wind A *continuous* plasma stream,

mainly of protons and electrons, emitted by the Sun that moves at *hypervelocities* (250–800 km/s) and pervades all interplanetary space. It affects the Earth's magnetic field and causes accelerations in the tails of comets.

solidity A measure of the effective area of a propeller, usually measured as the ratio of the sum of the blade chords to the circumference at a standard radius. See **chord line**.

solid propellant Rocket propellant in solid state, usually in caked, plastic-like form, comprising a fuel-burning compound of fuel and oxidizer.

solid-state Pertaining to a circuit, device or system, which depends on some combination of electrical, magnetic or optical phenomena within a material which is usually a crystalline semi-conductor. Loosely applied to all active devices or circuits which do not rely on valves or tubes.

solstice (1) One of the two instants in the year when the Sun reaches its greatest excursion north or south of the equator. (2) One of the two points on the **ecliptic** midway between the **equinoxes**.

sonar Abbrev. for *SOund NAvigation and Ranging*. See **asdic**.

sone Unit of loudness equal to a tone of 1 kHz at a level of 40 dB above the threshold of the listener.

sonde Small telemetering system in satellite, rocket or balloon.

sonic boom Noise phenomenon due to the shockwaves projected outwards and backwards through the atmosphere from leading and trailing edges of an aircraft travelling at **supersonic speed**. The waves are discontinuities of atmospheric pressure with an N-shaped pressure time history having a sharp peak followed by a similar negative pressure. They are heard as a characteristic double report which may be of sufficient intensity to cause damage to buildings, etc.

sonic fatigue The deterioration (cracks) and failure which are caused in materials (metal, concrete etc.) by strong stress fluctuations of sound waves of high intensity. Sometimes happens in mechanical systems if the eigenfrequency is excited. Important for spacecraft and aircraft.

sonobuoy Equipment dropped and floated on the sea to pick up aqueous noise, and transmit a bearing of it to aircraft; three of such bearings enable the aircraft to 'fix' the source of underwater noise, e.g. from submarines.

sonogram Three-dimensional representation of a sound signal, the three co-ordinates being frequency, time and intensity. Intensity is often represented by shading.

sound The periodic mechanical vibrations and waves in gases, liquids and solid elastic media. Specifically the sensation felt when the eardrum is acted upon by air vibrations within a limited frequency range, i.e. 20 Hz to 20 kHz. Sound of a frequency below 20 Hz is called infrasound and above 20 kHz ultrasound.

sound absorption factor See **acoustic absorption factor**.

sound analyser One which measures each frequency, amplitude or phase of sound.

sound energy density See **energy density of sound**.

sound field Volume filled with a medium in which sound waves propagate.

sounding balloon A small free balloon carrying a meteorograph, used for obtaining records of temperature, pressure and humidity in the upper atmosphere.

sounding rocket Unmanned rocket-powered vehicle used for research purposes (atmosphere, astronomy, microgravity etc.) which does not go into Earth orbit, but follows a suborbital trajectory.

sound intensity Flux of sound power through unit area normal to the direction of propagation. If p is the acoustic pressure and v the velocity of the medium particles in the direction of propagation, the intensity is the time average of the product pv.

sound intensity level At any audio-frequency, the intensity of a sound, expressed in decibels above an arbitrary level, 10^{12} W/m^2, which is equivalent, in air, to a pressure of 20 μPa.

sound level Loosely used for **sound pressure level** and **sound intensity level**.

sound-level meter Microphone-amplifier-indicator assembly which indicates total intensity in decibels above an arbitrary zero. Also includes suitable weighting networks and time averaging. See **fast** and **slow**.

sound locator Apparatus for determining the direction of arrival of sound waves, particularly the noise from aircraft and submarines. See **asdic**.

sound pressure Fluctuating mean component of the pressure in the medium containing a sound wave, as opposed to the constant component, e.g. atmospheric or hydrostatic pressure. Also *acoustic pressure*.

sound pressure level Sound pressure which is expressed in **decibels** relative to a reference pressure which is taken as 20 μPa or 2×10^{-5} N/m^2.

sound probe Usually a very small

microphone to minimize the disturbance of the sound field which is being measured. It is often equipped with a fine tube which is inserted in the field.

sound ranging Determination of locality of a source of sound, e.g. from guns, by simultaneously recording through spaced microphones and making deductions from the differences of times of arrival.

sound reduction index Same as **transmission loss.**

sound reflection Return of sound waves from discontinuities (surface, tube end etc.). See **reflection co-efficient.**

sound-reflection factor The percentage of energy reflected from a discontinuity; proportional to the square of the **reflection co-efficient.**

sound spectrograph An electronic instrument which makes **sonograms.**

Southern Cross A striking constellation of the southern hemisphere, visible only in latitudes below 30° N. It is a cruciform group of 4 stars, having the 2 bright stars α and β Centauri some way to the east, which makes it easy to identify.

southern oscillation A slow fluctuating exchange of air between the eastern tropical Pacific on the one hand, and the Indian Ocean and Indonesia on the other, with a corresponding negative correlation between annual mean pressure values over the two areas. The interval between corresponding points in successive cycles varies from 1 to 5 years, and the oscillation is linked to variations in sea-surface temperature and the pattern of rainfall.

Soyuz A three-man vehicle used by the USSR for ferrying crews to and from its orbital stations. Part of the spacecraft is a re-entry module used for the return to Earth, employing parachutes and rockets for the actual landing.

space See panel on p. 166.

space commercialization Exploitation of space and space-produced products by industry for profit.

spacecraft See panel on p. 167.

spacecraft stability For spacecraft, attitude control is important. Normally a spacecraft is spin-axis or 3-axis stabilized. Attitude can be corrected by **momentum wheels**, reaction wheels, gyroscopes, magnetic torques, **nutation** dampers or by reaction control thrusters. Attitude can be determined by Earth horizon, **limb**, Sun, star and magnetometer sensors. The interface between sensor and actuator is made by the *attitude control system electronics.*

space environment The extra-terrestrial

conditions existing in a particular region between Earth and distant bodies. Specifically, it involves the phenomena of vacuum, fields, particles and related effects.

Spacelab Reusable orbital research laboratory which extends the Space Shuttle capabilities and is carried in its cargo bay; modular in construction it comprises a manned module and unmanned pallets, permitting module-only, pallet-only or mixed modes to be used. The module is cylindrical, of diameter 4 m and consists of either one or two 2.7 m long sections enclosed between end cones. Each of five pallets is 3 m long and 4 m wide.

space parallax The difference in bearing between a moving object, such as a machine in flight, and the direction of arrival of the sound waves emitted by it. This arises from the comparable velocity of flight with that of the propagation of sound waves.

space programmes See panel on p. 168.

space qualified Used of space systems, subsystems and components to denote that they meet the specifications relevant to their use in space.

space research Investigation of the space environment, its effect on things and its use as a vantage point for viewing the Earth and deep space. Generally, the research is performed with the aid of a space system.

Space Shuttle See panel on p. 171.

space station See panel on p. 172.

space suit Specially designed suit which, when donned, allows an astronaut to operate in a space environment. The design includes the provision of a pressurized oxygen supply, and provides for temperature control and the purification of exhaled gases. Sometimes *pressure suit.*

space system The total assembly of space-related hardware and software; it can refer to the launch system plus payload or a spacecraft plus payload and includes the related ground segment.

space velocity The rate and direction of a star's motion in space of three dimensions, as deduced from its observable components (*a*) in the line of sight by the spectroscope, and (*b*) perpendicular to the line of sight by proper motions.

span The distance between the wing tips of an aircraft.

span loading The gross weight of an aircraft or glider divided by the square of the span.

spar A main spanwise member of an aerofoil or control surface. The term can be applied either to individual beam(s) designed to resist bending, or to the box structure of

---

space

That region of near-vacuum surrounding all bodies in the Universe. In practice, one talks of *near-Earth space* as that extra-atmospheric region just surrounding our planet, *interplanetary space* between planets, *interstellar space* between stars and *intergalactic space* between galaxies. Although normally regarded as void, space is not entirely empty. Interplanetary space is permeated by electromagnetic radiation and cosmic rays, as well as force and magnetic fields, and contains electrons, protons, neutral hydrogen and small particles of dust. Heavier atoms and molecules such as sodium and formaldehyde have been detected in interstellar space. The density of this matter is extremely low and it is less for each region as one progresses outward from near-Earth space (the density for interplanetary space is about 100 particles per cc and for interstellar space ten times less). Also, it varies considerably with the local conditions. The vacuum of space is much greater than that obtainable on Earth.

Two areas of human *space activity* can be recognized: exploration and exploitation. In the former, the physical nature of the Universe and its contents are investigated, and the properties of space and its constituents defined. Space exploration is performed with the aid of satellites and probes, suitably instrumented for measurements *in situ* or the observation of more distant objects. The exploitation of space involves the use of space and space-related phenomena in our everyday lives. This includes using man-made objects like weather, observation and communications satellites. Later, the exploitation of space could well lead, among other things, to the harnessing of solar energy and obtaining material from extra-terrestrial bodies.

Space flight can be achieved using unmanned or manned spacecraft. To date, unmanned vehicles have been put into various low-Earth orbits (LEO), or geosynchronous orbits (GEO) and deep-space elliptical orbits on near-Earth space missions. They have been used to explore the planets of our solar system and interplanetary space itself. Manned exploration has been confined to LEO and the Moon, although one can envisage future missions involving manned interplanetary flights and a human presence in GEO. These two modes of space investigation, conducted on a purely national or international basis, can only lead to greater knowledge of our Universe, a better use of its resources and a more harmonious existence of mankind on Earth.

---

spanwise vertical webs, transverse ribs, and skin which form a torsion box.

spar frame A specially strong transverse fuselage, or hull, frame to which a wing spar is attached.

spatial filtering The removal of part of the optical diffraction pattern by opaque masks so that the high frequency components in the image formation are removed or enhanced. Removal rounds sharp edges in the image; enhancement sharpens edges.

special rules zone A 3-dimensional space, under **air-traffic control**, wherein aircraft must obey special instructions.

specific excess power Thrust power available to an aircraft in excess of that required to fly at a particular constant height and speed, thus being usable for climbing, accelerating or turning.

specific heat capacities of gases Gases have two values of specific heat capacity: c_p, the s.h.c. when the gas is heated and allowed to expand against a constant pressure, and c_v, the s.h.c. when the gas is heated while enclosed within a constant volume.

specific heat capacity The quantity of heat which unit mass of a substance requires to raise its temperature by one degree. This definition is true for any system of units, including SI, but whereas in all earlier systems a unit of heat was defined by putting the specific heat capacity ($s.h.c$) of water equal to unity, SI employs a single unit, the joule, for all forms of energy including heat, which makes the s.h.c. of water 4.1868 kJ/kg K. See **mechanical equivalent of heat**.

specific humidity The ratio of the mass of water vapour in a sample of moist air to the total mass of the air.

spacecraft

A vehicle containing all the necessary subsystems to support a payload for the performance of a particular space mission. By its very nature, it must be self-supporting at all times and mission events are controlled by on-board software programs or ground intervention. The subsystems provide for such items as the electrical power for the operation of the individual subsystems themselves and for the payload; communications to and from the ground; thermal control of the spacecraft, its subsystem elements and its payload; data handling of input data and commands; propulsion for course changes and reaction control thrusters; attitude control and guidance, and control for executing the necessary trajectory and pointing manoeuvres to satisfy mission requirements. The subsystem equipment is mounted on and/or contained within structures designed to cope with launch and mission eventualities. A spacecraft may be manned or unmanned. In the former, additional subsystems, such as life support and specific crew-related items, are also needed.

Micro-electronics, computers, new high-strength materials and similar highly sophisticated devices have allowed the production of relatively small, lightweight and efficient spacecraft, in the design of which safety and reliability must be paramount. The overall mass must also be controlled during the design process so as to be consistent with the available launch system and mission objectives.

The spacecraft requires some support from the ground for mission control and data dissemination, and this is provided by a dedicated ground segment.

Today spacecraft are commonplace in the exploration and exploitation of space. The following table gives some typical examples of those which have been used or are in use at the present time.

Spacecraft name	Origin	Purpose	Size (m)	Mass (kg)	Power (kw)	Orbit
Meteosat	Europe	Meteor-ology	2.1 dia × 3.2	320	0.20 (solar)	GEO
Voyager	US	Science (planetary)	1.8 dia bus 3.7 dia antenna	815	1.20 (radio-isotope)	Inter-planetary
Intelsat 6	Inter-national	Communi-cations	3.8 dia × 6	2250	2.0 (solar)	GEO
Landsat 5	US	Remote sensing	4 × 2 structure, 6 × 2.3 solar panels	2000	2.0 (solar)	LEO (Sun synchron-ous)
Suisei	Japan	Science (cometary)	1.4 dia × 0.7	125	0.10 (solar)	Inter-planetary
Buran	USSR	Manned/ unmanned space trans-portation.	36 length, 24 span, 5.6 dia of fuselage	105 000	20 (?) (fuel cells ?)	LEO

space programmes

Principal contributors. Major world space activities are centered in the US, USSR, ESA, Japan and China. These five account for over 95% of the world's investment in space. Each has its own particular reasons for involvement. The two superpowers, the US and USSR and, to some extent, China, spend a considerable amount of their space budgets on military applications. At the same time the US and USSR undertake space activities for utilization, commercial and prestige reasons. ESA, applying the well-established technical base of Europe, seeks an independent approach to space, confining its involvement to strictly peaceful uses. Similarly, the Japanese efforts reflect a growing desire for a non-military national space programme based solely on Japanese technology. Chinese interests are linked to the need of an emerging China.

The US civilian efforts are centered on NASA which has been responsible for many advances in all fields of space research. The unmanned programme has resulted in a complete range of launchers, communications satellites (e.g. the early SYNCOM), remote sensing systems (e.g. LANDSAT) and considerable advances in the space sciences, as evidenced by the spectacular Voyager mission. Currently, many of the products of this basic work have been taken over by other agencies (e.g. NOAA) and commercial organizations, while NASA remains the main centre for new systems and technology (e.g. the NIMBUS meteorological instruments). NASA's manned space programme started with Mercury and we have witnessed the Apollo Moon landings, the Space Shuttle and now preparation for a Space Station with its international partners ESA, Japan and Canada. The Space Shuttle Challenger disaster, after the euphoria of Apollo, affected the whole US space programme. This led to some indecision in policy and a resurgence of expendable launchers. Recently there has been a noticable shift in space funding to the military field, of which little is known, but a considerable effort is devoted to surveillance and the much-vaunted Strategic Defence Initiative or Star Wars.

Less is known of the activities of the other great space power, the USSR, but recent developments (*glasnost*) permit a considerably more transparent view. Much work has been performed under the guise of the military but GLAVCOSMOS has now been identified as a major civilian commercial agency. Developments have paralleled those of NASA in the fields of launch systems, communications and remote sensing. The USSR programme is particularly celebrated for its planetary science (Venus and Mars probes) and space stations (SALYUT and MIR). Its many 'firsts' include Sputnik and man-in-space, and over 2000 successful launches have been registered. Interest in the planets and commitment to long duration space missions has led to much speculation on a future manned Mars mission. International co-operation in such a programme is a distinct and attractive possibility.

ESA is dedicated to the peaceful exploitation of space. It co-ordinates European space activities and executes programmes on behalf of its 13 member states. The major contributors are France, the Federal Republic of Germany, Italy and the United Kingdom. Its two types of programmes, mandatory (particularly science) and optional (e.g. Columbus), reflect both the need for a core programme and the indivdual desires of its Member States. Its present programmes cover science, telecommunications, remote sensing and space station activities and the Ariane family of launchers is well known. Present emphasis is on manned space flight with Spacelab operational and Ariane V, Hermes and Columbus elements for the international 'Freedom' space station and a European Data Relay Satellite system under devlopment.

continued on next page

space programmes (contd)

The Japanese space effort is through its two main agencies, NASDA and ISAS. A systematic and comprehensive approach will ensure autonomy in its own space affairs by the end of the century. Its present emphasis is on developing applied and scientific satellites, its own launcher H-II, the JEM laboratory for Freedom and a future spaceplane. China has only recently emerged as a space power, its policies being geared to military and economic considerations. Its main strength is in launchers as evidenced by the Cheng Zen, or CZ, family with a capability for low and geosynchronous Earth orbit. China is particulaly experienced in recoverable capsules.

An encouraging philosophy pervades all the strategies of the major space powers. This is a strongly expressed desire for international co-operation which bodes well for the future.

Other main contributors. Two other groups of nations wishing to pursue space research can be identified: industrialized countries with relatively small available budgets and emerging nations. The former are mainly Western European countries which possess capabilities in advanced technology and which, as well as contributing jointly through ESA, also conduct their own national space activities. Eastern European countries are closely tied to the USSR through INTERCOSMOS which was created to co-ordinate purely research efforts between the Soviets and their neighbours. The emerging countries interested in the possibilities of space tend to be financially poor but potentially rich through latent resources. The development of such countries can be considerably aided by the use of space for communications and remote sensing, for example. Both groups are usually engaged in their own, purely national, programmes and in joint programmes with other countries, normally on a bilateral basis. The bias towards the latter tends to depend on the financial resources available.

In Western Europe, France is the premier space country with a total investment (ESA and national programme) of about 0.12% of its GNP. This compares with about 0.5% in the US and about 0.03% in the UK. France has pioneered many space developments in diverse areas, such as launch vehicles (ARIANE), remote sensing (SPOT) and manned space transportation (Hermes). Next in order of financial commitment come the Federal Republic of Germany, Italy and the UK. As examples of their achievement, one can cite the German D.1 Spacelab mission, the Italian Tethered Satellite System and the British involvement in the INTELSAT programme. Other ESA Member States have relatively small national programmes. The Swedish-led Tele-X communications satellite for Nordic regions is particularly noteworthy. In addition many European countries support international bodies like EUMETSAT and EUTELSAT for the operation of regional satellites.

Canada is a country whose sheer size and richness in natural resources have been instrumental in the development of communications (e.g. ANIK) and remote sensing (e.g. RADARSAT) space programmes which have been achieved through co-operation with the US and ESA.

India leads the emerging countries with a space budget in excess of those for the UK and Canada, its interest stemming from its large size and population with poor roads and many village communities. It is a fine example of how a country with a relatively small GNP can build up a meaningful space programme through co-operation with other countries, in this case particularly the US, USSR and ESA. India's space programmes mainly under the aegis of ISRO, include the INSAT series of communications satellites and the IRS remote sensing satellites. Non-Indian launchers have been used (Space Shuttle,

continued on next page

space programmes (contd)
Ariane and Vostok) but every attempt is being made to develop an indigenous launch capability.

Another example of how a country's size and political and social conditions can lead to the use of space-based systems, is Brazil. Although similar activities are carried out by other South American states, Brazil is the only one with a fully developed space policy. It is presently developing a national launcher, its own Earth observation satellite and a domestic telecommunications network based on BRASILSAT satellites.

Other nations throughout the World have some involvement in space. For example, some Arab States use the ARABSAT communications network, Australia has its AUSAT and Indonesia its Palapa for similar uses. Very recently, Israel has entered the space field with its own satellite, OFFEQ-1. Also, many countries use the INTELSAT, INTERSPUTNIK and INMARSAT communications networks for various applications.

specific impulse A measure of rocket performance, defined as the thrust in kg (force) obtained from one kg (mass) of propellant burned for one second. It is denoted by I_{sp} and has the units of seconds. It is related to the exhaust velocity (V_e) by: $I_{sp} = V_e/g_0$, where g_0 is the acceleration due to gravity at the Earth's surface. A high value of I_{sp} and, hence, V_e is desirable.

speckle interferometry A technique using the principle of interference of light which enables very small angles such as the diameters of stars, to be measured directly.

spectral distribution curve The curve showing the relation between the radiant energy and the wavelength of the radiation from a light source.

spectral line Component consisting of a very narrow band of frequencies isolated in a spectrum. These are due to similar quanta produced by corresponding electron transitions in atoms. The lines are broadened into bands when the equivalent process takes place in molecules.

spectral types The Harvard classification of stars according to their spectra, giving a graded list represented by the letters (W) O B A F G K M S (R N), which represents a sequence (called the *main sequence*) of descending temperature, the O type stars being hot, white and gaseous, while the cooler M type show molecular band spectra.

spectre of the Brocken The shadow of an observer cast by the Sun on to a bank of mist. The phenomenon, often seen from a hilltop, may present the illusion that the shadow is a gigantic form seen through the mist. Also *Brockenspectre*.

spectrograph Normally used of spectroscope designed for use over wide range of frequencies (well beyond visible spectrum) and recording the spectrum photographically. The *mass spectrograph* separates particles of different specific charge in an analogous manner to the separation of spectrum lines of an optical spectrum.

spectroheliogram The recorded result of an exposure on the Sun by the spectroheliograph.

spectroheliograph An instrument for photographing the Sun in monochromatic light. It consists essentially of a direct-vision spectroscope, with a second slit instead of an eyepiece, which can be set so that only light of a desired wavelength passes through it on to a photographic plate.

spectrohelioscope An instrument in principle the same as the spectroheliograph, but adapted for visual use by the employment of a rapidly oscillating slit which, by the persistency of vision, enables an image of the whole solar disk to be viewed in light of one wavelength; it also detects the velocities of moving gases in the solar atmosphere by an adjustment called the *line-shifter*.

spectrometer Instrument used for measurements of wavelength or energy distribution in a heterogeneous beam of radiation.

spectrophotometer Instrument for measuring photometric intensity of each colour or wavelength present in an optical spectrum.

spectroradiometer A spectrometer for measurements in the infrared.

spectroscope General term for instrument (spectrograph, spectrometer etc.) used in spectroscopy. The basic features are a slit and collimator for producing a parallel beam of radiation, a prism or grating for 'dispersing' different wavelengths through differing angles of deviation, and a telescope,

---

space shuttle

A ground-to-orbit and return transportation system capable of lifting payloads into orbit and used for easy access to and departure from a permanent station or platform in space. In the latter role it can be employed for construction, crew transfer and logistic support. Two shuttles are presently in operation, the US Space Shuttle and the Soviet Energia-launched Shuttle.

The manned part of the US Space Shuttle (carrying up to 7 persons) is the Orbiter and is launched by rocket propulsion using its own engines which burn propellants (liquid oxygen and hydrogen) stored in a large external tank together with two solid-fuel boosters. The latter drop away after two minutes and the tank is then jettisoned just before orbit is reached. The orbital manoeuvering system (OMS), burning a **hypergolic** mixture of monomethyl hydrazine and nitrogen tetroxide, is used to insert the Orbiter into orbit, where the desired orientation is achieved by small rocket engines using the same hypergolic mixture. On completion of the mission, a retrorocket firing causes the reusable Orbiter to re-enter the atmosphere and it lands horizontally like a glider. The Space Shuttle can also be used for in-orbit experiments (when Spacelab may be employed), for the capture and repair of satellites in orbit and the return of payloads to Earth. It will be used particularly for the building and resupply of the international Space Station Freedom.

The Soviet Shuttle launches its orbiter (the first is called Buran) on an Energia heavy-lift booster. It is very similar to the US version but has no main engines of its own, relying on Energia which burns liquid hydrogen and oxygen. In particular the sizes of the cargo bay are identical (18.3 m long and dia 4.6 m) and the payload capabilities are similar at around 30 000 kg to low-Earth orbit. The orbiting vehicle has OMS engines for final orbit insertion, re-entry is initiated by retrorocket and it lands horizontally. The Energia booster, like the US large tank, is burned up in the atmosphere but the strap-on boosters are recovered by parachute. A crew of up to 10 persons can be carried; however, it was first flown fully automatically.

---

camera or counter tube for observing the dispersed radiation.

spectroscopic binary A binary star whose components are too close to be resolved visually, but which is detected by the mutual shift of their spectral lines owing to their varying velocity in the line of sight.

spectroscopic parallax The name given to the indirect method of deducing the distances of stars too far away to have detectable annual parallaxes; it involves the inferring of their absolute magnitudes from spectroscopic evidence which then, combined with the observed apparent magnitudes, gives their distances.

spectroscopy The practical side of the study of spectra, including the excitation of the spectrum, its visual or photographic observation, and the precise determination of wavelengths.

spectrum Arrangement of components of a complex colour or sound in order of frequency or energy, thereby showing distribution of energy or stimulus among the components. A mass spectrum is one which shows the distribution in mass, or in mass-to-charge ratio of ionized atoms or molecules. The mass spectrum of an element will show the relative abundances of the isotopes of the element.

spectrum analyzer (1) Electronic spectrometer usually working at microwave frequencies and displaying energy distribution in spectrum visually on a cathode-ray tube. (2) Pulse-height analyser for use with radiation detector.

spectrum colours When split up into a spectrum, white light is shown to be composed of a continuous range of merging colours; red, orange, yellow, green, blue, indigo, violet.

spectrum line Isolated component of a spectrum formed by radiation of almost uniform frequency. Due to photons of fixed energy radiated as the result of a definite electron transition in an atom of a particular element.

spectrum locus Curved line on the CIE chromaticity diagram representing the monochromatic hues.

space station

Several manned modules and/or unmanned platforms, launched separately but joined together in orbit to form a base which permits a permanent presence in space for exploration and exploitation of the environment, and as a staging post and refurbishment centre for other space activities. Two space stations are currently in operation or planned: the Soviet Mir and the International Space Station Freedom. Salyut (USSR, first use 1971) and Skylab (US, first use 1973) are examples of previous long duration manned orbital stations.

MIR

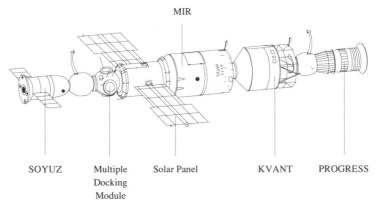

SOYUZ   Multiple   Solar Panel   KVANT   PROGRESS
           Docking
           Module

Soviet Space Station, MIR.

The Soviet Mir has a crew of 3–6 and has been developed from the earlier Salyut stations. It consists (see drawing above) of a central living module of length 13.15 m and max dia of 4.2 m with access to 6 docking ports, of which 5 are in the multiple docking module, and which allow specialized modules such as the Kvant astrophysics laboratory to be added. The station can grow, therefore, by adding more modules. Power is provided by photovoltaic cells on panels which can also be augmented as necessary. Crew rotation is by Soyuz spacecraft and re-supply by the unmanned logistics vehicle, Progress. The Soviet Shuttle will also be used for crew/logistic purposes when it is fully operational. Mir operates between 300 and 400 km in an orbit inclined at 51°. It is likely that a larger version of Mir will be launched by Energia in the mid-nineties.

The international Space Station Freedom (see drawing on next page), to be operational in the late 90s, consists of four manned modules attached to a single truss structure and interconnected by resource nodes. Solar panels attached to the 150 m truss can generate 75 kw of electrical power, although more use of solar dynamic generators will be made as the technology becomes available. The module diameters are 4 m with lengths up to 12 m. The habitation and laboratory modules are provided by the US, the Japanese Experimental Module (JEM) by Japan and the Columbus Attached Laboratory by the European Space Agency (ESA), with Canada providing servicing and manipulating devices. Up to 8 persons can live in the habitation module and the three laboratories are equipped for space-related experiments, particularly in microgravity science. JEM has its own logistics module and the Columbus Attached Laboratory, based on Spacelab experience, is part of ESA's Columbus programme. The latter also includes a man-tended Free-Flying Laboratory and a Polar Platform. The Columbus Polar Platform is unmanned and will work in conjunction with a similar US platform, complementing the Space Station. The US will ensure logistic supplies for the Space Station, via the Space Shuttle and its logistic modules. Freedom will operate in a circular orbit at an altitude of 450 km and an inclination of 28.5°.

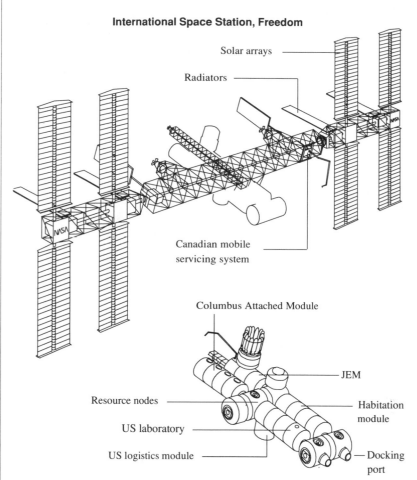

International Space Station, Freedom

Solar arrays

Radiators

Canadian mobile
servicing system

Columbus Attached Module

JEM

Resource nodes

Habitation
module

US laboratory

US logistics module

Docking
port

The pressurized modules (lower drawing) are slung below the main boom at the position indicated by the outline in the upper drawing. JEM is the Japanese Experimental Module. External payloads can also be attached to the main boom.

Freedom, as described here, is subject to adequate funding appropriations by the US government. This is by no means sure and the configuration and participation by the non-US partners is subject to change.

monochromatic hues.

specular reflectance Quotient of reflected to incident luminous flux for a polished surface.

specular reflection General conception of wave motion in which the wavefront is diverted from a polished surface, so that the angle of the incident wave to the normal at the point of reflection is the same as that of the reflected wave. Applicable to heat, light,

radio and acoustic waves.

specular transmittance See **transmittance**.

speed (1) The rate of change of distance with time of a body moving in a straight line or in a continuous curve (cf. *velocity*, a vector expressing both magnitude and direction). Units or speed are metres per second ($m\ s^{-1}$), feet per second ($ft\ s^{-1}$), miles per hour (mph), kilometres per hour expressed in revolutions per minute, radians per second etc.

speed bulges Streamlined bulges on the fuselage, or nacelles near the trailing edge of the wing, which meet the requirements of **area rule** for a smooth area distribution where it is impractical to give the fuselage a wasp-waist to reduce transonic **wave drag**.

speed of light The constancy and universality of the speed of light *in vacuo* is recognized by *defining* it (1983) to be exactly $2.997\ 924\ 58 \times 10^8$ ms^{-1}. This enables the SI fundamental unit of *length*, the metre, to be defined in terms of this value.

speed of rotation In a rotating body, the number of rotations about the axis of rotation divided by the time (see **speed**). Units are revolutions/second, minute or hour, or radians/ second, minute or hour. The axis of rotation may have a translatory speed of its own. See **moment of inertia**.

SPF/DB Abbrev. for *Super Plastic Forming/Diffusion Bonding*. See **super plastic forming**.

spherical astronomy The branch of astronomy concerned with the position of heavenly bodies regarded as points on the observer's celestial sphere. It comprises all diurnal and seasonal phenomena and the precise assignment of co-ordinates to the heavenly bodies. See **astrometry**.

spike Initial rise in excess of the main pulse in transmission.

spill burner A gas turbine burner wherein a portion of the fuel is recirculated instead of being injected into the combustion chamber.

spill door Auxiliary door mounted in an engine nacelle which opens to spill excess air provided by the intake but not needed by the engine. Designed to minimize drag; often spring loaded.

spin A continuous, but not necessarily even, spiral descent with the mean **angle of incidence** to the relative airflow above the **stalling** angle. In a *flat spin* the mean angle of incidence is nearer the horizontal than the vertical, while in an *inverted spin* the aircraft is actually upside down.

spin avoidance system One designed to detect the onset of a spin (primarily high angles of attack) and warn the pilot or make an angular correction of pitch.

spin chute See **antispin parachute**.

spinner A streamlined fairing covering the hub of a propeller and rotating with it.

spinning tunnel See **vertical wind tunnel**.

spin stabilized Refers to a spacecraft whose attitude is stabilized by causing it to spin about a rotationally symmetric axis.

spiral galaxy The second commonest morphological type, characterized by a large nuclear bulge of stars, surrounded by a pair of conspicuous spiral arms. These arms contain gas, dust and newly-formed star clusters. Our own Galaxy is a spiral. See **galaxy**.

spiral instability That form of lateral instability which causes an aircraft to develop a combination of side-slipping and banking, the latter being increasingly too great for the turn. This causes the machine to follow a spiral path.

split compressor An **axial-flow turbine** compressor in which front and rear sections are mounted on separate concentric shafts (being powered by separate turbines) as a means of increasing the pressure ratio without incurring difficulties with *surge*. Also *two-spool compressor*.

split flap A trailing edge flap in which only the lower surface of the aerofoil is lowered.

spoiler A device for changing the airflow round an aerofoil to reduce, or destroy, the lift, of which there are 3 principal types: (1) a small fixed spanwise ridge on the wing-root leading edge along the line of the **stagnation point** which improves lateral stability at the stall by ensuring that it starts at the root; (2) controllable devices at, or near, both wing tips which, by destroying lift on the side raised, impart a rolling moment to the aircraft; (3) small-chord spanwise flaps on top of the wing of a sailplane which can be raised to destroy a large part of the lift so as to make landing of the lightly-loaded aircraft more positive. Similar devices are often now fitted to jet aircraft so that by destroying lift at touchdown better braking can be achieved. See **thrust spoiler**

sponson A short, wing-like projection from a flying-boat hull to give lateral stability on the water.

spot beam A concentration of radio waves by an antenna-reflector so that a particular area is highly illuminated with radiation, resulting in the concentration of power over a small area and a high signal strength.

spring tab A *balance tab* connected so that its angular movement is geared to the compression or extension of a spring incorporated in the main control circuit. Its primary purpose is to reduce the effort required by the pilot to overcome the air loads on the main control surface resulting from high airspeeds. Cf. **servo tab**, **trimming tab**.

spring tides Those high tides occurring when the Moon is new or full, at which times the Sun and Moon are acting together to produce a maximum tide.

Sputnik A series of USSR artificial satellites; Sputnik I, launched in October 1957, was the first ever man-made object to orbit about the Earth.

squall A temporary sharp increase in the wind speed, lasting for some minutes.

squid A dynamically stable condition of a fully-deployed parachute canopy which will not fully distend.

SST Abbrev. for *SuperSonic Transport aircraft.*

stabilator See all-moving tail.

stability and control in aircraft See panel on p. 176. For spacecraft see **spacecraft stability.**

stability derivatives Quantitative expressions for the variation of forces and moments on an aircraft due to disturbances from steady motion.

stability limits (1) The extreme angles of incidence (maximum or minimum) to which a taxiing, taking-off or landing seaplane can be trimmed withour porpoising. (2) Now refers to the range of centre of gravity positions between which an aircraft can fly with acceptable safety.

stabilizer In US, *horizontal stabilizer* is *tailplane,* and *vertical stabilizer* is *fin.* See **automatic-.**

stable equilibrium The state of equilibrium of a body when any slight displacement increases its potential energy. A body in stable equilibrium will return to its original position after a slight displacement.

stage A section or part of a **launch system** which fires for a certain time only and then is separated from the main system; when more than one stage is used, the technique is termed *staging.*

stagger The horizontal distance between the leading edges of the wings of a multiplane as projected vertically. If the upper plane is ahead of the lower, stagger is positive; if behind it is negative.

staging The principle of increasing the velocity achieved by a **launch system** and its payload by using more than one propulsive stage. Tandem (nose-to-tail) or parallel (side-by-side) staging may be employed, each stage being jettisoned after the fuel has been expended, thus increasing the mass ratio and therefore, the efficiency of the whole system.

stagnation point The point at or near the nose of a body in motion in a fluid where the flow divides and where fluid pressure is at a maximum, and the fluid is at rest. Theoretically there is another stagnation point near the trailing edge.

stagnation temperature The temperature which would be reached if a flowing fluid were brought to rest adiabatically, which is

almost applicable in supersonic flight for the leading edges and air intakes, where the air in the boundary layer of a body is drastically and rapidly decelerated. Also *total temperature.*

stall The progressive breakdown of the lift-producing airflow over an aerofoil, which occurs near the angle of maximum lift.

stalling speed The airspeed of an aircraft at which the wing airflow breaks down.

stall-warning indicator A device fitted to those aircraft which do not give any positive warning of the approach of the stall by **buffeting.** Usually operated by the change of pressure and movement near the **stagnation point** near the stall, warning may be audible, visual or by a *stick-shaker* (or forward-pushing electric motor). See **stick pusher.**

standard atmosphere (1) The unit of pressure, defined as 101 325 N/m², equivalent to that exerted by a column of mercury 760 mm high at 0°C. Symbol *atm.* (2) See **International Standard Atmosphere** which is a hypothetical atmosphere used as the basis for assessing the performance of altimeters, aircraft etc.

standard beam approach A system of radio navigation which provides an aircraft with lateral guidance and marker-beacon indications at specific points during its approach. Abbrev. *SBA.*

standard mean chord The average chord, i.e. gross wing area divided by the span.

standard time The civil time in any of the *time zones* established by international agreement. These are about 15° of longitude wide, equal to one hour. Within a zone all civil clocks are set the same standard time or rather local *solar time.* Zones usually differ by a whole hour, but there are a few cases of half-hour zones (e.g. South Australia). See **time.**

stand-off bomb A small fast powered unmanned aircraft or rocket containing a nuclear warhead, released from a bomber to fly many hundreds of miles to the target. It is automatically piloted and navigated, usually by a **Doppler navigator** and/or **inertial navigation system.** Propulsion can be by ramjet, rocket, turbojet or in combination.

star See panel on p. 177.

star chart The name given to a systematic and accurately made map of the heavens in which the star positions are generally plotted according to equatorial co-ordinates.

star cluster See **galactic cluster, globular cluster, open clusters.**

star magnitude See **magnitude.**

star-streaming A phenomenon, discovered from analysis of observed stellar motions

stability and control in aircraft

Stability is the quality whereby any deviation from steady motion tends to decrease. Thus a given type of steady motion is *statically* stable if an aircraft will return to that state of motion after a disturbance without intervention from the pilot. *Dynamic* stability deals with the history of the motion and the rate at which it dies out. An aircraft has three axes about which its stability is defined with three associated degrees of freedom: *angular, normal displacement* and *change of velocity*.

A *control* is the device through which a human command is transmitted across the man/machine interface. It is the means, therefore, which cen also be automatic by which the orientation and path may be changed. Stability and control are considered together as they have strong interactions. For example, the more stable an aircraft the more control force is needed to produce a desired change in flight and *vice versa*. In advanced automatic flight control systems stability and control are inherently treated together.

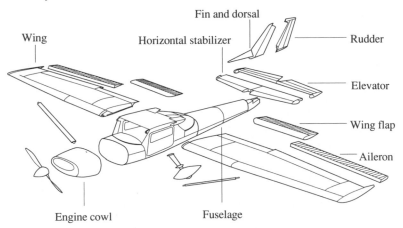

The diagram shows an exploded view of a light aircraft together with its control surfaces. The *ailerons* move differentially to give a rolling motion about the aircraft's horizontal axis. The wing *flaps* move together and can be adjusted to alter the lift of the aircraft as a whole. The *elevator* governs the pitching motion of the aircraft and the *rudder* the movement in the horizontal plane about the vertical axis.

There are a number of other control surfaces which are sometimes used: a *butterfly tail* has only two upwardly inclined surfaces which can be rotated in unison to perform the function of an elevator or differentially to create a yawing moment like that of a conventional rudder; *flaperons* are flaps which create a rolling moment by differential movement; a *foreplane* has hinged control surfaces on either side of the front fuselage which produce trimming pitching moments; *spoilers* are normally-retracted spanwise devices fitted to the upper wing surface which, in use, rise perpendicular to the wing surface to destroy lift and increase drag; a *taileron* is an all-moving two-piece tailplane which can produce a rolling moment by differential movement like that of an aileron; a *trim control surface* is an auxiliary winglet arranged to give a pitching moment for longitudinal control purposes.

See **active control systems, artificial feel, artificial stability, control column, control configured vehicle, Dutch roll, fly-by-wire, manoeuvre demand system, neutral point, phugoid oscillation, short period, spiral stability, stability derivatives, stability limits.**

star

Stars are self-luminous spheres of gas which produce energy through nuclear fusion reactions in their cores. Most of the mass we see in the universe resides in stars, and the study of their properties and evolution is central to modern astronomy. The nearest star, the **Sun**, is average and typical, being a **main sequence** dwarf star, burning hydrogen. It is only 150 million km away and can thus be studied in more detail than any other star.

Beyond the solar system, the nearest star is *Proxima Centauri*, about 4.3 light-years away. This is 27 million solar diameters: space is very empty, and within our vicinity only one part in 10^{22} is occupied by stellar material. About 5780 stars are visible to the naked eye and most of these are 100–1500 light-years from Earth. There are about 10^{11} stars in our Galaxy.

Stars are frequently found in characteristic configurations. *Double stars* can be chance alignments of two stars at different distances along the line of sight. Of greater interest are the *binary stars* in which the two orbit around a common centre of gravity. These can be: *visual*, if resolvable by telescope; *spectroscopic*, if only resolvable from their spectra; *eclipsing*, if they orbit across our line of sight. These latter are very important as they are the only systems where star masses can be measured. Close doubles can evolve into *interacting binaries* in which one member becomes a *red giant* and transfers mass to its dwarf companion. These can be observed as **nova** explosions and *X-ray stars*. A handful of exotic binaries include collapsed **neutron stars** and perhaps **black holes**.

Sometimes stars are found together in clusters. *Open clusters* (or *galactic clusters*) have a few hundred members, are concentrated in the galactic plane and are relatively young. *Globular clusters* have a striking spherical appearance, 10^5–10^6 stars and are up to 10^{10} years old.

New stars form by gravitational contraction within clouds of interstellar matter and during this phase the protostar may be a strong source of infrared radiation. Eventually the central pressure and temperature are high enough to sustain thermonuclear reactions in the core, and the star proper begins its evolution on the zero-age *main sequence*.

Star names follow several systems, many of the brighter being known by their Arabic designations. A more systematic method, developed from 1600, assigns Greek letters in order of decreasing brightness to the star in each constellation. Thus Deneb (Arabic) is also Alpha Cygni, the brightest star in Cygnus. Variable stars have two Roman letters followed by the constellation, as in RR Lyrae. For fainter stars the number assigned in a large catalogue is used and for the faintest objects positional co-ordinates alone suffice. See **Harvard classification, Hertzsprung-Russell diagram, magnitude, redshift-distance relation.**

(after removing the effects of the observer's own motions), by which the stars are found to have two preferential directions of motion, one towards the point RA 90°, declination 15° south, and the other towards RA 285°, declination 64° south; the first stream contains about 60% of the observed stars. The effect is due to the rotation of the Galaxy.

static discharge wick Wicks, usually of cotton impregnated with metallic silver, or of nichrome wire, fitted at the trailing edges of an airplane's flight control surfaces, by which static electricity is discharged into the atmosphere.

static instability *Hydrostatic instability.* Atmospheric state such that a parcel of air moved from its initial level experiences a hydrostatic force tending to remove it further from this level.

static inverter A non-rotating device for converting d.c. current to a.c. supply, usually of high voltage, for radio and instrument services.

static jet thrust See **static thrust**.

static line A cable joining a parachute pack

to the aircraft, so that when the wearer jumps, the parachute is automatically deployed.

static margin See panel on **stability and control.**

static pressure The pressure at any point on a body moving freely with a fluid in motion; in practice, the pressure normal to the surface of a body moving through a fluid.

static-pressure tube A tube with openings placed so that when the air is moving past it the pressure inside is that of still air. See **Pitot static tube, static vent.**

static stability Positive static stability in an aircraft means that if it is disturbed from a trimmed speed, the disturbance will be reduced.

static thrust The net thrust (kN or lb.s.t) of a jet engine at *International Standard Atmosphere* sea level and without translational motion.

static vent An opening, usually in the fuselage, found by experiment, where there is minimum **position error** and which is used instead of the **static-pressure tube.**

stationary orbit See **geostationary.**

stationary points Those points in the apparent path of a planet where its direct motion in right ascension changes to retrograde motion or *vice versa.*

station keeping The manoeuvres necessary to adjust a geostationary satellite's orbit so that its position in space is correct, its ground coverage does not vary and data transmission is optimized; the manoeuvres are usually effected by small jets or rockets.

stator The row of fixed, radially disposed aerofoils which forms an essential part of the dynamics of an axial compressor or axial turbine.

stator blade A small fixed aerofoil, usually of thin highly-cambered section, and of approximately parallel chord, mounted in the outer case of an axial compressor or turbine. See **exhaust-.**

steady state One in dynamic equilibrium, with entropy at its maximum.

steady-state cosmology See panel on p. 179.

stealth The technology of reducing the observable characteristics of military aircraft and missiles. Means include reducing size, noise, IR emissions from engines and from hot surfaces, radar reflections from intakes and between surfaces.

Stefan-Boltzmann law The total radiated energy from a black body per unit area per unit time is proportional to the fourth power of its absolute temperature, i.e. $E = \sigma T^4$ where σ (Stefan-Boltzmann constant) is

equal to 5.6696×10^{-8} W m^{-2} K^{-4}.

stellar energy See **carbon cycle, proton-proton chain.**

stellar evolution The sequence of events and changes covering the entire life cycle of a star. The principal stage of evolution is the nuclear burning of hydrogen to form helium, with a consequential release of energy. Eventually the hydrogen in the core is exhausted, and the star becomes a *red giant.* In the final stages of evolution there are several paths; the formation of a **white dwarf**, a **neutron star** or a **supernova.** See **Hertzsprung-Russell diagram, main sequence.**

stellar interferometer The device, developed by Michelson, by means of which, when fitted to a telescope, it is possible to measure the angular diameters of certain giant stars (all of which are below the limit of resolution of even the largest telescopes) by observations of interference fringes at the focus of the telescope.

stellar magnitude See **magnitude.**

stellar population See **population types.**

stellar wind Radial outflow of material from the atmosphere of a very hot star, analogous to the **solar wind.**

step The step discontinuity in the bottom of a flying-boat hull, to facilitate take-off from the water surface by allowing the forebody to plane and the afterbody to be clear of the forebody wake. Typical values range from 4% to 12% of the maximum beam. Low values are likely to induce **porpoising.**

Stevenson screen A form of housing for meteorological instruments consisting of a wooden cupboard having a double roof and louvred walls, these serving to protect the instruments from the Sun and wind while permitting free ventilation. The base of the screen should be about 1 metre above the ground.

stick force The force exerted on the **control column** by the pilot when applying aileron or elevator control.

stick-force recorder A device attached to the control column of an aircraft by which the pilot's effort is measured and transmitted to a recording instrument.

stick pusher A device fitted to the control column of some high-performance aircrafts with swept-back wings which moves the column sharply forward to prevent a stall. See **stall-warning indicator.**

stick shaker See **stall-warning indicator.**

stiffener A member attached to a sheet for the purpose of restraining movement normal to the surface. Usually of thin drawn or extruded light-alloy, L, Z or U section, attached by riveting, metal-bonding or spot welding.

steady-state cosmology

Cosmology aims to describe what the universe is 'really' like. To do so it builds models. In historical times, folklore, theology or philosophy provided acceptable cosmologies. Modern cosmology combines astronomical observations with the **general theory of relativity** to give a mathematical picture of the structure and evolution of the universe. There are two major classes of cosmological model: those in which the universe evolves, so that its appearance will change with time, and those in which the universe remains forever as a static entity. The first class has given rise to **Big Bang** cosmology and the second to *steady-state cosmology*. Currently almost all astrophysicists choose the Big Bang picture to interpret their results, but steady-state theories were once widely accepted.

Einstein was the first steady-state cosmologist. Although the general theory of relativity predicted a dynamic universe because of the interplay of gravitational forces, Einstein inserted a completely arbitrary cosmological repulsion factor into the field equations in order to stabilize the universe. Once the **redshift-distance relation** was established (1929) this idea was dropped and static models of the universe faded into the background.

In 1948, Herman Bondi and Thomas Gold revived the idea of a steady-state universe by postulating that matter is continuously created, in the expanding universe, to fill the voids left as the galaxies receded from each other. This notion of spontaneous creation was not new: it was suggested by James Jeans (1929) and Pascual Jordan showed (1939) how to modify general relativity so that matter could be created. Fred Hoyle extended this work by constructing a model in which the creation rate matches the dilution caused by expansion. The actual form assumed by the newly-created matter does not emerge from the theory.

The steady-state theory takes the view that the universe is infinitely old and will continue into the infinite future. The continuous creation of new matter everywhere means that this expanding universe looks more or less the same from all vantage points and at all times. New galaxies form alongside older ones. This has a crucial consequence which is that the universe far away should be more or less like the local universe and thus provides the key to test the theory. The great controversy in the 1960s between Hoyle and radio-astronomer Martin Ryle over the interpretation of the number of galaxies seen at different distances led to considerable public awareness of these rival theories of cosmology. Moreover the discovery of *cosmic background radiation* or **microwave background**, whose presence contradicts the steady-state theory, has not yet eliminated the attractions of the latter in the public mind.

An important outcome of the research on the steady-state theory is frequently overlooked. In the absence of a big bang, the only place where elements heavier than helium can be manufactured is inside stars. Hoyle's brilliant work on **nucleosynthesis**, motivated by the need to explain the origin of the elements in the steady-state universe, has survived intact as a major contribution to modern astrophysics.

See **integral-, stringer.**

stiffness control In a mechanically vibrating system, the condition in which the motion is mainly determined by the stiffness of the retaining springs and negligibly by the resistance and mass of the system.

stiffness criterion The relationship between the stiffness, strength and other structural properties which will prevent *flutter* or dangerous aero-elastic effects.

still air range The theoretical ultimate range of an aircraft without wind and with allowances only for take-off, climb to cruising altitude, descent and alighting.

stochastic noise See **random noise.**

Stokes' Law (1) The resisting force offered by a fluid of dynamic viscosity η to a sphere of radius r, moving through it at steady velocity V is given by

$$R = 6\pi\eta rV$$

whence it can be shown that the **terminal velocity** of a sphere of density ρ, falling under gravitational acceleration g through fluid of density ρ_0 is given by

$$v = \frac{2gr^2}{9\eta}(\rho - \rho_0).$$

Applies only for viscous flow with Reynolds number less than 0.2. (2) Incident radiation is at a higher frequency and shorter wavelength than the re-radiation emitted by an absorber of that incident radiation.

Stokes layer Very thin boundary layer along an interface between a fluid and a solid in which the velocity and temperature fluctuations in a sound wave are reduced because of friction and thermal conductivity respectively. Important for sound absorption. Also *AC-boundary layer*.

STOL *Short Take-Off and Landing*, a term applied to aircraft with high-lift devices and/or deflected engine thrust enabling them to operate from small airstrips, 1000 ft (300 m) or less being the criterion.

storm-centre The position of lowest pressure in a cyclonic storm.

STOVL See **short take-off and vertical landing**.

straighteners See **honeycomb**.

strain When a material is distorted by forces acting on it, it is said to be in a state of *strain*, or *strained*. Strain is the ratio

$$\frac{\text{deflection}}{\text{dimension of material}}$$

and thus has no units. The main types of strain are direct (tensile or compressive) strain:

$$\frac{\text{elongation or contraction}}{\text{original length}}$$

shear strain:

$$\frac{\text{deflection in direction of shear force}}{\text{distance between shear forces}}$$

volumetric (or bulk) strain:

$$\frac{\text{change of volume}}{\text{original volume}}$$

strapdown Any device mounted to an aircraft so that its attitudes change with those of the aircraft. Describes a class of navigation system like those using a **fibre-optic gyro** in contrast to ordinary gyroscopes which maintain a constant attitude.

Strategic Defence Initiative Abbrev.

SDI. A military programme commonly referred to as *Star Wars*. The intent of SDI is to provide a defensive shield based on satellite, laser and high-energy particle technology for the destruction of hostile ballistic missiles when the latter are still in the atmosphere after launch.

stratocumulus Grey and/or whitish patch, *sheet* or *layer* of *cloud* which almost always has dark parts, composed of *tessellations*, rounded masses, rolls etc., which are non-fibrous (except for **virga**) and which may or may not be merged; most of the irregularly arranged small elements have an apparent width of more than 5°. Abbrev. *Sc*.

stratopause Top of the **stratosphere**, at about 50 to 55 km above the surface of the Earth.

stratosphere Region of the atmosphere between the **tropopause** and the **stratopause**, in which temperature generally increases with height.

stratus Generally grey *cloud layer* with a fairly uniform base, which may give drizzle, ice prisms or snow grains. When the Sun is visible through the cloud, its outline is clearly discernable. Stratus does not produce halo phenomena, except possibly, at very low temperatures and sometimes it appears in the form of ragged patches. Abbrev. *St*.

streamline A line in a fluid such that the tangent at any point follows the direction of the velocity of the fluid particle at the point, at a given instant. When the streamlines follow closely the contours of a solid object in a moving fluid, the object is said to be of streamline form. Thus *streamlines* on a chart show the direction of the horizontal wind at some particular level.

streamline flow Path taken by fluid molecules or minute suspended particles. Usually qualified as **laminar flow** or **turbulent flow**. See **viscous flow**.

streamline motion The steady motion of a fluid in **laminar flow** past a body with neither abrupt changes in direction nor close curves.

streamline wire High-tensile steel wire of elliptical, not true streamline, cross-section, used to reduce the drag of external bracing wires. Fitted principally to biplanes and some early types of monoplane.

stress The force per unit area acting on a material and tending to change its dimensions, i.e. cause a *strain*. The *stress* in the material is the ratio of force applied to the area of material resisting the force i.e.

$$\frac{\text{force}}{\text{area}}$$

The two main types of stress are *direct* or *normal* (i.e. *tensile* or *compressive*) *stress* (symbols σ or *f*) and *shear stress* (symbols τ, *f* or *q*). The usual units are kPa, MPa, lbf/in^2, tonf/in^2, kN/m^2, MN/m^2, bar or hbar. (See **newton, bar.**) *To stress* may also mean 'to calculate stresses'; thus a structure has been *stressed* when its strength has been verified by calculation.

stressed-skin construction The general term for aircraft structures in which the skin (usually light alloy, formerly plywood, occasionally plastic such as Fibreglass, and in supersonic aircraft, titanium alloy or steel) carries a large proportion of the loads. In the more elementary forms, the framework may take bending and shear, with a thin skin transmitting torsion, but when of developed form, the skin is thick enough to support bending loads in the form of tension and compression in the respective surfaces. Since 1950 the principle of thick skin has been developed in the form of 'sculpturing' to vary the thickness to suit the local loads and to incorporate integral stiffeners, either by machining or by acid etching. See **monocoque.**

stringer A light auxiliary member parallel with the main structural members of a wing, fuselage, float or hull, mainly for bracing the transverse frames and stabilizing the skin material. See **stiffener.**

string theory A theory in fundamental physics that attempts to construct a model of elementary particles from one-dimensional entities rather than the zero-dimensional 'points' of conventional particle physics. See **superstring theory.**

strobe lighting An anti-collision lighting system based on the principle of a capacitor-discharge flash tube. A capacitor is charged to a very high voltage which is then discharged in a controlled sequence as a high-intensity flash of light, usually blue-white, through xenon-filled tubes located at wing tips and tail of an aircraft.

structural damping See **damping.**

structure (1) Framework or ensemble of rigid elements which is designed to withstand a variety of mechanical and thermal influences (e.g. thrust forces, bending moments, aerodynamic heating effects) during launch and flight of a spacecraft, and provide protective support for its subsystems and payload. (2) In aircraft, the rigid construction wich is designed to house the crewe, payload, fuel and subsystems and made to withstand the forces of aerodynamics, inertia, propulsion and landing, together with the reactions arising from the

release of armament.

structure-borne sound Sound in solid bodies, as opposed to sound in gases (e.g. air-borne sound) and sound in liquids (e.g. water sound).

STS Abbrev. for *Space Transportation System.* Usually refers to the Space Shuttle but actually includes **Spacelab**, inter-orbit stages carried by the Orbiter and the Tracking and Data Relay Satellite System (*TDRSS*), together with the supporting ground segment.

stub plane A short length of wing projecting from the fuselage, or hull, of some types of aircraft to which the main planes are attached.

Stüve diagram An **aerological diagram** with rectangular axes temperature and (pressure)$^{(\gamma-1)/\gamma}$ where γ is the ratio of the specific heats of a perfect gas.

subjective noise meter A noise meter for assessing noise levels on the phon scale, the loudness of the noise level being measured by ear with the adjusted reference tone, 1000 Hz. See **objective noise meter.**

subsonic Said of an object or flow which moves with a speed less than that of sound.

subsonic speed Any speed of an aircraft where the airflow round it is everywhere below Mach 1. See **Mach number.**

substellar point The point on the Earth's surface, regarded as spherical, where it is cut by a line from the centre of the Earth to a given star; hence the point where the star would be vertically overhead, the point whose latitude is equal to the star's declination. Applied also to the Sun and Moon as *subsolar point* and *sublunar point* respectively.

subsystem Constituent part of a system which performs a particular function; thus, *electrical power subsystem, data handling subsystem* etc. It is the sum of the coherent subsystem performances which provides a certain system capability.

sudden warming A rapid rise of temperature of the polar **stratosphere** of up to 50 K in a few days occurring in winter or early spring. It is associated with a breakdown of the winter polar stratospheric vortex and may be either temporary or, in spring, permanent. See **general circulation of the middle atmosphere.**

summer solstice See **solstice.**

Sun as a star See panel on p. 182.

sun pillar A vertical column of light passing through the Sun, seen at sunset or sunrise. It is caused by reflection of sunlight by horizontal ice crystals.

sunseeker A photo-electric device mounted

Sun as a star

The Sun is 150 million km from Earth, so close that it can be studied in far greater detail than any other star (see solar constants below). The layer we see is the *photosphere*, the lowest layer of the solar atmosphere, with a temperature of about 5800 K. High resolution observations show individual convection cells, formed through the phenomenon of *granulation*. The upper photosphere oscillates with a 5-minute period and these *solar oscillations* give information about the deep interior. In the spectrum of the photosphere there are hundreds of thousands of **Fraunhofer lines**, caused by the absorption of the inner radiation by the upper cooler regions of the upper solar atmosphere. Their careful study provides information about the abundances of chemical elements.

Above the photosphere lies the *chromosphere*, which is about 10 000 km thick and is visible as a pinkish glow during a total solar eclipse. Much of its matter is arranged in spiky cylinders called *spicules*. The temperature is about 15 000 K, high enough to excite emission lines, seen as the *flash spectrum*, during eclipse. The outermost layer is the *corona* where temperatures reach from 0.5 to 2 million K. At optical wavelengths plumes and streamers of matter extend far away from the Sun and X-ray telescopes have shown cool and dark zones in the corona, termed *coronal holes*.

The solar interior reaches 15 000 000 K at the core (see diagram), high enough for the proton-proton reaction in which protons fuse to form helium-4 nuclei, the source of solar energy, releasing enormous quantities of neutrinos. A major problem in solar astronomy is that there are far fewer neutrinos than any plausible model of solar evolution can explain. Because of the presence of free electrons the interior is opaque to photons; one consequence of this is the long diffusion time, about 1 million years, for the energy released today to work its way to the photosphere. The present age of the Sun, 5 billion years, means that about half of the available protons have already been processed in nuclear fusion reactions.

Much activity is superimposed on the basic structure of the Sun of which **sunspots** are the most obvious signs. The *solar wind* is a steady flow of matter from the corona, analogous to evaporation, and is responsible for keeping the tails of comets always pointing away from the Sun.

The Sun is located about 8 kiloparsec from the centre of the Galaxy and is about 10 parsec above the galactic plane. Its orbital velocity of 250 km/sec implies that it travels round the galaxy in about 220 million years. The table gives the main solar constants.

Solar constants:

Radius	6.96×10^5 km
Mass	1.989×10^{30} kg
Mean density	1.409×10^{32} kg m^{-3}
Surface gravity	2.74×10^2 m s^{-1}
Escape velocity	617.7 km s^{-1}
Apparent magnitude	-26.74
Absolute magnitude	$+4.83$
Luminosity	3.83×10^{26} W
Effective temperature	5770 K
Central temperature	1.5×10^7 K
Spectral class	G2V

Cross-section of the Sun's interior

in rockets or space vehicles, in which an instrument such as a spectrograph can be directed constantly to the Sun.

sunshine recorder The *Campbell-Stokes recorder* consists of a glass sphere arranged to focus the Sun's image on to a bent strip of card, on which the hours are marked. The focused heat burns through the card and the duration of sunshine is read off from the length of the burnt track.

sunspot A disturbance of the solar surface which appears as a relatively dark centre (*umbra*), surrounded by a less dark area (*penumbra*); spots occur generally in groups, are relatively shortlived and with few exceptions are found in regions between 30° N. and S. latitude. They are cooler regions (about 4000 K) in the photosphere where an intense magnetic field (up to one tesla) emerges from the interior. The number of spots visible varies over the 11-year *solar cycle*. About two years after the peak of each cycle, the polarity of the Sun's dipole field reverses. The seat of solar activity is the Sun's differential rotation; it rotates about once per month but the equatorial regions take three days less than the poles. The resultant twisting amplifies the internal magnetic field which erupts through the spots. Violent *flares* are associated with the largest spots which eject particles: protons as solar **cosmic rays**, as well as electrons and ions that cause the *aurora* and *magnetic storms* when they strike the Earth's upper atmosphere. See **Sun as a star**.

Sun-synchronous orbit A near-polar orbit synchronized to the motion of the Sun over the Earth's surface and used particularly for Earth observations; it ensures repeated passages over the same point on the Earth's surface at the same local time and with the same lighting conditions.

superaerodynamics Aerodynamics that occur at very low air densities above 100 000 ft, i.e. for spacecraft on ascending and re-entry trajectories. The mean free path of the molecules is long compared with vehicle length. The physics is described by *free-molecule-flow* and Newtonian aerodynamics. *Magnetohydrodynamic* features are likely to be significant.

supercharging (1) In aero-engines, maintenance of ground-level pressure in the inlet pipe up to the rated altitude by means of a centrifugal or other blower. Necessary for flying at heights at which the air pressure is low and normal aspiration would be insufficient. (2) In other IC engines, the term is used synonymously with *boosting*.

super-circulation A form of boundary layer control, creating and controlling high lift in which air (usually derived from a jet engine compressor) is blown supersonically over the leading edge of a plain flap so that it carries the main airflow downward below the actual surface as an invisible extension. May also involve blowing over the leading edge. See **jet flap**.

superconducting gyroscope Frictionless gyroscope supported in vacuum through magnetic field produced by currents in superconductor.

supercooled, subcooled, water Water which continues to exist as a liquid at temperatures below 0°C.

supergiant star Star of late type and abnormal luminosity, such as Betelgeuse and Antares; they are of enormous size and low density.

supergravity A particular version of a **supersymmetry** theory which postulates *gravitons* and *gravitinos* as carriers of the gravitational force.

superheat The increase (positive) or decrease of the temperature of the gas in a gas-bag as compared with the temperature of the surrounding air. Similarly, *superpressure*.

supernova Novae with an absolute magnitude −14 to −16; 3 have been recorded in our own Galaxy, and about 50 more in spiral nebular. The violent outburst results from the gravitational collapse of a massive star, the outer layers being ejected, while the core is left as a **neutron star**.

super plastic forming *Diffusion bonding.* Method of manufacturing by joining parts of structures together at high temperature and pressure. Abbrev. *SPF/DB*.

superpressure See **superheat**.

super-refraction Refraction greater than standard refraction.

supersonic Faster than the speed of sound in that medium. Erroneously used for ultrasonic. See **Mach number, ultrasonic**.

supersonic boom Shock wave produced by an object moving supersonically. At a large distance from the object the time history of the pressure has the shape of an N and is therefore called *N-wave*.

supersonic speed Applies to aircraft when its speed exceeds that of local sound. Applies to airflow anywhere when local speed exceeds that of sound.

supersonic wind tunnel A wind tunnel in which the stream velocity in the working section exceeds the local speed of sound.

super stall This phenomenon appeared with the adoption of high tailplanes for swept-wing jet aircraft. When the disturbed airflow from a stalled wing renders the tail controls inoperative, the aircraft will remain in a stable, substantially level attitude, while descending very rapidly. Recovery is by releasing the tail parachute to raise the tail clear of the wing wake so that the elevators again become operative.

superstring theory A version of **string theory** that incorporates ideas of supersymmetry in which all classes of elementary particles are placed on an equal footing. The astronomical context is that these classes of theory may have applied to matter in the very early universe. See **Big Bang**.

supersymmetry Theory which attempts to link all four fundamental forces, and postulates that each force emerged separately during the expansion of the very early universe.

surface acoustic wave Acoustic wave, which may have frequencies corresponding to the microwave bands, travelling along the optically polished surfaces of a piezoelectric substrate, at a velocity about 10^{-5} that of light. Used in microwave components and amplifiers. Abbrev. *SAW*.

surface boundary layer The atmospheric layer, extending to a height of about 100 m, in which the motion is controlled predominantly by the presence of the Earth's surface. It forms the lowest part of the **friction layer**.

surface-friction drag That part of the drag represented by the components of the pressures at points on the surface of an aerofoil, resolved tangential to the surface. Also *skin friction*.

surface loading The average force per unit area, normal to the surface, on an aerofoil under specified aerodynamic conditions.

surface wave Wave on the surface of liquids or solid bodies which have dimensions large compared with the wavelength. The amplitude is maximal at the surface and decays exponentially towards the interior of the body. The displacement of the medium particles is both longitudinal and transversal.

surface wind The wind at a standard height of 10 m (33 ft) above ground. Differs from the **geostrophic wind** and the **gradient wind** because of friction with the Earth's surface.

surge Unstable airflow condition in the compressor of a gas turbine due to a sudden increase (or decrease) in mass airflow without a compensating change in pressure ratio.

surveillance radar A plan position indicator radar showing the position of aircraft within an air traffic control area or zone.

sweat cooling Cooling of a component by the evaporation of a fluid through a porous surface layer; used for high-performance gas turbine blades or hypersonic vehicles.

sweep The angle, in plan, between the normal to the plane of symmetry and a specified spanwise line on an aerofoil. Most commonly, the quarter-chord line is used, but leading and trailing edges are sometimes stipulated. Sweep increases longitudinal stability by extending the centre of pressure and delays compressibility drag by reducing the chordwise component of the airflow. *Sweepback*, the more usual, is the aft displacement of the wings and *forward sweep* the opposite.

sweepback Aircraft wings making an acute angle with the fuselage, the wing tips being towards the tail.

swing The involuntary deviation from a straight course of an aircraft while taxiing, taking-off or alighting.

swing-wing See **variable sweep**.

swirl sprayers Fuel injectors in a gas turbine which impart a swirling motion to the fuel.

swirl vanes Vanes which impart a swirling motion to the air entering the flame tube of a gas-turbine combustion chamber.

swivelling propeller See **propeller**.

symmetrical flutter See **flutter**.

synchronous orbit Circular orbit of a satellite of a body, moving in the same direction and with the same period as the parent body. If the latter is the Earth, the term *geosynchronous* is used.

synchrotron Machine for accelerating charged particles to very high energies. The particles move in an orbit of constant radius guided by a magnetic field. The acceleration is provided at one point in their orbit by a high frequency electric field whose frequency increases to insure that particles of increasing velocity arrive at the correct instant to be further accelerated. Proton synchrotrons can produce energies greater than 200 GeV. Electron synchrotrons give energies up to 12 GeV. See **betatron, cyclotron**.

synchrotron radiation (1) Electrons accelerated in a *synchrotron* produce a very intense, highly collimated, polarized beam of electromagnetic radiation, whose wavelength ranges continuously from 10^{-2} mm to 10^{-2} nm. Used with a monochromator, it is an important source for research purposes. (2) A theory of the origin of cosmic radio waves; it is suggested that the electrons moving in an orbit in a magnetic field are accelerated as in a synchrotron, but on a vastly larger scale.

synodic month The interval (amounting to 29.530 59 days) between two successive passages of the Moon through conjunction or opposition respectively; therefore, the period of the phases. Also *lunation*.

synodic period An interval of time between two similar positions of the Moon or a planet, relative to the line joining the Earth and Sun; hence the length of time from one conjunction or opposition to another, and the period of the phases of the Moon or a planet.

synoptic chart See **weather map**.

synoptic meteorology That part of the science of meteorology which deals with the preparation of a *synoptic chart* of the observed *meteorological elements* and, from consideration of this chart, the production of a weather forecast.

synthetic aperture radar An instrument for the all-weather sensing of the Earth which uses microwaves. An effective increase in the antenna aperture is obtained by integrating individual 'looks' at the same area as the spacecraft or aircraft moves over it. The radar emits pulses continuously and at a precisely controlled frequency such that the transmitted power is coherent. All the echoes are recorded and processed off-line and extremely fine resolution is attainable.

system engineering A logical process of activities which transforms a set of **requirements** arising from a specific mission objective into a full description of a system which fulfills the objective in an optimum way. It ensures that all aspects of a project have been considered and integrated into a consistent whole.

syzygy A word applied to the Moon when in conjunction or opposition.

T

tab Hinged rear portion of a flight control surface. See **balance-, servo-, spring-,** and **trimming-**.

tail See **tail unit.**

tail boom One or more horizontal beams which support the tail unit where the fuselage is truncated; commonly used for cargo aircraft to facilitate loading trucks and bulky freight through full-width rear doors.

tail chute A parachute mounted in an aircraft tail. Cf. **antispin parachute** and **brake parachute.**

tail cone The tapered streamline **fairing** which completes a fuselage or **tail boom.**

tailerons Two-piece tailplane whose two halves can operate either together, performing the function of an elevator or differentially, causing rolling moments as does an aileron.

tail-first aircraft An aircraft in which the horizontal stabilizer (i.e. tailplane) is mounted ahead of the main plane; common on pioneer aircraft and reintroduced for supersonic flight; sometimes *canard* because of their similarity to a planform of a duck in flight.

tail heaviness That state in which the combination of forces acting upon an aircraft in flight is such that it tends to pitch up.

tailless aircraft An aircraft, or glider, in which longitudinal stability and control in flight is achieved without a separate balancing horizontal aerofoil. This balance is achieved by **sweep,** and many *delta-wing* aircraft are tailless because their sharp angle of sweepback renders a tailplane unnecessary.

tailplane A horizontal surface, fixed or adjustable, providing longitudinal stability of an aircraft or glider. See **stabilizer.**

tail rotor See **auxiliary rotor.**

tail slide A difficult aerobatic manoeuvre in which an aircraft is pulled up into a zoom and allowed to slide backward along its longitudinal axis after the vertical speed drops to zero. Flying surfaces are esp. strengthened to withstand the reverse airflow encountered.

tail unit Hindmost parts of an aircraft, the horizontal *tailplane,* fin rudder and any strakes of an aircraft. Would include oblique (V) surfaces as in butterfly tail. Also *empennage.*

tail wheel landing gear That part of the alighting gear taking the weight of the rear of the plane when on the ground. It consists of a shock-absorber carrying a wheel (*tail wheel*) or a shoe (*tail skid*).

take-off rocket A rocket, usually jettisonable, used to assist the acceleration of an aircraft. Cordite rockets were introduced for naval aircraft during the Second World War and replenishable liquid-fuel rockets, some with controllable thrust, thereafter. Sometimes referred to as a *booster rocket,* although strictly this is for the acceleration of missiles. Abbrev. *RATOG* (Rocket-Assisted Take-Off Gear). US abbrev. *JATO* (Jet-Assisted Take-Off).

talk-down See **ground-controlled approach.**

tangential wave path That of a direct wave, tangential to the surface of the Earth and which is curved by atmospheric refraction.

tank vent pipe The pipe leading from the air space in an aircraft fuel, or oil, tank to atmosphere, for equalizing changes in pressure due to alterations in altitude; in *aerobatic* aircraft a non-return valve is fitted to prevent liquid escaping when inverted.

target Reflecting object which returns a minute portion of radiated pulse energy to the receiver of a radar system.

target strength $T.$ In radar it is defined in dB by $T = E-S+2H,$ where E = echo level, S = source level and $2H$ = transmission loss.

TAS Abbrev. for **True AirSpeed.**

taxi-channel markers See **airport markers.**

taxi track A specially prepared track on an aerodrome used for the ground movement of aircraft. See **perimeter track.**

taxi-track lights Lights so placed as to define manoeuvring areas and tracks.

TBO, tbo See **time between overhauls.**

TDRSS Abbrev. for *Tracking and Data Relay Satellite System.* A satellite communication system for relaying data (at rates of hundreds of megabits per sec.) from Space Shuttle Orbiter, the Space Station and related orbital payloads to the ground.

tehp Abbrev. for **Total Equivalent brake HorsePower.**

telecommunication The transmission and reception of data-carrying signals usually between two widely separated points with the aid of a *communications satellite.* More generally any communication between two distant points by electrical means.

teleguided missile A small subsonic missile for attacking surface targets, e.g. tanks and ships, controlled by command guidance from an operator, or automatic device, by signals transmitted through fine wires connected to the control box and un-coiled in flight from the missile. Also *wire-guided*.

telescience A fully interactive mode of scientific operations where the experiment is performed remotely, using data presented to the experimenter (e.g. by television or on-board operator) who can remotely control elements of the experimental equipment and, thus, iterate the conduct of the experiment and its results.

telescopic star Star whose apparent magnitude is numerically greater than the 6th and which is too faint to be seen with the naked eye.

Telesto The thirteenth natural satellite of **Saturn**, discovered in 1980.

telluric line Absorption line or band in stellar and planetary spectra, caused by absorption in the Earth's atmosphere, mainly by water vapour and oxygen.

temperature A measure of whether two systems are relatively hot or cold with respect to one another. Two systems brought into contact will, after sufficient time, be in thermal equilibrium and will have the same *temperature*. A *thermometer* using a temperature scale established with respect to an arbitrary zero (e.g. **Celsius scale**) or to absolute zero (Kelvin thermodynamic scale) is required to establish the relative temperatures of two systems. See Zeroth law of **thermodynamics**.

temperature co-efficient The fractional change in any particular physical quantity per degree rise of temperature.

temperature inversion Anomalous increase in temperature with height in the troposphere.

temperature lapse rate The rate of decrease of temperature with height.

tephigram Aerological diagram in which the principal rectangular axes are temperature (T) and entropy (Φ): hence TΦ-gram. Equal area represents equal energy at all points.

tercom *TERrain COMparison, TERrain COntour Matching*. Stored digital data of ground contours are compared with those detected below the aircraft in flight and used to identify the aircraft's present position and track.

terminal velocity The maximum **limiting velocity** attainable by an aircraft as determined by its total **drag** and thrust.

terminal-velocity dive A nose dive to the greatest obtainable velocity of the machine at that altitude.

terminator The border between the illuminated and dark hemispheres of the

Moon or planets. Its apparent shape is an ellipse and it marks the regions where the Sun is rising or setting.

terprom *TERrain PROfile Matching*. See **tercom**.

terrain-avoidance system A system providing the pilot with a situation display of the ground or obstacles which project above a plane containing the aircraft so that the pilot can avoid the obstacles. Cf. **terrain-clearance system**.

terrain-clearance system A fully automatic system for sensing ground obstructions and guiding the aircraft away from them without pilot intervention. Cf. **terrain-avoidance system**.

terrestrial magnetism The magnetic properties exhibited within, on and outside the Earth's surface. There is a nominal (magnetic) North pole in Canada and a nominal South pole opposite, the positions varying cyclically with time. The direction indicated by a compass needle at any one point is that of the horizontal component of the field at the point. Having the characteristics of flux from a permanent magnet, the Earth's magnetic field probably depends on currents within the Earth and also on those arising from ionization in the upper atmosphere, interaction being exhibited by the Aurora Borealis.

terrestrial radiation At night the Earth loses heat by radiation to the sky, the maximum cooling occurring when the sky is cloudless and the air dry. Dew and hoar frost are the result of such cooling.

tesla SI unit of magnetic flux density or magnetic induction equal to 1 weber m^{-2}. Equivalent definition; the magnetic induction for which the maximum force it produces on a current of unit strength is 1 newton. Symbol T.

test vehicles Aircraft for aerodynamic, control and other tests in guided weapon development. They may simply be for gathering basic information or they may be actual missiles without a warhead. They are known by their initials: CTV, command (control) test vehicle; GPV, general-purpose vehicle; MTV, missile test vehicle; RJTV, ramjet test vehicle; RTV, rocket test vehicle.

tethered satellite A term applied to a satellite which is deployed from a spacecraft and attached to it by a wire or tether which may measure over 100 km.

Tethys The third natural satellite of **Saturn**, 1000 km in diameter.

TGT See **gas temperature**.

Thebe A tiny natural satellite of **Jupiter**, discovered in 1979 by the Voyager 2 mission.

theories of light Interference and diffraction phenomena are explained by the *wave theory*, but when light interacts with matter, the energy of the light appears to be concentrated in *quanta*, called photons. The *quantum* and *wave theories* are supplementary to each other.

thermal An ascending current due to local heating of air, e.g. by reflection of the Sun's rays from a beach.

thermal conductivity A measure of the rate of flow of thermal energy through a material in the presence of a temperature gradient. If (dQ/dt) is the rate at which heat is transmitted in a direction normal to a cross-sectional area A when a temperature gradient (dT/dx) is applied, then the thermal conductivity (k) is

$$- \frac{(dQ/dt)}{A(dT/dx)}.$$

SI unit is $W\ m^{-1}\ K^{-1}$. Materials with high electrical conductivities tend to have high thermal conductivities.

thermal cueing unit Visual display presented to the pilot showing likely targets detected by a **FLIR** system. Targets can be classified by temperature 'signature' and after selection the co-ordinates can be fed to the attack system.

thermal imaging Imaging based on the detection of weak infrared radiation from objects. Applications include the mapping of the Earth's surface from the air, weather mapping and medical thermography (thermal contours on the surface of the human body). See **optical-electronic devices**.

thermal noise That arising from random (Brownian) movements of electrons in conductors and semiconductors, and which limits the sensitivity of electronic amplifiers and detectors. If δf is the frequency bandwidth, R the resistance of the source, k Boltzman's constant and T the absolute temperature, then the noise voltage V is given by

$$V = \sqrt{4RkT}.\delta f\ .$$

Also *circuit noise, Johnson noise*.

thermal wind The vector difference between the geostrophic wind at some level in the upper air and the wind at some lower level.

thermodynamics The mathematical treatment of the relation of heat to mechanical and other forms of energy. Its chief applications are to heat engines (steam engines and IC engines) and to chemical reactions. *Laws of thermodynamics*: *Zeroth law*: if two systems are each in thermal equilibrium with a third system then they are in thermal equilibrium with each other. This statement is tacitly assumed in every measurement of temperature. *First law*: the total energy of a thermodynamic system remains constant although it may be transformed from one form to another. This is a statement of the principle of the conservation of energy. *Second law*: heat can never pass spontaneously from a body at a lower temperature to one at a higher temperature (Clausius) *or* no process is possible whose only result is the abstraction of heat from a single heat reservoir and the performance of an equivalent amount of work (Kelvin-Planck). *Third law*: the entropy of a substance approaches zero as its temperature approaches absolute zero.

thermograph A continuously recording thermometer. In the commonest forms the record is made by the movement of a bimetallic spiral, or by means of the out-of--balance current in a Wheatstone bridge containing a resistance thermometer in one of its arms.

thermometer An instrument for measuring temperature. A thermometer can be based on any property of a substance which varies predictably with change of temperature. For instance, the *constant volume gas thermometer* is based on the pressure change of a fixed mass of gas with temperature, while the *platinum resistance thermometer* is based on a change of electrical resistance. The commonest form relies on the expansion of mercury or other suitable fluid with increase in temperature.

thermonuclear energy Energy released by a *nuclear fusion* reaction that occurs because of the high thermal energy of the interacting particles. The rate of reaction increases rapidly with temperature. The energy of most stars is believed to be acquired from exothermic thermonuclear reactions. In the hydrogen-bomb, a fission bomb is used to obtain the initial high temperature required to produce the fusion reactions. Many different containment procedures have been designed to study the release of thermonuclear energy.

thermosphere The region of the Earth's atmosphere, above the **mesosphere**, in which the temperature rises steadily with height.

thickness chart A chart of upper air showing the difference in geopotential between two particular pressure levels. Centres of low thickness are *cold pools*. See **constant-pressure chart**.

thickness-chord ratio The ratio of the

maximum depth of an *aerofoil*, measured perpendicular to the *chord line*, to the *chord length*; usually expressed as a percentage.

three-axis stabilized Refers to a spacecraft which can be held in any position by the application of small torques provided by reaction control thrusters about three orthogonal axes. See **gravity stabilized, spin stabilized**.

three-body problem The problem of the behaviour of three bodies which mutually attract each other; no general solution is possible but certain particular solutions are known. See **Trojan group**.

three-point landing The landing of an aircraft so equipped on the 2 wheels and tail skid (or wheel) simultaneously; the normal 'perfect landing'.

threshold lights A line of lights across the ends of a runway, strip or landing area to indicate the usable limits.

threshold of hearing Minimum r.m.s. pressure of a sound wave which an the average listener can just detect at any given frequency. It is 2.10^{-5} Pa at a frequency of 1000 Hz. This value is often used as a reference pressure so that the threshold of hearing is at 0 dB (see **decibel**) at 1000 Hz.

threshold of pain Minimum intensity or pressure of sound wave which causes sensation of discomfort or pain in average human listener. It is between 130 and 140 dB.

threshold of sound audibility Minimum intensity or pressure of sound wave which average normal human listener can just detect at any given frequency. Commonly expressed in decibels relative to 2×10^{-5} Pascal (Pa).

thrust Reaction force produced on a rocket or other vehicle as a result of the expulsion of a high velocity exhaust gas. It is related to the acceleration (*a*) produced by Newton's second law: $T = ma$, where T is the thrust and *m* the mass of the rocket vehicle. Thrust is measured in newtons but is often quoted in kg.

thrust chamber The compartment in a rocket where the propulsive forces are developed before ejection; usually, but not necessarily, the *reaction chamber*. See **rocket engine**.

thrust deflector A device, usually a combination of doors closing the jet pipe and a cascade of guide vanes, for deflecting the efflux of a turbojet downward to provide upward thrust for *STOL* or *VTOL*.

thrust deflexion The direction of the efflux from a **turbojet**, **ramjet** or **rocket** in a direction other then along its axis, for the purpose

of obtaining a thrust component normal to this axis. Generally used for the guidance of rockets and for **STOL** and **VTOL** aircraft. See **jet deflection**.

thrust loading The sea-level static thrust of the engine(s) of a jet-propelled aircraft divided by its gross weight.

thrust reverser A device for deflecting the efflux of a turbojet forward in order to apply a positive braking thrust after landing. There are two basic types: mechanical ones in which the jet is blocked by hinged doors, which also direct the gases forward; and aerodynamic ones wherein high-pressure air injected into the centre of the jet causes it to impinge upon peripheral louvres that turn it forward.

thrust spoiler A controllable device mounted on, or just behind, the nozzle of a jet-propulsion engine to deflect and thus negate the thrust. See **thrust reverser**.

thrust-to-weight ratio (1) The ratio of thrust over weight which must be greater than 1 (both factors expressed in kg) if the rocket system is to leave the ground. (2) Similarly, the thrust of an aircraft's power plant(s) divided by the gross weight at takeoff.

thunder The crackling, booming or rumbling noise which accompanies a flash of lightning. The noise has its origins in the violent thermal changes accompanying the discharge, which cause non-periodic wave disturbances in the air. Its reverberatory characteristic arises mainly from the continuous arrival of the brief noise from sections of the discharge at increasingly remote locations, since the spark may be many kilometres long. Claps of thunder occur when the spark is, roughly, normal to the line of observation. The time interval between lightning and the corresponding thunder (seconds divided by three) gives the distance of the storm centre (in kilometres).

thunderstorm A storm in which **lightning** and **thunder** occur, usually associated with **cumulonimbus** cloud. The mechanism by which the cloud becomes electrically charged is not fully understood.

tidal friction See panel on p. 190.

tide See panel on p. 190.

tilt wing **VTOL** aircraft whose wing, complete with propulsion units, propellers or nozzles can be rotated through 90° about a transverse axis, so that thrust acts vertically. Such arrangements have flown but have proved impractical.

time See panel on p. 191.

time between overhauls The period in hours of running time between complete

tides

The distortion of the surface layers (whether liquid or solid) of a planet or natural satellite resulting from differences between the gravitational forces acting on its various parts. The most familiar is the ocean tide on the Earth. This is causd by the gravitational fields of the Moon and the Sun. They result from the *differences* between the gravitational attraction of the Sun or Moon at different points on the Earth. Simplistically we might imagine that the Moon pulls the oceans away from the Earth; if that were the explanation there would be only one tide per day as the Earth rotated under the bulging ocean beneath the Moon. In fact there are two tides, separated by about 13 hours. One of these is indeed on the side near the Moon. The tide on the opposite side occurs because the Earth also is attracted by the Moon and pulled away from the water on the side opposite the Moon. The tides are separated by 13 hours, rather than 12 hours, because the Moon is in orbit round the Earth. The Sun raises tides about one half as great as the Moon.

At full Moon and new Moon the lunar and solar tides are in phase; high *spring tides* result. At first and last quarter the attractions are orthogonal, causing the moderate *neap tides* when the tide is smallest.

Tidal friction is that caused by the ebb and flow of the tides which has over geological time slowed the rotation of the Earth, through friction along coasts and in estuaries. The day length is currently increasing by 2 milliseconds per century as a result. The total angular momentum of the Earth-Moon system is conserved and, as a result, the Moon is gradually spiralling away from the Earth. In the long run these tidal effects will lead to the day and the month being of equal length, about 47 of the present days.

---

dismantling of an aero-engine. Abbrev. *TBO* or *tbo*. Also *overhaul period*.

time dilation The time interval between two events appears to be longer when they occur in a reference frame which is moving relative to the observer's reference frame, than when they occur at rest relative to the observer. Time dilation is a consequence of the **Lorentz transformations** in the *special theory of relativity* and can be expressed as:

$$t = t_0 \sqrt{1 - \left(\frac{v}{c}\right)^2}$$

where t is the time of a clock moving with velocity v, t_0 is the elapsed time of a stationary clock and c is the velocity of light.

time-division multiple access A technique used extensively in satellite communications in which a stream of *time slots* in a time-division multiplex system is allocated to users in accord with the demands they are making at any time.

time-division multiplex Form of multiplex transmission which follows logically from the adoption of *pulse modulation* and processes involving *sampling*. The gaps between pulses which constitute a signal allow other pulses to be interleaved; extraction of the desired signal

at the receiver requires a system operating in synchrony with the transmitter. Cf. **frequency-division multiplex**.

tip-path plane The plane of rotation of the tips of a rotorcraft's blades, which is higher than the rotor hub in flight. See **coning angle**.

Titan Saturn's largest satellite (diameter 5150 km), and the second largest Moon in the solar system. It is the only satellite with a substantial atmosphere, principally composed of nitrogen and methane.

Titius-Bode law See **Bode's Law**.

TL See **transmission loss**.

tone Sound signal of a single frequency. In the terminology of music, tone is often used to specify a complex note having a constant fundamental frequency. To avoid misunderstanding, a single frequency sound is termed a pure tone.

torch igniter A combination igniter plug and fuel atomizer for lighting-up gas turbines.

tornado An intensely destructive, advancing whirlwind formed from strongly ascending currents. When over the sea, the apparent drawing up of water arises from the condensing of water vapour in the vacuous core; also, in West Africa, the squall

time

Originally measured by the **hour angle** of a selected point of reference on the celestial sphere with respect to the observer's meridian, four distinct time scales are now used in astronomy. Each has different application.

Sidereal time is the *hour angle* of the **equinox**. It measures the rotation of the Earth relative to the distant stars, and is the time base used in planning and making astronomical observations. It has irregularities, on account of the irregular rotation of the Earth and there is an unevenness in the passage of sidereal time.

Solar time is defined by reference to the passage of the Sun across the local meridian. *Apparent solar time* is that shown by the true Sun. However, the Earth's elliptical orbit leads to gross changes in the Sun's progress through our sky, and so a fictitious *mean Sun* is used to define a uniformly progressing *mean solar time*. *Universal time* corresponds closely with solar time and is also defined in terms of the Sun. It serves as the basis of all civil time keeping.

Dynamical time is the time factor that occurs in gravitational equations of motion. It is the time that is the fourth dimension in the general theory of relativity. Since 1984 it has replaced **ephemeris time** which was defined in terms of an ephemeris mean Sun moving uniformly around the mean equator. The differences between dynamical, or ephemeris, time and universal time cannot be predicted: they depend on the Earth's rotation.

International atomic time, introduced in 1972, is a fundamental system based on the *second*. In the SI system, a second is defined in terms of the hyperfine transition in ^{133}Cs (caesium) atoms. If the fundamental physical constants are absolutely unchanging, then there is a fixed relationship between dynamical time and atomic time. If, however, the gravitational constant (for example) varies over cosmological time scales, then these two time systems would differ. Dynamical time currently runs 32.184 seconds ahead of atomic time.

Leap seconds are used to ensure that univeral time and atomic time do not differ by more than 0.9 seconds. This is accomplished by adding a leap second, usually at the end of June or December, whenever wobbles in the Earth's motion make it necessary to bring both systems more closely in line.

Time units:

Anomalistic year, (1 revolution of Earth from perihelion to perihelion) 365.259 641 34 ephemeris days.

Sidereal year, (1 revolution of Earth relative to stars) 365.256 365 56 ephemeris days.

Tropical year, (1 revolution of Earth around Sun) 365.242 198 78 ephemeris days.

Ephemeris second, 1/31 556 925.9747 of the tropical year 1900.0.

---

following thunderstorms between the wet and dry seasons.

toroidal-intake guide vanes The flared annular guide vanes which guide the air evenly into the intake of a centrifugal *impeller*.

torque Propeller torque is the measure of the total air forces on the airscrew blades, expressed as a moment about its axis.

torque limiter Any device which prevents a safe torque value from being exceeded, but specifically one which is used on a constant-speed **turboprop** to prevent it from delivering excess power to its airscrew.

torque link A mechanical linkage, usually of simple scissor form, which prevents relative rotation of the telescopic members of an aircraft **landing gear** shock absorber.

torquemeter A device for measuring the torque of a reciprocating aero-engine or turboprop, the indication of which is used by the pilot, together with r.p.m. and other readings, to establish any required power rating.

torr Unit of low pressure equal to head of

1 mm of mercury or 133.3 N/m^2.

Torricellian vacuum   See **mercury barometer**.

toss-bombing A manoeuvre for the release of a bomb (usually with a nuclear warhead) which allows the pilot to evade the blast. A special computing sight enables the pilot to loop before reaching the target, when the bomb is lobbed forward; or the pilot can overfly the target and loop to toss the bomb 'over-the-shoulder'.

total equivalent brake horsepower The brake horsepower at the propeller shaft plus the bhp equivalent of the residual jet thrust of a turboprop. Abbrevs. *tehp*, *ehp*.

total head In fluid flow, the algebraic sum of the **dynamic pressure** and the **static pressure**.

total impulse That available from a self-contained rocket expressed as the product of the mean thrust, in newtons (*N*) and the firing time, in seconds (*s*), expressed as *Ns*.

total temperature See **stagnation temperature**.

Townend ring A cowling for radial engines consisting of an aerofoil section ring, which ducts the air on to the engine cylinders and directs a streamlined flow on to the fuselage or nacelle, thus reducing drag; now obsolete.

TRACE (1) Abbrev. for *Test-equipment for Rapid Automatic Check-out and Evaluation*. A computerized general purpose diagnostic testing rig for aircraft electrical and electronic systems. (2) Trace: visible record of instrument reading.

track (1) The distance between the outer points of contact of port and starboard main wheels of an aircraft. (2) The distance between the vertical centrelines of port and starboard undercarriages where the wheels are paired. (3) More generally, the projection of a flight path upon the Earth's surface.

tracking Automatic holding of a radar beam on to a target through the operation of return signals. More generally, the continuous process of following, from a distance, an object to determine its position in space.

track-while-scan Electronic process for detecting a radar target, computing its velocity and predicting its future position without interfering with the process of continuous scanning.

tractor A propeller which is in front of the engine and the structure of the aircraft, as contrasted with a *pusher*, which is behind the engine and pushes the aircraft forward. See **propeller**.

trade-off study A logical evaluation during the preliminary design process of the pros and cons of alternative concepts or approaches and/or parameters which leads to the choice of the preferred ones; typical criteria for the analysis are: performance, schedule, risk and cost.

tradewinds Persistent winds blowing from the N.E. in the northern hemisphere and from the S.E. in the southern hemisphere between the *horse latitudes* (calm belts at 30°N. and S. of the equator) towards the **doldrums**.

trailing edge The rear edge of an aerofoil, or of a strut, wire etc.

trailing flap A *flap* which is mounted below and behind the wing trailing edge so that it normally trails at neutral incidence and is rotated to various positive angles of incidence to increase lift, there always being a gap between the wing undersurface and the flap leading edge.

trailing vortex The vortex passing from the tips of the main surfaces of an aircraft and extending downstream and behind it.

trajectory (1) The path of a rocket or space vehicle. The word *trajectory* is used when the path is limited in length, e.g. a *trajectory* to the Moon, whereas *orbit* usually refers to a path which is closed and repetitive. (2) In meteorology, the actual path along which a small quantity, or parcel, of air travels during a definite time interval.

transatmospheric vehicle Aircraft capable of normal wingborne flight through the atmosphere and also of travelling into space orbits. See **aerospaceplane** and **HOTOL**.

transfer ellipse The trajectory (part of an ellipse) by which a space vehicle may transfer from an orbit about one body (e.g. the Earth) into an orbit about another (e.g. the Sun).

transient A sound of short period and irregular non-repeating waveform, which implies a continuous spectrum of sound-energy contributions.

transit (1) The apparent passage of a heavenly body across the meridian of a place, due to the Earth's diurnal rotation. See **culmination**. (2) The passage of a smaller body across the disk of a larger body as seen by an observer on the Earth, e.g. of Venus or Mercury across the Sun's disk, or of a satellite across the disk of its parent planet.

transit circle See **meridian circle**.

transition In VTOL aircraft flight, the action of changing to or from the vertical lift mode (jet, fan or rotor) to forward flight with wing lift.

transition point The point where the flow in a **boundary layer** changes abruptly from *laminar* to *turbulent*.

transmission loss Ten times the logarithm to base 10 of a power ratio, describing the transmission of sound through walls, windows etc. The nominator is the power of the transmitted sound and the denominator the power of the incident sound. Abbrev. *TL* or *TR* in continental Europe.

transmit-receive tube Switch which does or does not (*anti-transmit-receive tube*) permit flow of high-energy radar pulses. It is a vacuum tube containing argon for low striking, and water vapour to assist recovery after the passage of a pulse. Used to protect a radar receiver from direct connection to the output of the transmitter when both are used with the same scanning aerial through a common waveguide system. Abbrev. *TR tube*.

transmittance Ratio of energy transmitted by a body to that incident on it. If scattered emergent energy is included in the ratio, it is termed *diffuse transmittance*, otherwise *specular transmittance*. Also *transmission*.

transonic range The range of airspeed in which both **subsonic** and **supersonic** airflow conditions exist round a body. Largely dependent upon body shape, curvature and **thickness-chord ratio**, it can be broadly taken as Mach 0.8 to Mach 1.4.

transpiration The flow of gas along relatively long passages, the flow being determined by the pressure difference and the viscosity of the gas, surface friction being negligible.

transponder A form of transmitter-receiver which transmits signals automatically when the correct interrogation is received. An example is a radar beacon mounted on a flight vehicle (or missile), which comprises a receiver tuned to the radar frequency and a transmitter which radiates the received signal at an intensity appreciably higher than that of the reflected signal. The radiated signal may be coded for identification.

transverse frame The outer-ring members of a rigid airship frame. It may be of a stiff-jointed type, or braced with taut radial members to a central fitting. It connects the main longitudinal girders together.

trapping region Three-dimensional space in which particles from the Sun are guided into paths towards the magnetic poles, giving rise to **aurora**, and otherwise forming ionized shells high above the ionosphere. Also *magnetic tube*.

tribology The science and technology of interacting surfaces in relative motion (and the practices related thereto), including the subjects of friction, lubrication and wear.

tricycle landing gear A landing gear with a nose-wheel unit.

trim Adjustment of an aircraft's controls to achieve stability in a desired condition of flight. See **trimming strip, trimming tab**.

trimmer See **trimming tab**.

trimming strip A metal strip, or a cord of wire doped in place with fabric, on the trailing edge of a control surface to modify its balance or trim; it is adjustable only on the ground.

trimming tab A *tab*, which can be adjusted in flight by the pilot, for trimming out control forces; coll. *trim tab, trimmer*.

trim tab See **trimming tab**.

Triton The principal natural satellite of **Neptune** with a diameter of 3700 km.

Trojan group A number of minor planets, named after the heroes of the Trojan War, which have the same mean motion as Jupiter and travel in the same orbit. They are divided into 2 clusters, one of which is 60° of longitude ahead of Jupiter, the other 60° behind; each planet oscillates about a point which forms an equilateral triangle with Jupiter and the Sun. These are 2 particular solutions of the **three-body problem**.

tropical month The period of lunar revolution with respect to the equinox (27.321 58 days).

tropical revolving storm A small intense cyclonic depression originating over tropical oceans. Also *cyclone, hurricane, typhoon*, depending on the locality.

tropical year The interval between 2 successive passages of the Sun in its apparent motion through the **First Point of Aries**; hence the interval between 2 similar equinoxes or solstices and the period of the seasons; its length is 365.242 194 mean solar days.

tropopause The upper limit of the **troposphere**, where the lapse rate of temperature becomes small ($\leq 2°C/km$). Sometimes a single unique tropopause cannot be defined and there is a *multiple tropopause* structure.

troposphere The lower part of the atmosphere extending from the surface up to a height varying from about 9 km at the poles to 17 km at the equator, in which the temperature decreases fairly regularly with height.

trough A V-shaped extension of the isobars from a centre of low pressure.

true airspeed The actual speed of an aircraft through the air, computed by correcting the indicated airspeed for altitude, temperature, position error and compressibility effect. Abbrev. *TAS*

true altitude Altitude of a heavenly body as deduced from the *apparent altitude* by applying corrections for atmospheric refraction, for instrumental errors, and where necessary for geocentric parallax, Sun's semi-diameter, and dip of horizon.

T-tail A *tail unit* characterized by positioning the horizontal stabilizer at or near the top of the vertical stabilizer. The unit is employed on aircraft having rear-mounted engines, and has variable incidence. See **all-moving tail**.

tuned rate gyro An advanced gyro used in high-performance inertial platforms in **inertial navigation systems**. A rotor within its casing is spun by an electric motor; applied angular rotations cause the rotor to oscillate which is nullified by pivot springs, so reducing random errors. Simpler and more reliable than **floated rate integrating gyro**.

turbine See **axial-flow turbine**.

turbine aero-engine See **aero-engine**.

turbine blade temperature The temperature of the metal blades caused by the hot gases in a gas turbine.

turbine entry temperature The gas temperature at the entry to a gas turbine system.

turbofan, turbojet, turboprop See **aero-engine**.

turbopump A combination **ram-air turbine** and hydraulic, or fuel, pump for a guided weapon or aircraft in emergency.

turboramjet, turborocket See **aero-engine**.

turbo-starter An aero-engine starter in which rotation is imparted by a turbine motivated either by compressed air, a gas source, or the decomposition by catalysis of an unstable chemical, such as hydrogen peroxide.

turbo-supercharger See **exhaust-driven supercharger**.

turbulence See **turbulent flow**.

turbulent flow Fluid flow in which the particle motion at any point varies rapidly in magnitude and direction. This irregular eddying motion is characteristic of fluid mo-

tion at high Reynold's numbers. Gives rise to high drag, particularly in the **boundary layer** of aircraft. Also *turbulence*. See **streamline flow**.

turn-and-slip indicator A pilot's instrument for blind flying which indicates the rate of turn and sideslip, or error, in banking. Also *turn-and-bank indicator*.

turn indicator Any instrument that indicates the departure of an aircraft from its set course in a horizontal plane. Necessary for flying in clouds or at night.

21 centimetre line A line in the radio spectrum of neutral hydrogen at 21.105 cm. It is caused by the spontaneous reversal of direction of spin of the electron in the magnetic field of the hydrogen nucleus, but it may be detected only in the vast hydrogen clouds of the Galaxy.

twilight The period after sunset, or before sunrise, when the sky is not completely dark. Astronomical twilight is defined as beginning (or ending) when the Sun is 18° below the horizon; hence twilight will last all night for a period in the summer months in all latitudes greater than about 48°. See **civil-**, **nautical-**.

twin-shaft turbine See **split compressor**.

twist and steer The control of a guided weapon or *drone* about the pitch and roll axes only, turns being achieved by rolling into a bank so that the elevator can provide the required turning moment. The system simplifies the autopilot and power requirements and is sometimes used with differentially-mounted variable-incidence wings.

twister Plate with slats giving double reflection of a radar wave, one being half-wave retarded, to give a twist in direction of polarization of electric component of wave.

two-spool compressor See **split compressor**.

typhoon A **tropical revolving storm**, in the China Sea and western North Pacific.

U V

UFO Abbrev. for *Unidentified Flying Object*. Applied to any sighting in the sky which the observer is unable to account for in terms of known phenomena.

Uhuru The name of the first X-ray astronomy satellite. Launched from Kenya in 1970, it made the first good map of the X-ray sky.

ultimate load The maximum load which a structure is designed to withstand without a failure. See **load factor, limit load, proof load**.

ultrasonic Said of frequencies above the upper limit of the normal range of hearing, at or about 20 kHz. *Ultrasonics* is the general term for the study and application of ultrasonic sound and vibrations.

ultraviolet astronomy The detection and analysis of radiation from cosmic sources at wavelengths between 25 and 350 nm. The hottest stars emit the bulk of their radiation in this waveband.

ultraviolet radiation Electromagnetic radiation in a wavelength range from 400 nm to 10 nm approximately, i.e. between the visible and X-ray regions of the spectrum. The *near* ultraviolet is from 400 to 300 nm, the *middle* from 300 to 200 nm and the *extreme* from 200 to 190 nm.

ultraviolet spectrometer An instrument similar to an optical spectrometer but employing non-visual detection and designed for use with ultraviolet radiation.

umbilical cord The term (frequently *umbilical*) applied to any flexible and easily-disconnectable cable, e.g. for conveying information, power or oxygen to a missile or spacecraft before launching, for connecting an operational spacecraft with an external astronaut.

umbra Region of complete shadow of an illuminated object, e.g. dark central portion of the shadow of the Earth or Moon. Generally applied to eclipses of the Moon or of the Sun, the term is also applied to the dark central portion of a sunspot. The outer, less dark, shadow is known as the *penumbra*.

umkehr effect An effect used in meteorology to derive, on certain assumptions, the vertical distribution of ozone from a series of measurements of the relative intensities of two wavelengths in light scattered from the zenith sky; one wavelength is more, and the other less, strongly absorbed by ozone. As the Sun's zenith angle varies, a reversal (Ger. *Umkehr*) occurs in the variation of the ratio of the intensities.

undercarriage Each of the units (consisting of wheel(s), shock-absorber(s) and supporting struts) of an aircraft's alighting gear is an undercarriage, i.e. two main and either tail or nose undercarriages. Colloq. the whole *landing gear*.

unducted fan Recent term for **propeller** used in high-speed civil turbine engines. *Propellers* when mounted within a circular duct are called *fans* but when advanced gas turbines drive fans without surrounding ducts they are called *unducted fans*. This apparently illogical terminology may be excused as the blade shapes of the new *propellers* are closer to those of fans than to those of the 1960s.

unit charge, unit quantity of electricity In SI units, 1 coulomb. In unrationalized MKS units, the electric charge which experiences a repulsive force of 1 newton when placed 1 metre from a like charge *in vacuo*. Similarly, in the CGS electrostatic units, the force is 1 dyne when 1 centimetre apart.

universal time A name for *Greenwich Mean Time,* recommended in 1928 by the International Astronomical Union to avoid confusion with the pre-1925 GMT which began at noon, not midnight. Abbrev. *UT*. See **ephemeris time**.

universe In modern astronomy this term has the particular meaning: the totality of all that is in the cosmos and which can affect us by means of physical forces. The definition excludes anything which is in principle undetectable physically, such as regions of **spacetime** that have been irreversibly cut off from our own spacetime.

unpitched sound Any sound or noise which does not exhibit a definite pitch, but consists of components spread more or less continuously over the frequency spectrum.

unstable equilibrium The state of equilibrium of a body when any slight displacement decreases its potential energy. The instability is shown by the fact that, having been slightly displaced, the body moves farther away from its position of equilibrium.

unstick See **lift-off**.

up, down, locks Safety locks which hold the units of a retractable landing gear up in flight and down on the ground.

upper atmosphere A term used somewhat loosely for the region of the Earth's atmosphere above about 20 miles (30 km), which is not normally explored by sounding balloons, but can be studied by rockets and artificial satellites.

Uranus

The seventh major planet of the solar system in order from the Sun, discovered by William Herschel in 1781. It is just bright enough to be seen by the naked eye under good observing conditions. From the Earth, it appears as an almost featureless, greenish disk, in even the largest telescope. In 1986, the space probe **Voyager** 2 passed close to Uranus and its satellites, providing close-up images of them. Ten small satellites were discoved by Voyager 2. Five larger satellites were already known: *Miranda, Ariel, Umbriel, Titania* and *Oberon*.

Uranus is one of the four giant gaseous planets of the solar system. Its diameter is four times the Earth's and its mass 15 times greater, and it is composed almost entirely of hydrogen and helium. It is generally believed that there is a small rocky core at the centre of the planet, which is surrounded by a thick icy mantle of frozen water, methane and ammonia. The outermost layer is an atmosphere of hydrogen and helium, with small quantities of some molecular compounds.

Even in the Voyager close-ups, Uranus presents a bland, nearly featureless appearance, though there is some evidence for faint bands parallel to the equator. A curious feature of Uranus is that its rotation axis lies almost in the plane of the solar system, rather than being nearly perpendicular to it, as is the case for the other planets. The rotation period is 17 hours 12 minutes.

In 1977, a series of narrow rings was discovered, around Uranus in its equatorial plane. The rings are each only a few kilometres wide and not visible from Earth and were discovered when Uranus occulted at 8th magnitude star: the rings caused small dips in the observed brightness of the star just before and just after the occultation by the disk of the planet. Later occultations, by beta Scorpii and sigma Sagittarii, confirmed this result. The ring system was subsequently imaged by Voyager 2 in 1986, when two further rings were discovered, bringing the total to nine.

upper culmination See **culmination**.

upper transit Same as *upper culmination*.

upstream injection A gas turbine fuel system in which the fuel is injected towards the compressor in order to achieve maximum vaporization and turbulence.

Uranus See panel on this page.

useful load The gross weight of an aircraft, less the tare weight. Usually includes fuel, oil, crew, equipment not necessary for flight (such as parachutes) and payload.

UT Abbrev. for **universal time**.

v Symbol for (1) velocity, (2) specific volume of a gas.

V Symbol for volt.

V Symbol for (1) potential, (2) potential difference, (3) electromotive force, (4) volume.

V The subscripted symbols used in aircraft documentation. V_1: abbrev. for *critical speed*. V_{lo}; see **lift-off**. V_{ne}: abbrev. for the maximum permissible *indicated airspeed*: a safety limitation (the subscript means *Never Exceed*) because of strength or handling considerations. The symbol is used mainly in operational instructions. V_{no}:

normal operating speed, usually of an airliner or other civil aircraft; this term is used mainly in flight operation documents and may be quoted in *EAS, IAS* or *TAS*. V_r: abbrev. for *rotation speed*.

VAB Abbrev. for *Vehicle Assembly Building* at Kennedy Space Center, Florida, used for integrating large elements of a space system.

vacuum Literally, a space totally devoid of any matter. Does not exist, but is approached in inter-stellar regions. On Earth, the best vacuums produced have a pressure of about 10^{-8} N/m^2. Used loosely for any pressure lower than atmospheric, e.g. train braking systems, 'vacuum' cleaners etc.

vacuum evaporation Under normal conditions, there is an equilibrium of molecular exchange at the surface of a solid body – molecules leaving the surface and others captured by it. Under vacuum conditions and in space, these molecules are lost.

value engineering A total approach to engineering design that seeks to achieve required performance, reliability and quality at minimum cost by attention to simplicity, avoidance of unnecessary functions and integration of design and manufacturing techniques.

Van Allen radiation belts Two belts encircling the Earth within which electrically charged particles are trapped. The lower Van Allen belt extends from 1000 to 5000 km above the equator with the second at about 20 000 km. Within these zones electrons originally captured from the solar wind are trapped. The belts are named after the American space scientist who discovered them.

van der Waals' equation An equation of state which takes into account the effect of intermolecular attraction at high densities and the reduction in effective volume due to the actual volume of the molecules: $(P+a/v^2)(vt-b) = RT$, a and b being constant for a particular gas. See **gas laws**.

vanes See **inlet guide-, nozzle guide-, swirl-.**

vapour A gas which is at a temperature below its critical temperature and can therefore be liquefied by a suitable increase in pressure.

vapour concentration The ratio of the mass of water vapour in a sample of moist air to the volume of the sample. Also *absolute humidity.*

vapour pressure The pressure exerted by a vapour, either by itself or in a mixture of gases. The term is often taken to mean saturated vapour pressure, which is the vapour pressure of a vapour in contact with its liquid form. The saturated vapour pressure increases with rise of temperature. See **saturation of the air.**

variable-area propelling nozzle A turbojet *propelling nozzle* which can be varied in effective outlet area, either mechanically or aerodynamically, to match it to the optimum engine operating conditions (principally thrust), thereby improving fuel economy: essential for the efficient use of an *afterburner* (see **reheat**) and in supersonic flight.

variable cycle engine A gas turbine in which the gas path can be changed by diverters or valves so it can operate in different modes at different flight speeds. Examples are the tandem fan turbojet which operates with four nozzles in hovering flight but with only one as a straight turbojet at high speed. Similarly in hypersonic propulsion systems the airflow has to completely bypass the turbojet at Mach numbers over 4.

variable-density wind tunnel A closed-circuit wind tunnel wherein the air may be compressed to increase the **Reynolds number**. Also *compressed-air wind tunnel.*

variable geometry See **variable sweep.**

variable-inlet guide vanes See **inlet guide vanes.**

variable-pitch propeller See **propeller.**

variable star Any star with a luminosity not constant with time. The variation can be regular or irregular. The stars can vary in their apparent **magnitude** for a variety of reasons. (1) In an *eclipsing binary* the pair of stars periodically eclipse, as seen from the Earth, and the apparent magnitude of the pair falls when one member conceals the other. (2) Many stars pulsate, and the change in size and surface temperature leads to a change in luminosity. The principle types are **Cepheid variables,** *RR Lyrae stars* and the long-period *Mira variables.*

variable sweep An aircraft with wings so hinged that they can be moved backward and forward in flight to give high **sweepback** for low **drag** in supersonic flight and high **aspect ratio**, with good lifting properties, for take-off and landing. Colloq. *swing-wing.*

variation The name given to the fourth principal periodic term in the mathematical expression of the Moon's motion, caused by the variation of the residual attraction of the Sun on the Earth-Moon system during a synodic month; it has a maximum value of 39' and a period of 14.77 days.

variation of latitude A phenomenon, first detected in 1888 by Küstner, who showed that, owing to the spheroidal form and non-rigid consistency of the Earth, its axis of rotation does not remain constant in direction but varies in a regular manner about a mean position, so that the latitude of a place also undergoes periodic variations.

VASI Abbrev. for **Visual Approach Slope Indicator.**

V-beam radar One which uses two fan-shaped beams to determine the range, bearing and height of the target. One beam is vertical, the other inclined, intersecting at ground level. They rotate continuously about a vertical axis.

vector The course or track of an aircraft, missile etc. but generally a quantity possessing both magnitude and direction, e.g. wind velocity.

vectored thrust The deflexion of the thrust from turbojet(s) to provide a jet-lift component. Particularly applied to a system using a ducted fan engine with bifurcated nozzles on fan and jet pipe so that there are four sources of thrust, thereby contributing to the balance of the system. The swivelling nozzles which deflect the thrust at any angle from horizontally aft to several degrees forward of the vertical are under the control of the pilot. Also used in missile rocket-propulsion and control systems.

veering A clockwise change in the direction from which the wind comes. Cf. **backing**.

vee-tail An aircraft tail unit consisting of two surfaces on each side of the centre line, usually at about 45° to the horizontal, which serve both as tailplane and fin. The associated hinged control surfaces are so actuated that they move in unison up/down as elevators and left/right as rudders, following conventional movements of the control column and rudder bar respectively. Also *butterfly tail*.

velocity (1) The rate of change of displacement of a moving body with time; a vector expressing both magnitude and direction (cf. *speed* which is scalar). (2) For a wave, the distance travelled by a given phase divided by the time taken. Symbol *v*.

velocity amplitude Amplitude of the velocity of the volume elements oscillating with the sound wave.

velocity budget The sum of the **characteristic velocities** involved in a complete space mission.

velocity of light See **speed of light**.

velocity of sound In dry air at s.t.p. 331.4 m/s (750 mi/h). In fresh water, 1410 m/s, and in sea water, 1540 m/s. The above values are used for sonar ranging but do not apply to explosive shock waves. They must be corrected for variations of temperature, humidity etc.

velocity of sound A key factor in aircraft design and operation, *Mach* 1.0. Under *ISA* sea level conditions the speed of sound in air is 761.6 mi/h (1229 km/h), reducing with temperature, to 660.3 mi/h (1062 km/h) at the tropopause (ISA 36 090 ft, 11 000 m and temperature − 54.46°C or −69.64°F) above which height it remains constant. Also *speed of sound*.

vent (1) The opening (usually at the centre) in a parachute canopy which stabilizes it by allowing the air to escape at a controlled rate. (2) Opening to atmosphere from, e.g. a fuel tank.

ventilated wind tunnel A wind tunnel for **transonic** testing in which part of the walls in the working section are perforated, slotted or porous, to prevent choking by the presence of the model, which would otherwise render measurements unreliable in the range from Mach 0.9 to 1.4.

ventral fins Fins mounted under the rear fuselage to increase directional stability, usually under high incidence conditions when the main fin may be blanketed.

ventral tank An auxiliary *fuel tank*, fixed or jettisonable, mounted externally under the fuselage. Sometimes *belly tank*.

venturi A convergent-divergent duct in which the pressure energy of an air stream is converted into kinetic energy by the acceleration through the narrow part of the waspwaisted passage. It is a common method of accelerating the airflow at the working section of a supersonic wind tunnel. Small venturis are used on some aircraft to provide a suction source for vacuum-operated instruments, which are connected to the low-pressure neck of the duct.

Venus See panel on p. 199.

vernal equinox See **equinox**.

vertical gust A vertical air current, which can be of dangerous intensity, particularly when met by aircraft flying at high speed.

vertical-gust recorder An **accelerometer** which records graphically the intensity of accelerations due to vertical gusts and, simultaneously, the airspeed; used in assessment of aircraft fatigue life. Abbrev. *v.g. recorder*.

vertical separation See **separation**.

vertical speed indicator A sensitive form of differential pressure gauge which measures variations in pressure sensed at the **static-pressure tube** and indicates them in terms of rates of climb and descent. Mainly used in high performance gliding.

vertical take-off and landing See **VTOL**.

vertical wind tunnel A wind tunnel wherein the air flow is upward and which is used principally for testing freely spinning models. Also *spinning tunnel*.

VFR Abbrev. for **Visual Flight Rules**.

v.g. recorder Abbrev. for aircraft speed (*v*) and normal acceleration (*g*) in a vertical-gust recorder.

video map Electronic system for transferring a map of any chosen territory, which may be on a transparency or store in computer memory, on to a radar display. See **chart comparison unit**.

virga Slight rain or snow which evaporates before reaching the ground.

virtual temperature The virtual temperature of a sample of moist air is the temperature at which dry air of the same total pressure would have the same density as the sample. Use of the virtual temperature obviates the need for a variable *gas constant* in applying the usual equation of state to moist air.

viscosity The resistance of a fluid to shear forces, and hence to flow. Such shear resistance is proportional to the relative velocity between the two surfaces on either side of a layer of fluid, the area in shear, the **coefficient of viscosity** of the fluid and the

Venus

The second major planet of the solar system, in order from the Sun, is one of the terrestrial planets, similar in nature to Earth and only slightly smaller. Like the Earth, it is surrounded by a substantial atmosphere. Venus is the nearest planet to the Earth and can be the brightest object in the sky (apart from the Sun and Moon). Because its orbit lies inside the Earth's, its position in the sky can never be further than 47° away from the Sun. As a result, Venus can be viewed either in the western sky in the evening, or in the eastern sky in the morning.

As a further consequence of its location within the Earth's orbit, Venus appears to go through a cycle of phases, similar to the Moon. At its brightest and nearest, even a small telescope will show that Venus is actually a crescent. The surface is perpetually covered by dense, highly reflecting clouds, which show few features in visible light, though ultraviolet photographs reveal a banded structure. These clouds consist of droplets of dilute sulphuric acid, created by the action of sunlight on the carbon dioxide, sulphur compounds and water vapour present in the atmosphere.

This atmosphere is almost entirely of carbon dioxide and the surface pressure is more than 90 times that at the surface of the Earth. The exceptionally high surface temperature of 730 K is a result of a runaway greenhouse effect. Venus has been investigated by a number of Soviet and US probes, notably the Soviet Venera and Vega series and the US Pioneer Venus probes. The extremely high temperature and pressure present considerable difficulties. Nevertheless, it has proved possible to analyse the chemical composition of surface rocks and to return limited panoramic views of the surface terrain, showing rocky desert landscapes. The surface has been mapped in considerable detail by means of radar on spacecraft orbiting the planet. Most of the surface consists of vast planes, above which several large plateaus rise to heights of several kilometres. The two main highland areas are *Ishtar Terra* in the northern hemisphere and *Aphrodite Terra* in the equatorial region. The *Maxwell Montes* are the highest feature, rising to 11 kilometres above the mean level of the planetary surface.

Circular structures have been found, that could be impact craters. Features that could be of volcanic origin have also been identified. This, and the high level of sulphur compounds in the atmosphere and long-term changes in their concentrations, could be interpreted as evidence of recent volcanic activity, though there is no direct proof.

reciprocal of the thickness of the layer of fluid. For comparing the viscosities of liquids, various scales have been devised, e.g. *Redwood No. 1 seconds* (UK), *Saybolt Universal seconds* (US), *Engler degrees* (Germany). See **kinematic viscosity.**

viscous damping Opposing force, or torque, proportional to velocity, e.g. resulting from viscosity of oil or from eddy currents.

viscous flow A type of fluid flow in which fluid particles, considered to be aggregates of molecules, move along streamlines so that at any point in the fluid the velocity is constant or varies in a regular manner with respect to time, random motion being only of a molecular nature. The name is also used to describe *laminar flow* or *streamline flow*.

visibility The maximum distance at which a black object of sufficient size can be seen and recognized in normal daylight.

visibility meter A meter which attenuates visibility to a standardized value, and measures such visibility on a scale.

visible radiation Electromagnetic radiation which falls within the wavelength range of 780 to 380 nm, over which the normal eye is sensitive.

visual acuity A term used to express the spatial resolving power of the eye. Measured by determining the minimum angle of separation which has to be subtended at the eye between two points before they can be seen as two separate points.

visual approach slope indicator A luminous device for day and night use, consisting of red, green and amber light bars on each side of a runway which, by being directed through restricting visors, show a pilot if he is below, on or above, and in line with, the approach path for an accurate touchdown. Developed by the Royal Aircraft Establishment from a World War II night lighting system known as the *Visual Glide Path Indicator*. Abbrev. *VASI*.

visual binary A **double star** whose two components may be seen as separate in a telescope of sufficient resolving power.

visual flight rules The regulations set out by the controlling authority stating the conditions under which flights may be carried out without radio control and instructions. The regulations usually specify minimum horizontal visibility, cloud base, and precise instructions for the distance to be maintained below and away from cloud. Abbrev. *VFR*.

visual meteorological conditions T h o s e weather conditions in which an aircraft can fly under freedom from air-traffic control except in controlled air space. Abbrev. *VMC*.

vital actions Sequences of pilot actions to be performed in preparation for flight and learned as part of good airmanship practice. Learned by mnemonics such as 'BUMPF' standing for 'Check brakes, undercarriage, mixture, pitch, flaps'.

VLBI Abbrev. for *Very Long Baseline Interferometry*. A technique of **aperture synthesis** used in **radio astronomy** to link telescopes separated by thousands of kilometres.

VMC Abbrev. for **Visual Meteorological Conditions**.

V-n diagram See **flight envelope**.

volt The SI unit of *potential difference*, electrical potential, or e.m.f., such that the potential difference across a conductor is 1 volt when 1 ampere of current in it dissipates 1 watt of power. Equivalent definition: if, in taking a charge of 1 coulomb between two points in an electric field, the work done on or by the charge is 1 joule, the potential difference between the points is 1 volt. Named after Count Alessandro Volta (1745-1827).

volume A general term comprehending the general loudness of sounds, or the magnitudes of currents which give rise to them. Volume is measured by the occasional peak values of the amplitude, when integrated over a short period, corresponding to the time constant of the ear. See **volume indicator, volume unit**.

volume compression, expansion In a radio-telephone transmission, particularly of speech, the automatic compression of the volume range so that the envelope of the waveform is transmitted at a higher average level with respect to interfering noise levels. After expansion at receiving end, the resulting transmission is freer from noise.

volume compressor In those communication systems which depend on amplitude modulation, intelligence transmitted as a modulation is limited to 100%. So that this is not exceeded with very loud sounds in the modulation, the original transmission has to be compressed into a relatively small dynamic range to maintain a high signal/noise ratio.

volume indicator Instrument for measuring *volume, volume compression, volume expansion*.

volume range The difference between the maximum amplitude and the minimum useful amplitude of a transmitted signal, expressed in decibels. In speech it is generally taken to be 15–20 dB, and for a full orchestra 60–70 dB.

volume unit One used in measuring variations of modulation in a communication circuit, e.g. telephone or broadcasting. The unit is the decibel expressed relative to a reference level of 1 mW in 600 ohms, and standard *volume indicators* are calibrated in these units. Abbrev. *VU*.

vortex An eddy, or intense spiral motion in a limited region; a *vortex sheet* is a thin layer of fluid with intense vorticity; *tip vortices* are a form of *trailing vortex* from aerofoils, caused by shedding of lateral and line-of-flight airflows.

vortex generators Small aerofoils, mounted normal to the surface of a main aerofoil and at a slight angle of incidence to the main airflow, which re-energize the **boundary layer** by creating vortices. Used on the wings and tail surfaces of high-speed aircraft to reduce **buffeting** caused by compressibility effects, so raising the critical **Mach number**, and sometimes to improve the airflow over control surfaces near the stall, thereby improving controllability. Cf. **wing fence**.

vortex street Regular procession of vortices forming behind a bluff or rectangular body in two parallel rows. The vortices are staggered and each vortex in the opposite direction from its predecessor. Also *Kármán street*.

vorticity equation An equation for the rate of change of the vorticity, or curl of the velocity for atmospheric flow, esp. the vertical component which is the dominating one. If f is the **Coriolis parameter**, then the vorticity is $\zeta = \dfrac{\partial v}{\partial x} - \dfrac{\partial u}{\partial y}$ and the horizontal

divergence is $D = \dfrac{\partial u}{\partial x} + \dfrac{\partial v}{\partial y}$. If the small terms are neglected then

$$\frac{d}{dt}(\zeta + f) = -(\zeta + f)D.$$

Vostok A series of Russian manned Earth satellites, first used successfully in April 1961. Vostok I carried Yuri Gagarin, the first man to travel in space.

Voyager The name of two unmanned spacecraft, Voyager 1 and 2, designed for exploring the outer planets of the solar system. Spectacular images of Jupiter and its satellites, and Saturn and its rings were sent back to Earth. Launched in late 1977, Voyager II completed its grand tour of the outer planets with a fly-by of Neptune in August 1989. See **Neptune**.

VSI Abbrev. for **Vertical Speed Indicator**.

VTOL A general term for aircraft, other than conventional helicopters, capable of vertical take-off and landing: in Britain, the initials *VTO* were originally used, but the US expression *VTOL* has become general.

VU Abbrev. for **volume unit**.

Vulcan Hypothetical planet inside the orbit of Mercury.

VU meter Instrument calibrated to read intensity of electro-acoustic signals directly in **volume units**. See **volume indicator**.

W X Y Z

w Symbol for work.

W Symbol for watt.

W Symbol for (1) weight, (2) work.

wake The region behind an aircraft in which the **total head** of the air has been modified by its passage.

warm front The leading edge of a mass of advancing warm air as it rises over colder air. There is usually continuous rain in advance of it.

wash-in, wash-out Increase (wash-in) or decrease (wash-out) in the **angle of incidence** from the root toward the tip of an aerofoil, principally used on wings to ensure that the wing tips stall last so as to maintain aileron control.

WAT curves Complicated graphs relating the take-off and landing behaviour of an aircraft to its weight, airfield altitude and ambient temperature. Their preparation and use is mandatory for UK public transport aircraft. Many other countries use the *Weight/Altitude/Temperature* information in tabular form.

water A colourless, odourless, tasteless liquid, m.p. 0°C, bp 100°C. It is hydrogen oxide, H_2O, the liquid probably containing associated molecules, H_4O_2, H_6O_3 etc. On electrolysis it yields two volumes of hydrogen and one of oxygen. It forms a large proportion of the Earth's surface, occurs in all living organisms, and combines with many salts as water of crystallization. Water has its maximum density of 1000 kg/m³ at a temperature of 4°C. This fact has an important bearing on the freezing of ponds and lakes in winter, since the water at 4°C sinks to the bottom and ice at 0°C forms on the surface. Besides being essential for life, water has a unique combination of solvent power, thermal capacity, chemical stability, permittivity and abundance.

water bomber Aircraft, usually a flying-boat, designed to combat forest fires by collecting water whilst planing on, e.g. a lake and jettisoning its load from low altitude over the fire zone.

water channel An open channel in which the behaviour of the surface of water flowing past a stationary body gives a visual simulation of supersonic airflow.

water/methanol injection (1) The use of the *latent heat of evaporation* of water (the methanol is an antifreeze agent) injected into a piston engine intake to cool the charge, thereby permitting the use of greater power without detonation for take-off. (2) The injection of water into the airflow of the compressor of a **turbojet** or **turboprop** to restore take-off power by cooling the intake air at high ambient temperatures.

water recovery The recovery, principally by condensation, of the water in the exhaust gases of an aero-engine. Used in airships for ballast purposes, as a partial set-off against the loss of weight due to the consumption of fuel during flight.

waterspout See **tornado**.

water tunnel A tunnel in which water is circulated instead of air to obtain a visual representation of flow at high Reynolds numbers with low stream velocities.

water vapour pressure That part of the atmospheric pressure which is due to the water vapour in the atmosphere.

wave cloud A cloud that appears at the crest of a **lee wave** and thus remains more or less stationary relative to the ground. Wave clouds are usually rather smooth in appearance and often occur in regularly spaced bands demonstrating the lee waves causing them.

wave clutter See **sea clutter**.

wave drag The drag caused by the generation of **shock waves**, applied to the aircraft as a whole. See **compressibility drag**.

wave equation Linear partial differential equation of at least second order which describes the propagation of a wave in space and time.

waveguide Electromagnetic waves in the microwave region can be transmitted efficiently from a source to other parts of a circuit by means of hollow metal conductors called waveguides. The transmission can be described by the patterns of electric and magnetic fields produced inside the guide, different modes being characterized by different electric and magnetic field configurations. Dielectric guides operate similarly but generally have higher losses.

wavelength Symbol λ. (1) Distance, measured radially from the source, between two successive points in free space at which an electromagnetic or acoustic wave has the same phase; for an electromagnetic wave it is equal in metres to c/f where c is the velocity of light (in m s^{-1}) and f is the frequency (in Hz). (2) Distance between two similar and successive points on a harmonic (sinusoidal) wave, e.g. between successive maxima or minima. (3) For electrons, neutrons and other particles in motion when considered as a *wave train*, $\lambda = h/p$, p is the momentum of the particle and h is Planck's constant.

wavelength of light   The wavelength of visible light lies in the range from 400 to 700 nm approximately.

wave-particle duality   Light and other electromagnetic radiations behave like a wave motion when being propagated, and like particles when interacting with matter. Interference, diffraction and polarization effects can be described in terms of waves. The photoelectric effect and the Compton effect can be described in terms of *photons*, quanta of energy $E = h\nu$ where h is Planck's constant and ν is the frequency.

weapons system   The overall planned equipment and backing required to deliver a weapon to its target, including production, storage, transport, launchers, aircraft etc.

weathercock stability   The tendency for an aircraft to turn into the relative wind, due to the side areas aft of the cg exceeding the value for directional stability (as with aircraft designed for flying at low airspeeds); excessive weathercock stability causes an oscillating yawing motion when flying in a cross-wind.

weather forecast   See **forecast, weather map.**

weather map   A map on which are marked synchronous observations of atmospheric pressure, temperature, strength and direction of the wind, the state of the weather, cloud and visibility. Weather maps (also known as *synoptic charts*) are used as a basis for forecasting.

weather minima   The minimum horizontal visibility and cloud base stipulated (1) by the air-traffic authority and (2) by the standing orders of each airline, under which take-off and landing is permitted.

weather radar   A radar installation, either PPI or RHI, designed to be useful for the detection of **precipitation** and utilizing a wavelength of 3 to 20 cm. As the strength of the echo varies as the sixth power of the diameter, heavy showers and thunderstorms are much more conspicuous than widespread light rain or drizzle.

weber   The SI unit of magnetic flux. An e.m.f. of 1 volt is induced in a circuit through which the flux is changing at a rate of 1 weber per second. 1 weber equals 1 volt-second equals 1 joule per ampere. Equivalent definition: 1 weber is the magnetic flux through a surface over which the integral of the normal component of the magnetic induction is 1 tesla m^2.

wedge   See **ridge.**

wedge aerofoil   A supersonic aerofoil section (much used for missiles) comprising plane, instead of curved, surfaces tapering from a very sharp leading edge at an acute included angle to give a **thickness-chord ratio** of 5% or less; the aerofoil may have a blunt trailing edge, or it may have the section of a very elongated lozenge, or it may have a parallel mid-portion with leading- and trailing-edge wedges, the two latter cases being known as *double-wedge aerofoils.*

weight and mass   See panel on p. 204.

weightlessness   A condition obtained in **free fall** when reaction is absent; a body has then no 'weight' only inertia. See **microgravity.**

wet and dry bulb hygrometer   A pair of similar thermometers mounted side by side, one having its bulb wrapped in a damp wick dipping into water. The rate of evaporation of water from the wick and the consequent cooling of the 'wet bulb' is dependent on the relative humidity of the air; the latter can be obtained by means of a table from readings of the two thermometers. Also *psychrometer.*

wet-bulb potential temperature   The wet bulb potential temperature of a sample of moist air at any level may be found on an **aerological diagram** by following the **saturated adiabatic** curve through the **wet-bulb temperature** of the sample until it intersects the 1000 mb isobar and then reading off the temperature there. It is, for all practical purposes, a conservative quantity for such processes as evaporation, condensation and dry and saturated adiabatic temperature changes, and is thus a useful quantity for **air-mass** analysis.

wet-bulb temperature   The temperature at which pure water must be evaporated adiabatically at constant pressure into a given sample of air in order to saturate the air under steady-state conditions. It is approximated closely by the temperature indicated by a thermometer, freely exposed to the air (but shielded from radiation), whose bulb is covered by muslin wetted with pure water.

wet lease   Hire of commercial aircraft complete, with original crew, serviced by the original owner, but perhaps carrying the new operator's logo.

wetted area   Total surface area of body immersed in an airflow and over which a boundary develops.

whirling arm   An apparatus for making certain experiments in aerodynamics, the model or instrument being carried round the circumference of a circle, at the end of an arm

weight and mass

Weight is the gravitational force acting on a body at the Earth's surface. Units of measurement are the newton, dyne or pound-force. Symbol W. Weight equals mass times the **acceleration due to gravity**, and must therefore be distinguished from *mass*, which is determined by quantity of material and measured in pounds, kilograms etc.

The definition of weight in an *aircraft* is vital in all stages of its life, from the drawing board to operational service over many decades. At the design stage weight is first estimated from the drawings and from formulae derived from previous aircraft whose weights have been measured. There are national and international reporting formats for the manner in which the weight of the several hundred separate items, contained in an aircraft, must be tabulated so that there is no misunderstanding between the aircraft builder, the engine supplier, the customer (airline or air force) and the operating staff. There are some differences between national standards but in general the terms used are as follows.

Basic weight consists of the *structure* (wing, body, tail unit and landing gear), the *propulsion systems* and the *airframe sevices and equipment* (mechanical, avionics, fuel tanks and pipes). Basic weight includes residual oil and undrainable fuel but no operational equipment or *payload*, defined as that part of the general payload from which revenue is derived.

All-up (gross) weight is the basic weight plus, for *civil aircraft*, the fuel, crew, ballast, catering load, passengers, baggage and freight and, for *military aircraft*, the fuel, crew, pylons and armament.

There are a number of terms used in the US: *empty, empty minus engines, standard empty, standard aircraft and unusable fuel, basic empty, standard empty and operating equipment, useful load.*

The following abbreviations and definitions are also in general use: *OEW*, operating empty weight; *MTOW*, (also *max gross*) maximum take-off weight; *ZFW*, zero fuel weight; *MLW*, maximum landing weight; *MRW*, maximum ramp weight; *overload weight*, MTOW in unusual circumstances; *target bogey weight*, a project design target 4% less than specification.

For *guided missiles*, payload is the warhead section with contained and actuating devices.

In the design of *spacecraft*, weight control is again essential to ensure that the capabilities of the launch vehicle are met. The following terms are used: *launch weight* is the total (gross) weight at launch; *dry weight* is the total weight of a system (a launcher or an orbiting spacecraft) without fuel and consumables. Sometimes the terms *full* and *empty* are used for launch and dry respectively. *Initial* and *burn-out weights* refer to the weights of a rocket stage before and after fuel consumption respectively.

---

rotating in a horizontal plane.

whirlwind A small rotating windstorm which may extend upwards to a height of many hundred feet; a small tornado.

whistlers Atmospheric electric noises which produce relatively musical notes in a communication system.

white dwarf Small dim star in the final stages of its evolution. The masses of known white dwarfs do not exceed 1.4 solar masses.

They are defunct stars, collapsed to about the diameter of the Earth, at which time they stabilize with their electrons forming a **degenerate gas**, the pressure of which is sufficient to balance gravitational force. See **Hertzsprung-Russell diagram**.

white frost See **hoar frost**.

white noise Noise, which may be of the random or impulse variety, having a flat frequency spectrum over the range of interest.

white-out A situation where the horizon is indistinguishable, when the sky is overcast and the ground is snow-covered.

wide-cut fuel   Low octane petrol (gasoline) obtained from wide-cut distillation used in turbojets in order to conserve kerosine. Abbrev. *avtag*. See **aviation fuels**.

Wien's laws for radiation from a black body (1) *Displacement law*: $\lambda_\mu T$ =constant (0.0029 metre kelvin). (2) *Emissive power* (*El*): within the maximum intensity wavelength interval $d\lambda$, $El = CT^5 d\lambda$, where $C = 1.288 \times 10^{-5}$ W/m^2 K^5 (*T* is absolute temperature in kelvins). (3) *Emissive power* (*dE*) in the interval $d\lambda$ is $dE = A\lambda^{-5}$ exp($-B/\lambda T$)$d\lambda$. λ_μ is wavelength at E_{max}, A =4.992 Jm, B =0.0144 mK.

winch launch Launching a glider by towing it into the air by a cable on a motorized winch, or pulling through a pulley by a car.

wind The horizontal movement of air over the Earth's surface. The direction is that from which the wind blows. The speed may be given in metres per second, miles per hour, or knots.

wind axes Co-ordinate axes, having their origin within the aircraft, and directionally orientated by the relative airflow.

wind chill factor Assessment of the power of a cold wind to chill objects and (esp.) living beings which combines the wind speed with the temperature and **relative humidity** of the air.

wind frost An air frost where the cold air has been propelled by the wind.

window Strips of metallic foil of dimensions calculated to give radar reflections and hence confuse locations derived therefrom. Also *chaff, rope*.

wind rose A star-shaped diagram showing, for a given location, the relative frequencies of winds from different directions and of different strengths.

wind shear Rate of change of the vector wind in a direction (horizontal or vertical) normal to the wind.

wind sock A truncated conical fabric sleeve, on a 360° free pivot, which indicates local wind direction. Also *wind stocking*.

wind T A T-shaped device displayed at airfields to indicate the direction of the surface wind. The leg of the T corresponds to the wind direction, with wind arrowhead at top of T.

wind tunnel Apparatus for producing a steady airstream past a model for aerodynamic investigations. These and their related facilities have many different designs, e.g. continuous, intermittent, atmospheric, pressurized, blow-down, shock tunnels, shock

tubes etc depending on speed range and applications.

wing The main supporting surface(s) of an aeroplane or glider.

wing area See **gross wing area**, **net wing area**.

wing car See **car**.

wing fence A projection extending chordwise along a wing and projecting from its upper surface. It modifies the pressure distribution by preventing a spanwise flow of air which would otherwise cause a breakaway of the flow near the wing tips and lead to tip stalling. Also *boundary layer fence*. Cf. **vortex generators**.

wing loading The gross weight of an aeroplane or glider divided by its **gross wing area**.

wing-tip float A watertight float which gives stability and buoyancy on the water; placed at the extremities of the wings of a seaplane, flying-boat or amphibian.

winter solstice See **solstice**.

wire-guided missile See **teleguided missile**.

Wolf-Rayet star An abnormal class of stars, with spectra similar to those of novae, broad bright lines predominating, indicating violent motion in the stellar atmosphere.

work One manifestation of energy. The work done by a force is defined as the product of the force and the distance moved by its point of application along the line of action of the force. For example, a tensile force does work in increasing the length of a piece of wire; work is done by a gas when it expands against a hydrostatic pressure. As for all forms of energy, the SI unit of work is the *joule*, performed when a force of 1 newton moves its point of application through 1 metre along the line of action of the force. Alternatively, when 1 watt of power is expended for 1 second, 1 joule of work is done. See **joule**, **kilowatt-hour**.

work function The minimum energy that must be supplied to remove an electron so that it can just exist outside a material under vacuum conditions. The energy can be supplied by heating the material (*thermionic* work function) or by illuminating it with radiation of sufficiently high energy (*photoelectric* work function). Also *electron affinity*.

x-axis The longitudinal, or roll, axis of an aircraft. Cf. **axis**.

X-band Microwave band lying roughly between 8 and 12 GHz; slight discrepancy between US and UK band limits. Widely used for 3 cm radar which is now correctly designated *Cx-band.*

X-rays Electromagnetic waves of short wavelength (ca 10^{-3} to 10 nm) produced when high-speed electrons strike a solid target. Electrons passing near a nucleus in the target are accelerated and so emit a continuous spectrum of radiation (*bremsstrahlung*) ranging up from a *minimum wavelength.* In addition, the electrons may eject an electron from an inner shell of a target atom, and the resulting transition of an electron of a higher energy level to this level produces radiation of specific wavelengths. This is the *characteristic* X-ray spectrum of the target and is specific to the target element. X-rays may be detected photographically or by a counting device. They penetrate matter which is opaque to light; this makes X-rays a valuable tool for medical investigations. See **synchrotron radiation.**

X-ray sources There are several sources of cosmic X-rays. (1) The solar **corona.** (2) Interacting **binary stars,** in which one member is a **black hole** or **neutron star.** (3) **Supernova** remnants, such as the **Crab Nebula.** (4) Some **radio galaxies,** such as **Cygnus A,** some **Seyfert galaxies** and some **quasars.**

yaw Angular rotation of an aircraft or other vessel about a vertical axis. The *yaw angle* is measured between the relative wind and the axis of the vessel.

yaw damper See **damper.**

yawing moment The component about the normal axis of an aircraft due to the relative airflow.

yaw meter An instrument, usually on experimental aircraft or missiles, which detects changes in the direction of airflow by the pressure changes induced thereby, or by a weather recording vane transmitting to instruments.

yaw vane A small aerofoil on a pivoted arm at the end of a long boom or probe, attached to the nose of an aircraft or missile, which measures the angle of the relative airflow and transmits it to recording instruments.

y-axis The lateral, or pitch, axis of an aircraft. Cf. **axis.**

year The civil or calendar year as used in ordinary life, consisting of a whole number of days, 365 in ordinary years, and 366 in leap years, and beginning with January 1. See **anomalistic-, eclipse-, sidereal-, tropical-, leap years.**

Youngman flap A trailing-edge flap which is extended below the main aerofoil to form a slot before being traversed rearward to increase the wing area and before being deflected downward to increase lift and drag co-efficient.

z-axis The normal, or yaw, axis of an aircraft. Cf. **axis.**

zenith The point on the celestial sphere vertically above the observer's head; one of the two poles of the horizon, the other being the **nadir.**

zenith distance The angular distance from the zenith of a heavenly body, measured as the arc of a vertical great circle; hence the complement of the altitude of the body.

zenith telescope An instrument similar to the meridian circle, but fitted with an extremely sensitive level and a declination micrometer; used to determine latitude, by observing the difference in zenith distance of two stars whose meridian transit is at a small and equal distance from the zenith, one north and one south.

zero fuel weight See **weight and mass.**

zero-g The state of weightlessness or **free fall.**

Z marker beacon A form of marker beacon radiating a narrow conical beam along the vertical axis of the cone of silence of a radio range.

Zodiac A name, of Greek origin, given to the belt of stars, about 18° wide, through which the ecliptic passes centrally. The Zodiac forms the background of the motions of the Sun, Moon and planets.

zodiacal light A faint illumination of the sky, lenticular in form and elongated in the direction of the ecliptic on either side of the Sun, fading away at about 90° from it; best seen after sunset or before sunrise in the tropics, where the ecliptic is steeply inclined to the horizon; it is caused by small particles reflecting sunlight, and appears to be an extension of the solar corona to a distance well beyond the Earth's orbit.

zonal index Numerical index measuring the strength of the westerly zonal flow in middle latitudes, e.g. between 35° and 65°. A common type is the mean pressure or **geopotential height** difference between latitude circles. Indices may be defined for various levels in the atmosphere.

zone of audibility The hearing of explosions at great distances from the source, although, nearer, there is a **zone of silence.**

zone of silence Local region where sound or electromagnetic waves from a given source cannot be received at a useful intensity. Also *shadow source.*

zone time See **standard time.**

zoning (1) The specification of areas surrounding an airfield in which there is a known clearance above obstruction for the safe landing and take-off of aircraft. (2) The division of an aircraft's fuselage, wings and engine nacelles into specific areas for precise location of equipment, identification and fire protection purposes.

zooming Utilizing the kinetic energy of an aircraft in order to gain height. *Zoombombing* involves the release of a nuclear bomb during a zooming manoeuvre to give the aircraft time to excape the blast. See **toss-bombing**.

Appendices

The Radio and Radar frequencies used in Aircraft Navigation

Band	Application
KHz	
9–14	Omega long-range continuous-wave phase-comparison location
70–90	Decca continuous-wave hyperbolic system
110–130	
90–110	Loran-C long-range pulse time-comparison location system
190–526*	Aeronautical and maritime direction finding; fixed beacons
1600–2000*	Loran-A medium-range pulse time-comparison location system
MHz	
75	Instrument landing distance fan-markers
108–112	Instrument landing localiser beams
112–118	VHF omni range
328–336	Instrument landing glide-slope beams
150 & 400	Transit satellite position fixing
582–606*	Air-traffic-control primary radar
960–1215*	Aeronautical secondary radar. Aero distance measuring equipment. Tacan
1215–1240	Navstar satellite position fixing system
1300–1350	Primary and secondary aeronautical radar
1559–1626*	Navstar satellite system and radio altimeters
2700–3100*	Aeronautical and maritime radar
4200–4400	Radio altimeters
5000–5250	Future microwave landing system
5350–5650*	Airborne and shipborne radar, beacons and transponders
8750–8850	Airborne dopplet navigation
8850–9800	Aeronautical and maritime radar
GHz	
13.25–13.4	Airborne doppler navigation
14.0–14.3*	Docking radar for ships
15.4–15.7	Aeronautical navigation
24.25 25.25	Navigation generally
31.8–33.4	Airfield surface movement indicator (radar)

* There are restrictions on the use of some parts of these bands for some types of navigation aid.

The Letter Designations of the Frequency Bands

Frequency bands are identified by defence organisations by capital letters, and these are often found in radar documents. In 1972 the bands indicated by the letters were changed, leading to confusion. The system is not internationally standardised, but the lists given here are often used.

Old System		New System	
Letter	Frequency band	Letter	Frequency band
	Megahertz		Megahertz
P	80–390	A	0–250
		B	250–500
L	390–2500	C	500–1000
			Gigahertz
	Gigahertz	D	1–2
S	2.5–4.1	E	2–3
		F	3–4
C	4.1–7	G	4–6
		H	6–8
X	7–11.5	I	8–10
J	11.5–18	J	10–20
K	18–33	K	20–40
O	33–40		
O	40–60	L	40–60
V	60–90	M	60–100

More recently the following microwave frequency allocations have been made for communication in space:

L-band (0.39 to 1.55); S-band (1.55 to 5.2); C-band (3.70 to 6.20); X-band (5.20 to 10.9); Ku-band (15.35 to 17.25); K-band (10.9 to 36.0); Ka-band (33.0 to 36.0).

The Specifications for Jet Fuels

	Designation				Specification		
Fuel Type	UK (JSD)	NATO	US Mil.	US Civil	UK (DERD)	US Mil.	US Civil
Kerosine				Jet A			ASTM D1655
	Avtur	F-35		Jet A-1	2494		ASTM D1655
	Avtur/FSII	F-34	JP-8		2453	T-83133	
Wide Cut	Avtag			Jet B	2486		
	Avtag/FSII	F-40	JP-4		2454	T-5624M	
High Flash	Avcat	F-43			2498		
	Avcat/FSII	F-44	JP-5		2452	T-5624	
High Thermal			JP-6			F-25656	
Stability			JPTS			T-25524C	
			JP-7			T-38219	
High Density			JP-10			P-87107	

DERD = Directorate of Engine Research & Development (UK)
FSII = Fuel System Icing Inhibitor (UK)
JSD = Joint Services Designation (UK)

ASTM = American Society for Testing & Materials (US)
JP = Jet Propellant (US)
JPTS = Thermally Stable Jet Propellant (US)

Physical Concepts in SI Units

Concept	Symbol	Name of Unit	Abbreviation of Unit Name	Definition or Defining Equation	Explanations; Equivalent Units; Alternative Definitions; etc.
Length	l	metre	m		1 m = 1 650 763·73 wavelengths in vacuo of radiation ($2p_{10}$—$5d_5$) of Kr 86.
Mass	m	kilogramme	kg		International Prototype Kilogramme.
Time	t	second	s		1 s = 9 192 631 770 periods of the radiation corresponding to the transition between the two hyperfine levels of the ground state of the caesium-133 atom.
Electric current	I	ampere	A		An ampere in each of two infinitely long parallel conductors of negligible cross-section 1 metre apart in vacuo will produce on each a force of 2×10^{-7} N/m.
Thermodynamic temperature	T	kelvin	K		The kelvin is 1/273·16 of the thermodynamic temperature of the triple point of water.
Luminous intensity	I	candela	cd		The luminous intensity of a black body radiator at the temperature of freezing platinum at a pressure of 1 std. atm. viewed normal to the surface is 6×10^5 cd/m².
Amount of substance		mole	mol		The amount of substance of a system which contains as many elementary units as there are carbon atoms in 0·012 kg of ^{12}C. The elementary unit must be specified (atom, molecule, ion, etc.).

All the above are internationally agreed basic units except the mole, which, though recommended, awaits international acceptance.

Concept	Symbol	Name of Unit	Abbreviation of Unit Name	Definition or Defining Equation	Explanations; Equivalent Units; Alternative Definitions; etc.
Plane angle	α, β, θ, etc.	radian	rad		A radian is equal to the angle subtended at the centre of a circle by an arc equal in length to the radius.
Solid angle	Ω, ω	steradian	sr		A steradian is equal to the angle in three dimensions subtended at the centre of a sphere by an area on the surface equal to the radius squared.
Area	A, a	square metre	m²	$a = l^2$	
Volume	V, v	cubic metre	m³	$V = l^3$	
Velocity	v, u	metre/second	m s⁻¹	$v = dl/dt$	
Acceleration	a	metre/second²	m s⁻²	$a = d^2l/dt^2$	
Density	ϱ	kilogramme/metre³	kg m⁻³	$\varrho = m/V$	
Mass rate of flow	$\overset{\circ}{m}, \overset{\circ}{M}$	kilogramme/sec	kg s⁻¹	dm/dt	
Volume rate of flow	$\overset{\circ}{V}$	cubic metre/sec	m³ s⁻¹	dV/dt	

Concept	Symbol	Name of Unit	Abbreviation of Unit Name	Definition or Defining Equation	Explanations; Equivalent Units; Alternative Definitions; etc.
Moment of inertia	I	kilogramme metre^2	kg m^2	$I = Mk^2$	
Momentum	p	kilogramme metre/sec	kg m s^{-1}	$p = mv$	
Angular momentum	$I\omega$	kilogramme metre^2/sec	kg m^2 s^{-1}	$I\omega$	
Force	F	newton	N	$F = ma$	kg m s^{-2}
Torque (Moment of Force)	T, (M)	newton metre	N m	$T = Fl$	
Work (Energy, Heat)	W, (E)	joule	J	$W = \int Fdl$	$1\,\text{J} = 1\,\text{N m} = 1\,\text{kg m}^2\,\text{s}^{-2}$ by definition
Potential energy	V	joule	J	$V = \int Fdl$	
Kinetic energy	T, (W)	joule	J	$T = \tfrac{1}{2} mv^2$	
Heat (Enthalpy)	Q, (H)	joule	J	$H = U + pV$	Definition for a fluid U = Internal energy
Power	P	watt	W	$P = dW/dt$	$1\,\text{W} = 1\,\text{J s}^{-1}$ by definition
Pressure (Stress)	p (σ, f)	newton/metre^2	N m^{-2}	$p = F/A$	Usually pressure in fluid; stress in solids
Surface tension	γ (σ)	newton/metre	N m^{-1}	$\gamma = F/l$	Free surface energy
Viscosity, dynamic	η, μ		N s m^{-2}	$\dfrac{F}{A} = \eta\, dv/dl$	$1\,\text{N s m}^{-2} = 1000$ centipoise (cP)
Viscosity, kinematic	ν		m^2 s^{-1}	$\nu = \eta/\varrho$	$1\,\text{m}^2\,\text{s}^{-1} = 10^6$ centistokes (cSt)
Temperature	θ, T	degree Celsius, kelvin	°C, K	$T\,\text{K} = (\theta + 273.15)°\text{C}$	International Temperature Scale
Velocity of light	c	metre/second	m s^{-1}	Fundamental, measured, constant	
Permeability of vacuum	μ_0	henry/metre	H m^{-1}	$\mu_0 = 4\pi \times 10^{-7}$ H/m	Defined value to give coherent rationalized electrical units
Permittivity of vacuum	ϵ_0	farad/metre	F m^{-1}	$\epsilon_0 = 1/\mu_0 c^2$	Derived in Maxwell's theory of e.m. radiation
Electric charge	Q	coulomb	C	$F = (Q_1 Q_2)/(4\mu\epsilon_0 r^2)$ (Coulomb's Law)	A s Also $q = \int idt$
Electric potential (Potential difference)	V	volt	V	$V = \int_\infty Edl$ $(V_{ab} = -\int_b^a Edl)$	
Electric field strength (Electric force)	E	volt/metre	V m^{-1}	$E = -dV/dl$	N C^{-1} E = Force on unit point charge
Electric resistance	R	ohm	Ω	$R = V/I$	
Conductance	G	siemens	S	$G = \dfrac{1}{R}$	℧ 1 ℧(mho) = 1 S
Electric flux	Ψ	coulomb	$\Psi = Q$		

Physical Concepts in SI Units *(contd.)*

Concept	Symbol	Name of Unit	Abbreviation of Unit Name	Definition or Defining Equation	Explanations; Equivalent Units; Alternative Definitions; etc.
Electric flux density (Displacement)	D	coulomb/metre²	$C\,m^{-2}$	$D = d\Psi/dA$	
Frequency	f	hertz	Hz		s^{-1} or cycles per second
Permittivity	ϵ	farad/metre	$F\,m^{-1}$	$\epsilon = D/E$	
Relative permittivity	ϵ_r			$\epsilon_r = \epsilon/\epsilon_0$	a numeric
Magnetic field strength	H	amp. turn/metre	$A\,t\,m^{-1}$	$dH = idl\sin\theta/4\pi r^2$	The *turn* is a numeric not a unit
Magnetic flux	Φ	weber	Wb	$\Phi = -\int e\,dt$	V s Faraday Law
Magnetic flux density	B	tesla	T	$B = d\Phi/dA$	$V\,s\,m^{-2}\ Wb\,m^{-2}$
Permeability	μ	henry/metre	H/m	$\mu = B/H$	
Relative permeability	μ_r			$\mu_r = \mu/\mu_0$	a numeric
Mutual inductance	M	henry	H	$e_2 = M di_1/dt$	$Wb\,A^{-1}$
Self inductance	L	henry	H	$e = L di/dt$	$Wb\,A^{-1}$
Capacitance	C	farad	F	$C = Q/V$	$C\,V^{-1}$
Reactance	X	ohm	Ω	$X = \omega L \text{ or } \dfrac{1}{\omega C}$	$\omega = 2\pi f$ rad/s
Impedance	Z	ohm	Ω	$Z = \sqrt{R^2 + X^2}$	} Sinusoidal a.c.
Susceptance	B	siemens	S	$B = \dfrac{1}{X}$	℧
Admittance	Y	siemens	S	$Y = \dfrac{1}{Z}$	℧
Total voltamperes	S	volt amp	VA	$S^2 = P^2 + Q^2$	
Reactive voltamperes	Q	volt amp reactive	VAr		
Power factor	p.f.			$\text{p.f.} = \dfrac{\text{power}}{\text{total voltamperes}}$	
Luminous flux	Φ	lumen	lm	$lm = cd\,sr$	
Illumination	E	lux	lx	$lx = lm\,m^{-2}$	

Some of the electrical definitions may occur with capital letter symbols in place of lower case, and vice versa. In general, in electrical engineering, lower case symbols are used for the instantaneous value of time-dependent quantities, and the corresponding capital letter symbols for values which are not a function of time.

SI Conversion Factors

(Exact values are printed in bold type)

Quantity	Unit		Conversion factor
Length	1 in.	=	**25·4 mm**
	1 ft	=	**0·3048 m**
	1 yd	=	**0·9144 m**
	1 fathom	=	**1·8288 m**
	1 chain	=	**20·1168 m**
	1 mile	=	1·609 34 km
	1 International nautical mile	=	**1·852 km**
	1 UK nautical mile	=	1·853 18 km
Area	1 in.^{2}	=	**6·4516 cm^2**
	1 ft^2	=	0·092 903 m^2
	1 yd^2	=	0·836 127 m^2
	1 acre	=	4046·86 m^2 = 0·404 686 ha (hectare)
	1 sq. mile	=	2·589 99 km^2 = 258·999 ha
Volume	1 UK minim	=	0·059 193 8 cm^3
	1 UK fluid drachm	=	3·551 63 cm^3
	1 UK fluid ounce	=	28·4131 cm^3
	1 US fluid ounce	=	29·5735 cm^3
	1 US liquid pint	=	473·176 cm^3 = 0·4732 dm^3 (= litre)
	1 US dry pint	=	550·610 cm^3 = 0·5506 dm^3
	1 Imperial pint	=	568·261 cm^3 = 0·5682 dm^3
	1 UK gallon	=	1·201 US gallon
		=	4·546 09 dm^3
	1 US gallon	=	0·833 UK gallon
		=	3·785 41 dm^3
	1 UK bu (bushel)	=	0·036 368 7 m^3 = 36·3687 dm^3
	1 US bushel	=	0·035 239 1 m^3 = 35·2391 dm^3
	1 in.^{3}	=	16·3871 cm^3
	1 ft^3	=	0·028 316 8 m^3
	1 yd^3	=	0·764 555 m^3
	1 board foot (timber)	=	0·002 359 74 m^3 = 2·359 74 dm^3
	1 cord (timber)	=	3·624 56 m^3
2nd moment of area	1 in.^{4}	=	41·6231 cm^4
	1 ft^4	=	0·008 630 97 m^4 = 86·3097 dm^4
Moment of inertia	1 lb ft^2	=	0·042 140 1 kg m^2
	1 slug ft^2	=	1·355 82 kg m^2
Mass	1 grain	=	0·064 798 9 g = 64·7989 mg
	1 dram (avoir.)	=	1·771 85 g = 0·001 771 85 kg
	1 drachm (apoth.)	=	3·887 93 g = 0·003 887 93 kg
	1 ounce (troy or apoth.)	=	31·1035 g = 0·031 103 5 kg
	1 oz (avoir.)	=	28·3495 g
	1 lb	=	**0·453 592 37 kg**
	1 slug	=	14·5939 kg
	1 sh cwt (US hundredweight)	=	45·3592 kg
	1 cwt (UK hundredweight)	=	50·8023 kg
	1 UK ton	=	1016·05 kg
		=	1·016 05 tonne
	1 short ton	=	2000 lb
		=	907·185 kg
		=	0·907 tonne
Mass per unit length	1 lb/yd	=	0·496 055 kg/m
	1 UK ton/mile	=	0·631 342 kg/m
	1 UK ton/1000 yd	=	1·111 16 kg/m
	1 oz/in.	=	1.116 12 kg/m = 11·1612 g/cm
	1 lb/ft	=	1·488 16 kg/m
	1 lb/in	=	17·8580 kg/m

SI Conversion Factors *(contd.)*

(Exact values are printed in bold type)

Quantity	Unit		Conversion factor
Mass per unit area	1 lb/acre	=	$0{\cdot}112\,085\ \text{g/m}^2 = 1{\cdot}120\,85 \times 10^{-4}\ \text{kg/m}^2$
	1 UK cwt/acre	=	$0{\cdot}012\,553\,5\ \text{kg/m}^2$
	1 oz/yd^2	=	$0{\cdot}033\,905\,7\ \text{kg/m}^2$
	1 UK ton/acre	=	$0{\cdot}251\,071\ \text{kg/m}^2$
	1 oz/ft^2	=	$0{\cdot}305\,152\ \text{kg/m}^2$
	1 lb/ft^2	=	$4{\cdot}882\,43\ \text{kg/m}^2$
	1 lb/in.^{2}	=	$703{\cdot}070\ \text{kg/m}^2$
	1 UK ton/mile^2	=	$0{\cdot}392\,298\ \text{g/m}^2 = 3{\cdot}922\,98 \times 10^{-4}\ \text{kg/m}^2$
Density	1 lb/ft^3	=	$16{\cdot}0185\ \text{kg/m}^3$
	1 lb/UK gal	=	$99{\cdot}7763\ \text{kg/m}^3 = 0{\cdot}099\,78\ \text{kg/l}$
	1 lb/US gal	=	$119{\cdot}826\ \text{kg/m}^3 = 0{\cdot}1198\ \text{kg/l}$
	1 slug/ft^3	=	$515{\cdot}379\ \text{kg/m}^3$
	1 ton/yd^3	=	$1328{\cdot}94\ \text{kg/m}^3 = 1{\cdot}328\,94\ \text{tonne/m}^3$
	1 lb/in.^{3}	=	$27{\cdot}6799\ \text{Mg/M}^3 = 27{\cdot}6799\ \text{g/cm}^3$
Specific volume	1 in.^{3}/lb	=	$36{\cdot}1273\ \text{cm}^3/\text{kg}$
	1 ft^3/lb	=	$0{\cdot}062\,428\,0\ \text{m}^3/\text{kg} = 62{\cdot}4280\ \text{dm}^3/\text{kg}$
Velocity	1 in./min	=	$0{\cdot}042\,333\ \text{cm/s}$
	1 ft/min	=	$\mathbf{0{\cdot}005\,08}\ \text{m/s} = \mathbf{0{\cdot}3048}\ \text{m/min}$
	1 ft/s	=	$\mathbf{0{\cdot}3048}\ \text{m/s} = \mathbf{1{\cdot}097\,28}\ \text{km/h}$
	1 mile/h	=	$1{\cdot}609\,34\ \text{km/h} = \mathbf{0{\cdot}447\,04}\ \text{m/s}$
	1 UK knot	=	$1{\cdot}853\,18\ \text{km/h} = 0{\cdot}514\,773\ \text{m/s}$
	1 International knot	=	$\mathbf{1{\cdot}852}\ \text{km/h} = 0{\cdot}514\,444\ \text{m/s}$
Acceleration	1 ft/s^2	=	$\mathbf{0{\cdot}3048}\ \text{m/s}^2$
Mass flow rate	1 lb/h	=	$0{\cdot}125\,998\ \text{g/s} = 1{\cdot}259\,98 \times 10^{-4}\ \text{kg/s}$
	1 UK ton/h	=	$0{\cdot}282\,235\ \text{kg/s}$
Force or weight	1 dyne	=	$\mathbf{10^{-5}}\ \text{N}$
	1 pdl (poundal)	=	$0{\cdot}138\,255\ \text{N}$
	1 ozf (ounce)	=	$0{\cdot}278\,014\ \text{N}$
	1 lbf	=	$4{\cdot}448\,22\ \text{N}$
	1 kgf	=	$\mathbf{9{\cdot}806\,65}\ \text{N}$
	1 tonf	=	$9{\cdot}964\,02\ \text{kN}$
Force or weight per unit length	1 lb/ft	=	$14{\cdot}5939\ \text{N/m}$
	1 lbf/in.	=	$175{\cdot}127\ \text{N/m} = 0{\cdot}175\,127\ \text{N/mm}$
	1 tonf/ft	=	$32{\cdot}6903\ \text{kN/m}$
Force (weight) per unit area or pressure or stress	1 pdl/ft^2	=	$1{\cdot}488\,16\ \text{N/m}^2$
	1 lbf/ft^2	=	$47{\cdot}8803\ \text{N/m}^2$
	1 mm Hg	=	$133{\cdot}322\ \text{N/m}^2$
	1 in. H$_2$O	=	$249{\cdot}089\ \text{N/m}^2$
	1 ft H$_2$O	=	$2989{\cdot}07\ \text{N/m}^2 = 0{\cdot}029\,890\,7\ \text{bar}$
	1 in. Hg	=	$3386{\cdot}39\ \text{N/m}^2 = 0{\cdot}033\,863\,9\ \text{bar}$
	1 lbf/in.^{2}	=	$6{\cdot}894\,76\ \text{kN/m}^2 = 0{\cdot}068\,947\,6\ \text{bar}$
	1 bar	=	$10^5\ \text{N/m}^2$
	1 std. atmos.	=	$\mathbf{101{\cdot}325}\ \text{kN/m}^2 = \mathbf{1{\cdot}013\,25}\ \text{bar}$
	1 ton/ft^2	=	$107{\cdot}252\ \text{kN/m}^2$
	1 tonf/in.^{2}	=	$15{\cdot}4443\ \text{MN/m}^2 = 1{\cdot}544\,43\ \text{hectobar}$
Specific weight	1 lbf/ft^3	=	$157{\cdot}088\ \text{N/m}^3$
	1 lbf/UK gal	=	$978{\cdot}471\ \text{N m}^3$
	1 tonf/yd^3	=	$13{\cdot}0324\ \text{kN/m}^3$
	1 lbf/in.^{3}	=	$271{\cdot}447\ \text{kN/m}^3$

Quantity	Unit		Conversion factor
Moment, torque or couple	1 ozf in. (ounce-force inch)	=	0·007 061 55 N m
	1 pdl ft	=	0·042 140 1 N m
	1 lbf in	=	0·112 985 N m
	1 lbf ft	=	1·355 82 N m
	1 tonf ft	=	3037·03 N m = 3·037 03 kN m
Energy	1 erg	=	10^{-7} J
	1 hp h (horsepower hour)	=	2·684 52 MJ
	1 thermie = 10^6 cal $_{15}$	=	4·1855 MJ
	1 therm = 100 000 Btu	=	105·506 MJ
Power	1 hp = **550** ft lbf/s	=	0·745 700 kW
	1 metric horsepower (ch, PS)	=	735·499 W
Heat	1 cal$_{IT}$	=	**4·1868** J
	1 Btu	=	1·055 06 kJ
Specific heat	1 Btu/lb degF		
	1 Chu/lb degC	=	**4·1868** kJ/kg deg C
	1 cal/g deg C		
Heat flow rate	1 Btu/h	=	0·293 071 W
	1 kcal/h	=	**1·163** W
	1 cal/s	=	**4·1868** W
Intensity of heat flow rate	1 Btu/ft^2 h	=	3·154 59 W/m^2
Electric energy	1 kWh	=	3·6 MJ
Electric stress	1 kV/in.	=	0·039 370 1 kV/mm
Dynamic viscosity	1 lb/ft s	=	14·8816 poise = 1·488 16 kg/m s
Kinematic viscosity	1 ft^2/s	=	929·03 stokes = 0·092 903 m^2/s
Calorific value or specific enthalpy	1 Btu/ft^3	=	0·037 258 9 J/cm^3 = 37·2589 kJ/m^3
	1 Btu/lb	=	**2·326** kJ/kg
	1 cal/g	=	**4·1868** J/g
	1 kcal/m^3	=	**4·1868** kJ/m^3
Specific entropy	1 Btu/lb °R	=	**4·1868** kJ/kg K
Thermal conductivity	1 cal cm/cm^2 s degC	=	**4·1868** W cm/cm^2 degC
	1 Btu ft/ft^2 h degF	=	1·730 73 W m/m^2 degC
Gas constant	1 ft lbf/lb °R	=	0·005 380 32 kJ/kg K
Plane angle	1 rad (radian)	=	57·2958°
	1 degree	=	0·017 453 3 rad = **1·1111** grade
	1 minute	=	2·908 88 × 10^{-4} rad = 0·0185 grade
	1 second	=	4·848 14 × 10^{-6} rad = 0·0003 grade
Velocity of rotation	1 rev/min	=	0·104 720 rad/s

Physical Constants, Standard Values and Equivalents in SI Units

(Figures in parentheses are the standard deviation in last digit(s))

Velocity of light in vacuo:	c =	$2.997\,925(1) \times 10^8 \text{ m s}^{-1}$
Gravitational constant:	G =	$6.670(5) \times 10^{-11} \text{ N m}^2 \text{ kg}^{-2}$
Standard acceleration of gravity:	g =	$9.806\,65 \text{ m s}^{-2} (= 32.1740 \text{ ft s}^{-2})$
Acceleration of gravity at Greenwich:	g =	$9.818\,83 \text{ m s}^{-2}$
Standard atmosphere ($\equiv 760$ mm Hg to 1 in 7×10^6):	1 atm =	$101\,325 \text{ N m}^{-2}$
Electron charge:	e =	$1.602\,10(2) \times 10^{-19} \text{ C}$
Avogadro constant:	N_A =	$6.022\,52(9) \times 10^{26} \text{ kmol}^{-1}$
Mass unit:	u =	$1.660\,43(2) \times 10^{-27} \text{ kg}$
Electron rest mass:	m_e =	$9.109\,08(13) \times 10^{-31} \text{ kg}$
	=	$5.485\,97(3) \times 10^{-4} \text{ u}$
Proton rest mass:	m_p =	$1.672\,52\,(3) \times 10^{-27} \text{ kg}$
	=	$1.007\,276\,63(8) \text{ u}$
Neutron rest mass:	m_n =	$1.674\,82(3) \times 10^{-27} \text{ kg}$
	=	$1.008\,665\,4(4) \text{ u}$
Charge/mass ratio for electron:	$\dfrac{e}{m_e}$ =	$1.758\,796(6) \times 10^{11} \text{ C kg}^{-1}$
Faraday constant:	F =	$9.648\,70(5) \times 10^4 \text{ C mol}^{-1}$
Planck constant:	h =	$6.625\,59(16) \times 10^{-34} \text{ J s}$
Fine structure constant:	α =	$7.297\,20(3) \times 10^{-3}$
	$\dfrac{1}{\alpha}$ =	$137.038\,8(6)$
Rydberg constant:	R_∞ =	$1.097\,373\,1(1) \times 10^7 \text{ m}^{-1}$
Bohr radius:	a_0 =	$5.291\,67(2) \times 10^{-11} \text{ m}$
Compton wavelength of electron:	λ_{ce} =	$2.426\,21(2) \times 10^{-12} \text{ m}$
Electron radius:	r_e =	$2.817\,77(4) \times 10^{-15} \text{ m}$
Compton wavelength of proton:	λ_{cp} =	$1.321\,398(13) \times 10^{-15} \text{ m}$
Gyromagnetic ratio of proton:	γ =	$2.675\,192(7) \times 10^8 \text{ rad s}^{-1} \text{ T}^{-1}$
Bohr magneton:	μ_B =	$9.273\,2(2) \times 10^{-24} \text{ J T}^{-1}$
Gas constant:	R_0 =	$8.314\,34(35) \text{ J K}^{-1} \text{ mol}^{-1}$
Standard volume of ideal gas:	V_0 =	$2.241\,36 \times 10^{-2} \text{ m}^3 \text{ mol}^{-1}$
Boltzmann constant:	k =	$1.380\,54(6) \times 10^{-23} \text{ J K}^{-1}$
First radiation constant:	c_1 =	$3.741\,50(9) \times 10^{-16} \text{ W m}^2$
Second radiation constant:	c_2 =	$1.438\,79(6) \times 10^{-2} \text{ m K}$
Stefan-Boltzmann constant:	σ =	$5.669\,7(10) \times 10^{-8} \text{ W m}^{-2} \text{ K}^{-4}$

Solar year = 365 d 5 hr 48 min 45.5 s   Sideral year contains 365.256 360 42 mean solar days.
Mean solar second = 1/86400 mean solar day.

International Temperature Scale of 1948 (all at a pressure of one standard atmosphere)

b.p. Oxygen	-182.970 °C	b.p. Water	100.0 °C	f.p. Silver	960.8 °C
Triple point of water	0.0100 °C	b.p. Sulphur	444.600 °C	f.p. Gold	1063.0 °C

m.p. Ice on Kelvin scale.   0°C = 273.15 K   (one std. atmos.)

Velocity of sound at sea level at 0°C. $v = 1088 \text{ ft s}^{-1} = 331.7 \text{ m s}^{-1}$

SI Prefixes

The following prefixes are used to indicate decimal multiples and sub-multiples of SI units.

Symbol	Prefix	Factor	Symbol	Prefix	Factor
T	tera	10^{12}	d	deci	10^{-1}
G	giga	10^9	c	centi	10^{-2}
M	mega	10^6	m	milli	10^{-3}
k	kilo	10^3	μ	micro	10^{-6}
h	hecto	10^2	n	nano	10^{-9}
da	deda	10^1	p	pico	10^{-12}
			f	femto	10^{-15}
			a	atto	10^{-18}